# Shaping Tomorrow's Arctic

# Shaping Tomorrow's Arctic

Editors

**Stephanie Pfirman**
**Gail Fondahl**
**Grete K. Hovelsrud**
**Tero Mustonen**

MDPI • Basel • Beijing • Wuhan • Barcelona • Belgrade • Novi Sad • Cluj • Manchester

*Editors*
Stephanie Pfirman
College of Global Futures
Arizona State University
Tempe
United States

Gail Fondahl
Department of Geography, Earth and Environmental Sciences
University of Northern British Columbia
Prince George, BC
Canada

Grete K. Hovelsrud
Nordland Research Institute
Nord University
BODØ
Norway

Tero Mustonen
Snowchange Cooperative
University of Eastern Finland
Lehtoi
Finland

*Editorial Office*
MDPI
St. Alban-Anlage 66
4052 Basel, Switzerland

This is a reprint of articles from the Special Issue published online in the open access journal *Sustainability* (ISSN 2071-1050) (available at: www.mdpi.com/journal/sustainability/special_issues/shaping_arctic).

For citation purposes, cite each article independently as indicated on the article page online and as indicated below:

Lastname, A.A.; Lastname, B.B. Article Title. *Journal Name* **Year**, *Volume Number*, Page Range.

**ISBN 978-3-0365-6925-3 (Hbk)**
**ISBN 978-3-0365-6924-6 (PDF)**
**doi.org/10.3390/books978-3-0365-6924-6**

Cover image courtesy of Priscilla Frankson

# Contents

# About the Editors

**Stephanie Pfirman**

Stephanie Pfirman is Foundation Professor, College of Global Futures, and Senior Sustainability Scientist, Julie Ann Wrigley Global Institute of Sustainability and Innovation, Arizona State University. Pfirman focuses on understanding and responding to the changing Arctic, developing innovative approaches to formal and informal education, and exploring the intersection between diversity and interdisciplinarity. Her Arctic research addresses implications of changes in sea ice origin, drift, and melt patterns, including defining the Last Ice Area, recently established as a protected area by Canada. She co-led the National Research Council's analysis of the Arctic in the Anthropocene: Emerging Research Questions, and chaired the National Science Foundation's Office of Polar Program Advisory Committee, as well as NSF's Advisory Committee for Environmental Research and Education.

**Gail Fondahl**

Dr. Gail Fondahl is Adjunct Professor in the Department of Geography, Earth and Environmental Sciences at the University of Northern British Columbia, Canada. Her research focuses on legal geographies of Indigenous territorial rights in Russia, and cultural and governance dimensions of Arctic sustainability.

**Grete K. Hovelsrud**

Grete K. Hovelsrud is Professor of Environmental Sociology, at Nordland Research Institute and Nord University, Bodø, Norway. Grete is an anthropologist who has worked and travelled in the Arctic since 1980, as a research assistant and social scientist. She is President of the Norwegian Scientific Academy for Polar Research and of the International Arctic Social Sciences Association. She has conducted extensive field work in East and West Greenland, Svalbard, and Northern Norway. Her work focuses on interdisciplinary studies of adaptation to climate change, adaptive capacity of coupled social-ecological systems, and on transformation to a low-emission society. She considers the local context as a critical starting point for understanding impacts, transitions, and change, and continues to conduct fieldwork in Arctic communities in parallel with linking the results to national and international processes. She is also a spokesperson for including social science and socio-economic research questions in projects and processes addressing Arctic issues.

**Tero Mustonen**

Tero Mustonen, Adjunct Professor at Snowchange Cooperative has served as the Lead Author in the IPCC A6 WG2. He is also a professional winter seiner (fisher). Mustonen is a well-known scholar of Arctic Indigenous, biodiversity and climate issues and large-scale ecosystem restoration. He leads the award winning Landscape Rewilding Programme, and is recipient of several environmental and human rights awards globally. Currently he serves in the Climate Crisis Advisory Group led by Sir David King.

# Preface

The Special Issue "Shaping Tomorrow's Arctic," explores the past, present, and future of the Arctic. Decisions made now are fundamentally shaping the multiple Arctics of 2050. This collection of research articles, reviews, and commentaries address Arctic futures from local, national, regional, and international perspectives, spanning a wide range of disciplines and including perspectives from Indigenous and early career, as well as senior scholars. Special Issue authors address questions such as: In looking forward, what can we learn from past experiences, as well as from recent responses to change? What steps should we be taking now to lay the foundation for equitable, just and inclusive Arctics? How might futures differ for multiple Arctic populations, economies, cultures, and governance?

The resulting scope of "Shaping Tomorrow's Arctic" extends from the need to restructure approaches to societal and economic equity, to ways to build capacity for adaptation, knowledge co-production for resilient futures, and Arctic-global repercussions. The articles present enduring foundations for future research and decision-making that can guide us through this complexity, to think ahead, co-produce approaches, and try to make the better choices—not just for us and now—rather systematically, for the most vulnerable, and into the future. We know our prior paths have led us to the uncertain present and a perilous future. It is time to make intentional choices to shape thriving Arctic futures.

**Stephanie Pfirman, Gail Fondahl, Grete K. Hovelsrud, and Tero Mustonen**
*Editors*

*sustainability*

MDPI

*Editorial*

# Shaping Tomorrow's Arctic

**Stephanie Pfirman [1,*], Gail Fondahl [2], Grete K. Hovelsrud [3] and Tero Mustonen [4]**

1 College of Global Futures, Arizona State University, Tempe, AZ 85281, USA
2 Department of Geography, Earth and Environmental Sciences, University of Northern British Columbia, Prince George, BC V2N 4Z9, Canada
3 Nordland Research Institute and Nord University, 8049 Bodø, Norway
4 University of Eastern Finland (Snowchange), 70210 Kuopio, Finland
* Correspondence: spfirman@asu.edu

This Special Issue "Shaping Tomorrow's Arctic" explores the past, present and future of Arctic sustainability. Decisions made now are fundamentally shaping the multiple Arctics of 2050. In looking forward, what can we learn from past experiences, as well as from recent responses to change? What steps should we be taking now to lay the foundation for equitable, just and inclusive Arctics? How might futures differ for multiple Arctic populations, economies, cultures and governance (Figure 1)? These are some of the profound questions we ask in this Special Issue.

We invited contributions that advance our understanding of Arctic sustainability from all disciplines, and from local, national, regional, and international perspectives. We were eager to include commentaries as well as research articles and reviews. The community responded with a range of analyses and perspectives, including from Indigenous and early career, as well as senior, scholars. As a result, the scope of "Shaping Tomorrow's Arctic" extends from the need to restructure for societal and economic equity, to ways to build capacity for adaptation, knowledge co-production for resilient futures, and Arctic-global repercussions.

**Restructuring for Societal and Economic Equity**

Exploring differential national relationships to their Arctic regions and communities, Young [1] asks " ... do we need to acknowledge that the Arctic encompasses a number of different subregions whose futures may diverge more or less profoundly?" He continues: " ... some observers take the view that we need to think more about future Arctics than about Arctic futures". While people living in the northern regions of Finland, Norway, and Sweden are comparatively more integrated with their southern compatriots, the Arctic has historically been viewed as a region distinct from the rest of the country and more of a hinterland for Russia, Canada and especially the United States [1]. Huntington [2] reflects on the legacies of colonialism that are pervasive in many regions of the Arctic. The Sámi, the Indigenous people of Fennoscandia, along with the Indigenous peoples of the Russian North, the Inuit, Aleut, Gwich'in and Innu of the North American Arctic, and the Kalaallit of Greenland, have all been exposed to aggressive colonial practices that led to assimilation, loss of land, language, and culture. Colonization and the initiation of trade routes situated much of the Arctic at the "mercy of distant market forces, on which the Arctic had little or no influence" [2].

Both now and in the future, another outside force—climate change—is reshaping the Arctic, including the environment, societies, and development. Young [1] notes: "A striking feature of this development is the strengthening of linkages between the Arctic and the outside world". As we head towards those futures, Huntington [2] points out that: "The current path is one of incremental and reactive compromise, which leads not to sustainability, but to an inevitable decline". He proposes instead that " ... perhaps shaping tomorrow's Arctic should be reframed as creating the conditions that will allow

**Citation:** Pfirman, S.; Fondahl, G.; Hovelsrud, G.K.; Mustonen, T. Shaping Tomorrow's Arctic. *Sustainability* **2023**, *15*, 3732. https://doi.org/10.3390/su15043732

Received: 3 February 2023
Accepted: 7 February 2023
Published: 17 February 2023

tomorrow's Arctic the greatest scope for shaping itself". Huntington [2] calls for a core commitment to equity, "specifically a commitment to living equitably within the social and ecological bounds of the Arctic, from which specific policies and actions can be developed and adjusted as needed".

**Figure 1.** Sidney: *"My picture shows what I think it will look like in the future. It might be like a city". The visions of the children as painted tell of a future of multiple-coloured snow, sea level rise, urbanisation and houses of distinct colours—reflecting multiple pathways for this strong Norton Sound community might take in the future.* Sidney, an Indigenous Alaskan child from Unalakleet, drew the future of his home community as a part of Snowchange community work in 2020, showing the awareness and issues the youth of the Arctic have today. (c) Snowchange, used with permission. http://www.snowchange.org/2020/02/children-of-unalakleet-alaska-paint-their-future-2020/ (accessed on 23 September 2022).

Oddsdóttir et al. [3] stress the need to focus on equity, specifically as related to gender. They point out that " ... economic development throughout much of the region affects men and women differently. It is a cause for concern that future development in the North, for the most part, focuses on traditional male sectors such as oil and gas, mining, shipping, and tertiary industrial development". They raise the issue that if we are not sensitive to gender and other inequalities, we can inadvertently exacerbate vulnerability: "Adaptation to climate change, in Arctic research and policy, should thus be reframed to systematically account for health, education, food security, and Arctic economies, all of which are simultaneously differentiated by gender". They call for " ... evaluating the effects of all actions, policies, and programmes on all genders to ensure that decisions do not perpetuate existing inequalities and create new ones".

Rozanova-Smith [4] takes up the youth perspective, with a focus on two Russian cities, and with applications across the Arctic. Through her research she found that "the Arctic communities face tremendous risks associated with youth "flight", making their future social sustainability uncertain". She calls for " ... the engagement of young people in defining problems and drawing up policies is vital to allow younger generations to have control over their own futures in the Arctic and responsibility for the future and social sustainability of their communities" [4].

**Building Capacity for Adaptation**

How do we build capacity to adapt to changes in ways that will set the stage for both Arctic communities and environmental systems to thrive? National and international interest in Arctic renewable and non-renewable resources continues to grow, posing challenges to local, regional and larger-scale adaptation strategies, while path-dependencies confound urgently needed transformations. Yet opportunities exist both for restoring time-honored traditional practices and fostering new ones that address socio-cultural and environmental adaptation requisites.

Tackling this question of capacity-building requires understanding and dealing with systemic issues caused by colonialist legacies. Tabata [5] considers contributions of natural resource extraction, focusing on mining, oil and diamonds, to the economic development of the Sakha Republic, in Russia. Mustonen and Van Dam [6] observe that in Alaska dependencies on cash economies and firearms, as well as epidemics and the takeover of resources and cultural practices, contribute to the loss of self-governance. The consequence is that "mining, energy and infrastructure projects have thoroughly altered the ... land and lifescapes" [7].

Focusing on the perspective of wild reindeer and their interconnections with biological, cultural and linguistic diversity as well as food security, Mustonen et al. [7] observe

> Traditional skills do not necessarily disappear during industrial modernization. Instead, in certain encouraging conditions, they can re-emerge and pave routes to endemic futures". The conditions they derive for successful emplacement: (1) surviving natural core ecosystems, (2) engagement with cultural landscape knowledge including social-ecological networks) and (3) agency to renew endemic links and direct adaptation, can be considered a framework for post-extractive futures across the Arctic. [7]

Furthermore, Mustonen and Van Dam [6] position "Indigenous resurgence, restoration and wisdom" as necessary to thriving futures in the face of climate change.

Fresco et al. [8] also focus on Alaska, but explore the potential for farming and gardening through the lens of "cultivating opportunities" in the face of climate change. They find

> ... many new possible avenues for expansion of Alaska's agricultural potential at the local scale".... "Especially in the context of climate-related agricultural uncertainty, challenges in other regions and possible climate-related needs for greater local autonomy (due to disruptions in supply chains and/or reduced use of fossil fuels for transportation of crops), and the need to diversify Alaska's economy [offer] new opportunities for farms and gardens ... [8]

**Knowledge Co-Production for Resilient Futures**

Wilson [9] observes that "Balancing community-based and community-driven research with the academic freedom to pursue curiosity-driven research will be a difficult and increasingly contentious task in the future ... There is both room and a need for both types of research". Historically, both types of Arctic research were controlled and dominated by scholars, funding, and publication venues partly outside the Arctic. With respect to community-relevant research, " ... we have seen a profound shift in attitudes on the part of academic researchers from being in control of the research process from start to finish, to co-producing research with local partners and conducting research that meets the needs of communities" [9].

Degai et al. [10] raise the critical question with respect to "securing a sustainable Arctic tomorrow. Ultimately, we call on individual researchers to ask themselves: what can I do to make this happen?" They call out the need for a more just understanding of co-production, stating that "Co-production should imply co-identification of research needs, co-creation of research ideas, co-design of research questions, co-definition of research objectives, co-development of research programs, co-authorship of research results, co-implementation of research projects and co-evaluation of research outcomes" [10].

Turning attention to the critical issue of broadening education about the Arctic, Wilson [9] observes that:

> One of the most important roles played by post-secondary educators and, indeed, one that is often overlooked is to inform post-secondary students and the broader public about the issues confronting the North. . . . It may seem like a very small and inconsequential act compared to the important work being done in the North by organizations and governments but raising awareness about the North among students, a group of people who will be the country's future political leaders, businesspeople, activists and citizens, will pay dividends in the years ahead. [9]

Perrin et al.'s [11] contribution focuses on integration of research advances into policies, actions and decision-making. "Engaging stakeholders that will be involved in incorporating results into decisions or programs during early stages of project development can promote knowledge transfer later in the project" [11].

**Arctic-Global Repercussions**

Schlosser et al. [12] and Bodansky and Pomerance [13] argue that while most people in the world outside the Arctic do not realize it, their current and future livelihoods and circumstances are now vulnerable to an increasingly forceful and erratic warming Arctic. "The unraveling of the Arctic is bad enough for the Arctic itself, but it will have enormous consequences for the entire planet since the Arctic is a crucial component of the global climate system" [13].

Both papers argue for aggressive action. As Schlosser et al. [12] put it "We are now in the decisive decade concerning the future we leave behind for the next generations. The Arctic's future depends on global action, and in turn, the Arctic plays a critical role in the global future". Schlosser et al. [12] focus on the critical need for investment in scaling up carbon capture and sequestration in order to reduce the rate and magnitude of warming.

Bodansky and Pomerance [13] are concerned that even scaling up decarbonation may not happen fast enough, and that we need more tools that we can deploy:

> Most of the emission pathways considered by the IPCC to achieve the 1.5 °C target assume that we will initially overshoot the target and then need negative emissions to bring temperature back down. The problem is that some of the harm from overshoot will be effectively irreversible in meaningful time frames, such as the release of carbon dioxide and methane from thawing permafrost and the disappearance of Greenland and other Arctic—as well as Antarctic and mid-latitude—glaciers . . .

Solar climate intervention is perhaps the most controversial proposal to address climate change but may be the only means of cooling the earth quickly enough to save the Arctic. [13]

**Looking Ahead**

The concept that we could 'Shape Tomorrow's Arctic' seems at odds with the extraordinary complexities facing the world today—global pandemic, Russian invasion of the Ukraine, as well as climate change breaking record after record for heat waves, floods, fires, etc.

Yet this is exactly what these articles are about. They present enduring foundations for future research and decision-making that can guide us through this complexity, to think ahead, co-produce approaches, and try to make the better choices—not just for us and now—rather systematically, for the most vulnerable, and into the future.

But things are changing far too slowly. We are literally losing ground, as the Arctic thaws and melts. Some attribute this to a mismatch between those who know and those who decide, others attribute it to timescales in decision-making processes, and others to inertia where the path of least resistance is the path that is easiest to follow. These all are true, and yet they don't have to be. We know our prior paths have led us to the uncertain present and a perilous future. It is time to make intentional choices to shape thriving Arctic futures.

**Funding:** Pfirman: This research was partially funded by the U.S. National Science Foundation award 1928235, ARCNAV: Arctic Robust Communities-Navigating Adaptation to Variability. Hovelsrud: This research was partially funded by the Research Council of Norway, Grant number: 302914; and FACE-IT which has received funding from the European Union's Horizon 2020 research and innovation programme under grant agreement No. 869154. www.face-it-project.eu.

**Conflicts of Interest:** The authors declare no conflict of interest.

## References

1. Young, O.R. Arctic Futures–Future Arctics? *Sustainability* **2021**, *13*, 9420. [CrossRef]
2. Huntington, H. What Does the Arctic's Unstable Past Say about a Sustainable Future? *Sustainability* **2021**, *13*, 8067. [CrossRef]
3. Oddsdóttir, E.E.; Ágústsson, H.; Svensson, E.-M.; Gjørv, G.H.; Kendall, S.S.; Smieszek, M.; Prior, T.; Hayfield, E.; Williamson, K.J.; Rozanova-Smith, M.; et al. Gender Equality for a Thriving, Sustainable Arctic. *Sustainability* **2021**, *13*, 10825. [CrossRef]
4. Rozanova-Smith, M. Stay or Leave? Arctic Youth Prospects and Sustainable Futures of the Russian Arctic Communities. *Sustainability* **2021**, *13*, 12058. [CrossRef]
5. Tabata, S. The Contribution of Natural Resource Producing Sectors to the Economic Development of the Sakha Republic. *Sustainability* **2021**, *13*, 10142. [CrossRef]
6. Mustonen, T.; Van Dam, B. Climate Change and Unalakleet: A Deep Analysis. *Sustainability* **2021**, *13*, 9971. [CrossRef]
7. Mustonen, T.; Andreeva, T.; Shadrin, V.; Mustonen, K. Return of Nimat?—Wild Reindeer as an Indicator of Evenki Biocultural Systems. *Sustainability* **2021**, *13*, 12107. [CrossRef]
8. Fresco, N.; Bennett, A.; Bieniek, P.; Rosner, C. Climate Change, Farming, and Gardening in Alaska: Cultivating Opportunities. *Sustainability* **2021**, *13*, 12713. [CrossRef]
9. Wilson, G.N. Developing a Sustainable and Inclusive Northern Knowledge Ecosystem in Canada. *Sustainability* **2021**, *13*, 9213. [CrossRef]
10. Degai, T.; Petrov, A.N.; Badhe, R.; Egede Dahl, P.P.; Döring, N.; Dudeck, S.; Herrmann, T.M.; Golovnev, A.; Mack, L.; Omma, E.M.; et al. Shaping Arctic's Tomorrow through Indigenous Knowledge Engagement and Knowledge Co-Production. *Sustainability* **2022**, *14*, 1331. [CrossRef]
11. Perrin, A.D.; Ljubicic, G.; Ogden, A. Northern Research Policy Contributions to Canadian Arctic Sustainability. *Sustainability* **2021**, *13*, 12035. [CrossRef]
12. Schlosser, P.; Eicken, H.; Metcalf, V.; Pfirman, S.; Murray, M.S.; Edwards, C. The Arctic Highlights Our Failure to Act in a Rapidly Changing World. *Sustainability* **2022**, *14*, 1882. [CrossRef]
13. Bodansky, D.; Pomerance, R. Sustaining the Arctic in Order to Sustain the Global Climate System. *Sustainability* **2021**, *13*, 10622. [CrossRef]

sustainability

MDPI

*Essay*

# Arctic Futures–Future Arctics?

**Oran R. Young**

Bren School of Environmental Science and Management, University of California Santa Barbara, Santa Barbara, CA 93106, USA; oran.young@gmail.com

**Abstract:** Is the Arctic sufficiently distinctive and uniform to justify adopting a holistic perspective in thinking about the future of the region? Or do we need to acknowledge that the Arctic encompasses a number of different subregions whose futures may diverge more or less profoundly? In the aftermath of the Cold War, a view of the Arctic as a distinctive region with a policy agenda of its own arose in many quarters and played a prominent role in shaping initiatives such as the launching of the Arctic Environmental Protection Strategy in 1991 and the creation of the Arctic Council in 1996. Yet not everyone found this perspective persuasive at the time, and more recent developments have raised new questions about the usefulness of this perspective as a basis for thinking about the future of the Arctic. As a result, some observers take the view that we need to think more about future Arctics than about Arctic futures. Yet, today, climate change provides a central thread tying together multiple perspectives on the Arctic. The dramatic onset of climate change has turned the Arctic into the frontline with regard to the challenges of adapting to a changing biophysical setting. Ironically, the impacts of climate change also have increased the accessibility of massive reserves of hydrocarbons located in the Arctic, contributing to a feedback loop accelerating climate change. This means that the future of the Arctic will reflect the interplay between efforts to address the biophysical and socioeconomic consequences of climate change on the one hand and the influence of the driving forces underlying the political economy of energy development on the other.

**Keywords:** adaptation; climate change; energy development; tipping elements

**Citation:** Young, O.R. Arctic Futures–Future Arctics? *Sustainability* **2021**, *13*, 9420. https://doi.org/10.3390/su13169420

Academic Editors: Stephanie Pfirman, Gail Fondahl, Grete K. Hovelsrud and Tero Mustonen

Received: 22 July 2021
Accepted: 20 August 2021
Published: 22 August 2021



Is the Arctic sufficiently distinctive and uniform to justify adopting a holistic perspective in thinking about the future of the region? Or do we need to recognize that the Arctic encompasses a number of different subregions that may follow different trajectories in the future? Are we concerned, in other words, with Arctic futures or with what may be thought of more accurately as future Arctics? In this reflective essay, I offer some responses to these questions based on my own engagement with Arctic affairs starting in the 1970s. What emerges is a view of the future of the region linked to the consequences of climate change in the high latitudes of the northern hemisphere.

## 1. The Arctic Region

Those of us who developed and promoted the idea of the "Age of the Arctic" during the 1980s grounded our thinking about the future of the northern hemisphere's high latitudes in two major premises [1]. We argued, to begin with, that it was reasonable to approach the Arctic in holistic terms, treating the entire circumpolar North as a spatially delimited area exhibiting enough common features to be treated as an international region in much the same way that we treat Africa, the Middle East, and Southeast Asia as regions. Obviously, such regions are not sealed off from the rest of the world, especially in an era marked by globalization and the rise of digital technologies. However, the prominence of commonalities justifies treating them as regions, especially when we think about issues arising in the realm of public policy that have a distinctive regional character. We asserted, in addition, that while the Arctic had been an important theater of operations for strategic weapons systems during the Cold War, the region was not itself a locus of serious conflicts between or among states. As the Cold War faded, we anticipated, the region would emerge as an area in which efforts to promote cooperation could unfold somewhat outside the main currents of world affairs.

Taken together, these premises led us to conclude that the Arctic had a policy agenda of its own featuring issues of environmental protection and, more broadly, sustainable development and that we could and should launch cooperative responses to these issues, regardless of the course of tensions or conflicts arising in other regions [2]. For various reasons, this perspective took root in the thinking of members of the policy communities of the Arctic states and especially among leaders in Canada, Finland, and Russia [3]. It evolved into a narrative underpinning the creation of the Arctic Environmental Protection Strategy in 1991 followed by the launching of the Arctic Council in 1996 [4]. Often characterized as the "Arctic zone of peace" narrative, this perspective shaped the practices of the Arctic Council; it continues to inform the efforts of many of those interested in Arctic affairs today.

Yet, others are raising critical questions about the validity of this perspective as a roadmap for the work of the Arctic Council and as a guide to thinking about the future of the Arctic region more generally. The most thoughtful critics are not only questioning the extent to which this narrative is helpful as a framing device in assessing the future of the region; they are also asking searching questions about the extent to which the narrative provided a realistic picture regarding Arctic affairs even in the 1990s. This has brought into focus a consideration of the role of social construction as a central feature of any effort to frame complex issues in ways that make them tractable for consideration in policy arenas and an assessment of alternative perspectives on the past, present, and future of the Arctic in particular [5]. What is the significance of these observations for those contemplating Arctic futures?

## 2. A Uniform Arctic?

Consider first the idea of treating the circumpolar North as an area with enough in common to justify treating it as a distinct region. No doubt, it is correct to say that the Arctic is composed in large measure of remote segments of states whose centers of gravity lie far to the South, that the region is a homeland for numerous groups of Indigenous peoples struggling to maintain the integrity of their cultures and communities in an era of rapid change, and that it is an object of interest to outsiders first and foremost as a source of the raw materials needed to fuel the engines of industrial societies. Nevertheless, differences among the various sectors of the Arctic are also apparent, and these differences have far-reaching implications for how we organize our thinking about the future of the relevant areas.

The Arctic segments of Finland, Norway, and Sweden enjoy a surprisingly moderate climate due to the impact of the North Atlantic Drift. They constitute the homeland of the Sami people, but they have sizable and settled populations of non-Indigenous residents, and they are fully integrated into the social welfare systems of Scandinavia. What is more, these are not recent developments; there have been strong ties between the northern and southern segments of these countries for centuries. As a result, it makes little sense in considering the future of this area to employ the concept of hinterlands as the term is used in discussions of core-periphery relations or internal colonialism. Life in Tromsø does not differ fundamentally from life in Oslo; the same is true with regard to Umeå and Stockholm or Rovaniemi and Helsinki. This accounts for the resistance of many Scandinavians to recent characterizations of the entire Arctic as a coherent and broadly homogeneous region. It explains the development of the idea of the "Old North" as a point of departure for an alternative narrative to employ in efforts to make sense of the past, present, and future of this sector of the Arctic [6].

The North and the Arctic (the two terms are sometimes differentiated in Russian thinking) have loomed large in Russian history and culture for centuries. The associations have not always been happy ones. Convict labor played a major role in Russian (and Soviet) efforts to develop the North. There is a long history of banishing dissidents to remote corners of Siberia; the gulags operating under Arctic conditions in the Russian Far

East were notorious. Nevertheless, the North looms large in most accounts of Russian history. Today, Russia is home to about half of the region's human residents, the majority of whom are settlers in contrast to Indigenous peoples. There is a sense in which Russia is the preeminent Arctic country, and the North remains prominent in Russian thinking, perhaps even more so following the disintegration of the Soviet Union. As a consequence, a number of former Soviet republics with little or no orientation toward the Arctic went their separate ways as independent countries, leaving Russia itself with a more pronounced northern orientation. In rebuilding its economy and reasserting its claim to great-power status, Russia has prioritized the development of the world-class reserves of natural gas in the Arctic, the promotion of the Northern Sea Route as a major artery for commercial shipping, and the strengthening of its armed forces based in the Arctic [7]. Nowhere else does the future of the Arctic figure as prominently in thinking about the future of an entire society as it does in Russia.

The Arctic sectors of Canada and the United States, by contrast, lend themselves more easily to consideration as hinterlands or peripheries. Arctic Canada is sparsely populated, provides a homeland for Aboriginal peoples, and contains raw materials of interest to southern-based corporations. In recent decades, the government of Canada has worked hard to develop enlightened policies in its dealings with the Aboriginal peoples of the Canadian Arctic. Still, the Arctic is a remote place that has little to do with the day-to-day lives of most ordinary Canadians residing in the country's southern reaches, despite their fondness for the idea of the "true North strong and free". For most Americans, Alaska is almost an afterthought. Often derided as "Seward's folly" at the time of its purchase from Russia in 1867 and of interest mainly to those concerned with military security, the extraction of raw materials, and the protection of wilderness areas, the American Arctic does not loom large in the national consciousness. None of this is to deny the legitimate claims of Canada and the United States to be acknowledged as Arctic states. Still, the contrast between the North American Arctic and the thriving social welfare systems of Scandinavia and the central place of the North in the culture and political economy of Russia is striking. Thus, it is not hard to see why many observers are skeptical about the practice of lumping these areas together as components of a relatively homogeneous or uniform Arctic region.

In this account of similarities and differences among the various parts of the Arctic, Greenland stands out as a special case [8]. As a longstanding Danish colony, Greenland has developed a Scandinavian social welfare system along with economic and political systems that are distinctly European in origin. Yet, most Greenlanders are ethnically Inuit and have strong ties to their Inuit brethren spread across the Canadian North, Alaska, and onward to Chukotka. Today, Greenland has developed a thriving political system of its own and is preoccupied with the issue of whether there is a way to develop the economic base needed to support severing its remaining ties with Denmark, establishing itself as an independent, wholly Arctic, and largely Indigenous polity. For its part, Denmark has acknowledged increasingly that its claim to the status of an Arctic state rests with the role of Greenland in Arctic affairs. Denmark announced recently, for example, that Greenland will take the lead henceforth in representing Denmark in the Arctic Council. As the turmoil stirred up by US President Trump's summer 2019 expression of interest in buying Greenland made clear, the future of Greenland itself is linked to shifting relationships among the great powers and the securitization of the general discourse regarding developments in the Arctic. Nevertheless, efforts to come to terms with the issues relating to Greenland's political and legal status dominate the discourse on public affairs within Greenland itself.

## 3. A Peaceful Arctic?

If the premise underlying the holistic view of the circumpolar Arctic as an international region is open to question when we start to look more closely at the circumstances of particular segments of the Arctic, so too is the premise regarding the Arctic as an oasis of peace in a turbulent world. Most disagreements about Arctic matters center on disputes about the delimitation of jurisdictional boundaries or the interpretation of the provisions of international agreements as applied to specific situations. Canada, Denmark, and Russia have submitted overlapping claims regarding jurisdiction over the seabed extending into the Arctic Ocean beyond the limits of their Exclusive Economic Zones to the UN Commission on the Limits of the Continental Shelf pursuant to the provisions of UNCLOS Article 76. Canada and the United States disagree about the delimitation of their boundary in the Beaufort Sea and about the application of the provisions of UNCLOS Article 37 regarding transit passage to ships plying the waters of the Northwest Passage. A number of signatories to the 1920 Treaty of Paris dealing with the Svalbard Archipelago take the view that Norway's Svalbard Fisheries Protection Zone is incompatible with the provisions of the treaty.

Yet, there is no reason to expect these disagreements to trigger armed clashes in the Arctic. In the 2008 Ilulissat Declaration, the five Arctic coastal states asserted their primacy in matters pertaining to the Arctic and pledged to address any conflicts regarding such matters in a peaceful manner. Canada, Denmark, and Russia have said repeatedly that they will deal with their disagreements about jurisdiction over the seabed of the Arctic Basin peacefully and in accordance with the provisions of UNCLOS Article 76. There is no reason to doubt the sincerity of these statements. Canada and the United States signed an agreement in 1988 agreeing to disagree with regard to their divergent views on the applicability of the provisions of Article 37 to the waters of the Northwest Passage. Nothing has happened that is likely to destabilize this agreement. Norway and Russia were able to negotiate a treaty in 2010 resolving their differences concerning jurisdiction in the Barents Sea and formalizing a suite of cooperative arrangements relating to issues of common interest in this area. No one expects the disagreements regarding the provisions of the Treaty of Paris to erupt into a major conflict over the waters surrounding the Svalbard Archipelago. On the contrary, key players are taking steps to strengthen international cooperation regarding emerging Arctic issues. For example, the 2018 Central Arctic Ocean Fisheries Agreement, a legally binding arrangement committing the five Arctic coastal states and five others to pursuing a precautionary approach to activities likely to affect marine areas lying beyond national jurisdiction, entered into force in June 2021. In contrast to regions such as the Middle East and Southeast Asia, the Arctic seems largely devoid of severe regional conflicts of the sort that could trigger armed clashes.

Still, there is another side to this story that has important implications for the future of the Arctic, especially in the minds of those who see efforts on the part of major actors to maximize relative power and the role of geopolitical factors as the principal drivers of world affairs. To begin with, the use of the Arctic as a theater of operations for strategic weapons systems has never ceased. As Russia seeks to reclaim its status as a great power, it has upgraded the capabilities of the Northern Fleet based on the Kola Peninsula with more sophisticated ships and weapons systems; it has reopened and in some cases expanded military installations abandoned or closed following the collapse of the Soviet Union in the 1990s. Pursuing its objectives mainly through economic initiatives and developing the idea of the Polar Silk Road as a component of its overarching Belt and Road Initiative, China has described itself as a "near-Arctic state" and signaled a clear interest in becoming a significant player in Arctic affairs. The United States has responded to these developments aggressively, deploying the reactivated 2nd Fleet to the Barents Sea, mobilizing war games with an Arctic focus, authorizing the construction of new icebreakers, and initiating plans for upgrading the capacity of its armed forces to operate under Arctic conditions. There are significant disagreements regarding the motivations underlying all these activities. Although no one interprets them as responses to conflicts arising in the Arctic itself, it is reasonable to take note of the increased danger of inadvertent or unintended clashes occurring in the region, an observation underlying the suggestions that there is a need to (re)open lines of communication among leaders of armed forces and to encourage efforts to develop informal codes of conduct governing the deployment of military assets in the Arctic [9].

When it comes to thinking about the future of the Arctic, a disturbing aspect of these developments is the rise of a neo-realist Arctic narrative emphasizing the return of great-power politics to the region and asserting that the Arctic is undergoing a transition form a zone of peace to a zone of conflict [10]. Such views are particularly prominent in the writings of international relations scholars with a newfound interest in the Arctic and the reports of journalists endeavoring to capture the attention of readers with provocative images. However, they also are showing up in the pronouncements of prominent public officials. In a major speech preceding the May 2019 Ministerial Meeting of the Arctic Council, for example, the US Secretary of State asserted that "the region has become an area of global power and competition". He went on to say that, in response, the U.S. is "hosting military exercises, strengthening our force presence, rebuilding our icebreaker fleet, expanding Coast Guard funding, and creating a new senior military post for Arctic affairs" [11].

This great-power politics narrative provides a rationale for advocates of acting vigorously to enhance the capacity to exercise hard power in the region in contrast to the emphasis on softer forms of influence associated with the Arctic Council's vision of the Arctic as an area of unique international cooperation [12–14]. Like all policy narratives, this one features an effort to construct a coherent story in which selected observations about the actions of various players are assembled around a few guiding premises. There are good reasons to adopt a skeptical view regarding the persuasiveness of the great-power politics narrative. However, to the extent that this narrative captures the attention of members of the policy community and of those who shape the content of the public discourse about Arctic affairs, the influence of this Arctic zone of conflict narrative will grow, whether or not the premises on which it is built are persuasive. The result will be a securitization of the discourse regarding Arctic affairs and the development of a perspective on the future of the Arctic that differs sharply from the perspective embedded in the practices of the Arctic Council.

## 4. A "New" Arctic?

Some may draw the inference from this account that we ought to be thinking about future Arctics or the globalization of the Arctic rather than about the future of the Arctic as a distinctive region in our efforts to think carefully about what lies ahead for the lands and waters of the high latitudes of the northern hemisphere and for the peoples who reside there. It is hard to find a central thread linking the concerns of those seeking to come to terms with the economic crisis of the State of Alaska, the complex political future of Greenland, the Atlantification of the Barents Sea, and the extraction of the massive deposits of natural gas in Northwestern Siberia and the adjacent waters of the Kara Sea. At the same time, some see the Arctic as an increasingly prominent arena in the global competition among China, Russia, and the USA. These are all legitimate concerns; they occupy the attention of many who have no particular concern about the consequences of their actions for the future of the Arctic as a distinctive international region. Yet there is another stream of developments now unfolding at an accelerating pace on a circumpolar basis that raise profound questions about the sustainability of both the biophysical systems and the human systems of the whole Arctic and that suggest an alternative perspective on the future of the Arctic. Many associate this perspective with the idea of a "new" Arctic [15].

Taken together, Arctic marine and terrestrial systems constitute ground zero with regard to the onset of climate change [16–19]. Surface temperatures in the Arctic are rising at a rate that has reached three times the rate of change in surface temperatures anywhere else on the planet. Sea ice in the Arctic is receding and thinning at a dramatic pace. Experts now expect that the Arctic Basin will be essentially ice free for some part of the year within two to three decades. Permafrost is thawing at an accelerating rate not only compromising all sorts of infrastructure in the Arctic but also introducing the prospect of large releases of carbon dioxide and methane sequestered in permafrost in the tundra and in methane clathrates in shallow coastal waters in the Arctic. Wildfires are raging unchecked on an unprecedented scale both in Siberia and in the North American Arctic; some of them burn year around and affect tens of thousands of square kilometers. These developments are changing large ecosystems in ways that are leading to major shifts in the spatial distribution of fish stocks and in the health of populations of marine and terrestrial wildlife. As many observers have concluded, we are now facing a global climate emergency, and the consequences of this looming crisis are nowhere more apparent than they are in the Arctic.

The resultant challenges to the human communities of the Arctic are diverse in some respects. Some coastal communities, for example, are facing an urgent need to relocate as a consequence of severe coastal erosion making their current locations untenable. Others are confronted with growing threats arising from unprecedented flooding and the devastation caused by massive fires. Still others have to find ways to cope with major changes in the abundance and distribution of living resources critical to their livelihoods. From another perspective, however, all Arctic communities face common challenges of adapting to an environment that is not only changing rapidly but also prone to non-linear changes that are difficult to anticipate with any precision. What this means is that the key to the future of human settlements in the Arctic is not an issue of sustaining arrangements that are already in place but rather a matter of introducing dramatic adaptive measures in a timely manner, without compromising the quality of life of their inhabitants [20]. Although the nature of the specific measures required to achieve this goal will differ from place to place, the cultural, economic, political, and social dimensions of adaptation as a societal process have much in common regardless of the distinctive features of the settings in which adaptations to the onset of climate change occur. This suggests that efforts to devise effective adaptation strategies will constitute a central theme with regard to the future of all sectors of the Arctic and that this will provide opportunities to explore the value of cooperative initiatives on a circumpolar basis. At a minimum, those concerned with adaptation to the impacts of climate change in the high latitudes will find it useful to compare notes about the effectiveness of specific responses on a regular basis and perhaps to develop a clearinghouse providing access to information on the results of measures that have been implemented in specific places and to advice for those dealing with similar challenges in other places. The Arctic Council may be able to play a constructive role in this realm [21].

Ironically, the impacts of climate change also have made the Arctic increasingly accessible to those motivated by the economic and political attractions of launching largescale projects aimed at exploiting the natural resources of the region. The most dramatic example so far is the development of the massive deposits of natural gas located in Northwestern Siberia and the adjacent sector of the Kara Sea. Already, large shipments of liquefied natural gas are moving from the new port of Sabetta on the Yamal Peninsula westward to European markets and eastward to Asian markets. And major players, including international investors such as France's TotalEnergies and China's CNPC, as well as the Russian companies Novatek and Gazprom, are taking vigorous steps to accelerate the exploitation of what we have come to realize are massive reserves of natural gas in northwestern Siberia and the adjacent marine areas. Realistic assessments now treat this sector of the Arctic as an area rivaling the Middle East or the Gulf of Mexico as a source of hydrocarbons. Enhanced interest in exploiting the raw materials of the Arctic is apparent in other areas as well, including Alaska, Canada, Norway, and the Russian Far East. A fierce debate over the pros and cons of developing major deposits of rare earths and uranium became the focus of the April 2021 election in Greenland, for example, leading to the fall of the incumbent government and its replacement by a coalition led by the former opposition party.

A striking feature of this development is the strengthening of linkages between the Arctic and the outside world. China has displayed a growing interest in the Arctic as an economic frontier, exploring opportunities to invest in Arctic projects, developing the idea of the Polar Silk Road, and incorporating the Arctic into its globe-spanning Belt and Road Initiative. European and Japanese investors have become prominent stakeholders in projects focused on hardrock minerals in the Arctic as well as Arctic gas and oil. Korean firms such as DSME have taken the lead in the construction of a new class of LNG tankers that are now delivering natural gas from the Russian Arctic to markets in both Europe and Asia. Commercial shipping in the Arctic is growing accordingly. While this development has focused so far on destinational shipping associated with the transport of natural resources such as natural gas to non-Arctic markets, some major players foresee opportunities arising from the continued impact of climate change to open sea routes in the Arctic linking markets in Asia and Europe.

Of course, Arctic developments constitute only one determinant of the trajectory of both climate change and other change agents operating on a global scale. However, it is hard to miss the disconnect between the spreading crisis of adaptation in the Arctic on the one hand and the sharp rise of energy development and related industrial activities in the Arctic made possible by the impacts of climate change on the other.

## 5. A Paradoxical Future?

With all due respect to differences among the forces shaping the future of specific segments of the Arctic, therefore, there is a sense in which the future of the region as a whole will be determined by how we come to terms with a striking paradox. Nowhere on Earth are the impacts of climate change more dramatic and intensifying more rapidly than they are in the Arctic. As a result, both the biophysical systems and the human systems of the region are experiencing transformative changes, with consequences extending outward to the rest of the Earth system and triggering feedback processes affecting the Arctic. One of these changes centers on the increasing accessibility of the Arctic's massive deposits of hydrocarbons that are attractive to powerful players driven both by economic incentives and by political interests. However, extracting the Arctic's hydrocarbons and delivering them to the industrial societies of the outside world will contribute to sustaining, perhaps even increasing, emissions of the principal greenhouse gases that are the drivers of climate change. A particular concern in this regard stems from the fact that the extraction and shipment of nature gas leads to the release of methane, a dangerous short-lived climate pollutant [22]. The result is the prospect of a future dominated by a powerful feedback loop in which climate change increases the accessibility of hydrocarbons whose exploitation contributes to a continuation and perhaps an acceleration of the pace of climate change. Under the circumstances, familiar narratives such as the Arctic as a zone of peace or as a zone of conflict may become outmoded, overwhelmed by the force of this juggernaut.

Notable in this regard is an apparent inability or at least unwillingness of those who think about the future of the Arctic to come to terms with the tension between the elements of this paradox. There is a sizable community of people who focus on the impacts of climate change in the Arctic and the consequences for the Earth's climate system. There is an equally large community of people who are concerned with the economics and politics of the extraction of the Arctic's energy resources and the options for shipping these resources to outside markets. However, there is a striking disconnect between these communities. There is little overlap in the membership of the two communities; the discourses arising from the deliberations of their members have almost nothing in common, and individual members seldom engage in a focused effort to explore, much less to come to terms with, this paradox. Yet the future of the Arctic may well depend on efforts to address the tension between the increasingly severe biophysical and socioeconomic impacts of climate change on the region on the one hand and the forces driving the political economy of largescale projects focused on extracting and shipping the Arctic's massive reserves of fossil fuels on the other. What is more, the way in which we come to terms with this tension in the high latitudes of the northern hemisphere will have profound consequences not only for the future of the Arctic as a region but also for the future of the Earth system as a whole.

**Funding:** This research received no external funding.

**Institutional Review Board Statement:** Not applicable.

**Informed Consent Statement:** Not applicable.

**Data Availability Statement:** Not applicable.

**Conflicts of Interest:** The author declares no conflict of interest.

## References


1. Young, O.R. The Age of the Arctic. *Foreign Policy* **1985**, *62*, 160–179. [CrossRef]
2. Young, O.R. The Work of the Working Group on Arctic International Relations. *North. Notes* **1996**, *IV*, 1–19.
3. Gorbachev, M. Speech in Murmansk on the Occasion of the Presentation of the Order of Lenin and the Gold Star to the City of Murmansk, 1 October 1987. Available online: https://www.barentsinfo.fi/docs/Gorbachev_speech.pdf (accessed on 15 August 2021).
4. English, J. *Ice and Water: Politics, Peoples, and the Arctic Council*; Allen Lane: Toronto, ON, Canada, 2013.
5. Young, O.R. Shifting Ground: Competing Policy Narratives and the Future of the Arctic. In *The Arctic and World Order*; Spohr, K., Hamilton, D.S., Eds.; Brookings Institution Press: Washington, DC, USA, 2020; pp. 47–62.
6. Keskitalo, E.C.H. (Ed.) *The Politics of Arctic Resources: Change and Continuity in the "Old North" of Northern Europe*; Routledge: London, UK, 2019.
7. Mitrova, T. Arctic Resource Development: Economics and Politics. In *The Arctic in World Affairs: A North-Pacific Dialogue on Global-Arctic Interactions*; Corell, R., Kim, J.D., Kim, Y.H., Moe, A., VanderZwaag, D.L., Young, O.R., Eds.; KMI and EWC: Busan, Korea; Honolulu, HI, USA, 2019; pp. 205–224.
8. Nuttall, M. Greenland Matters: In the Crosscurrents of Arctic Change. In *The Arctic in World Affairs: A North Pacific Dialogue on Global-Arctic Interactions: The Arctic Moves from Periphery to Center*; Corell, R., Kim, J.D., Kim, Y.H., Moe, A., VanderZwaag, D.L., Young, O.R., Eds.; KMI and EWC: Busan, Korea; Honolulu, HI, USA, 2019; pp. 89–107.
9. Østhagen, A. The Good, the Bad, and the Ugly: Three Levels of Arctic Geopolitics. In *The Arctic and World Order*; Spohr, K., Hamilton, D.S., Eds.; Brookings Institution Press: Washington, DC, USA, 2020; pp. 357–378.
10. Huebert, R. A New Cold War in the Arctic: The Old One Never Ended. In Arctic Yearbook 2019. Available online: https://arcticyearbook.com/arctic-yearbook/2019/2019-commentaries/325-a-new-cold-war-in-the-arctic-the-old-one-never-ended (accessed on 15 August 2021).
11. Pompeo, M.R. Looking North: Sharpening America's Arctic Focus. U.S. Department of State, 6 May 2019. Available online: https://www.state.gov/looking-north-sharpening-americas-arctic-focus/ (accessed on 15 August 2021).
12. Arctic Council. Vision for the Arctic. Adopted at the 15 May 2013 Ministerial Meeting. Available online: http://hdl.handle.net/11374/287 (accessed on 15 August 2021).
13. Pincus, R. Three-Way Power Dynamics in the Arctic. *Strateg. Stud. Q.* **2020**, *14*, 40–63.
14. Barry, T.; Daviðsdóttir, B.; Einarsson, N.; Young, O.R. The Arctic Council: An agent of change? *Glob. Environ. Chang.* **2020**, *63*, 102099. [CrossRef]
15. Anderson, A. *After the Ice: Life, Death, and Geopolitics in the New Arctic*; Smithsonian Books: New York, NY, USA, 2009.
16. Arctic Monitoring and Assessment Programme. *Snow, Water, Ice and Permafrost in the Arctic (SWIPA) 2017*; AMAP: Oslo, Norway, 2017.
17. Arctic Monitoring and Assessment Programme. *Arctic Climate Change Update 2021: Key Trends and Impacts—Summary for Policy-Makers*; AMAP: Tromsø, Norway, 2021.
18. IPCC. Climate Change 2021, Summary for Policymakers. Available online: http://ipcc.ch/report/ar6/wg1 (accessed on 15 August 2021).
19. National Oceanic and Atmospheric Administration. *Arctic Report Card: Update for 2020*; National Oceanic and Atmospheric Administration: Washington, DC, USA, 2020. Available online: arctic.noaa.gov/Report-Card-2020 (accessed on 15 August 2021).
20. Arctic Council. Adaptive Actions for a Changing Arctic (AACA): Overview Reports 2017. Available online: http://hdl.dandle.net/11374/ (accessed on 15 August 2021).
21. Maddox, M. *Climate-Fragility Risk Brief: The Arctic*; Adelphi Research gGmbH: Berlin, Germany, 2021.
22. Miller, A.; Zaelke, D.; Andersen, S.O. *Resetting Our Future: Cut Super Climate Pollutants Now*; Changemakers Books: Winchester, UK; Washington, DC, USA, 2021.

*sustainability*

MDPI

*Essay*

# What Does the Arctic's Unstable Past Say about a Sustainable Future?

**Henry P. Huntington**

Ocean Conservancy, Eagle River, AK 99577, USA; hhuntington@oceanconservancy.org

**Abstract:** Visions for tomorrow's Arctic include complementary and conflicting ideas such as sustainability, security, prosperity, biodiversity, Indigenous rights, and more. Implicit in many of these views is the assumption that the right combination of policy and action will create a stable configuration producing the intended outcome for the foreseeable future. Even a cursory review of Arctic history, however, shows that economic, political, cultural, ecological, climatic, and other forms of stability are unlikely. Instead, the lessons of the past suggest that local and global factors will continue to interact to create high variability. Individual policies and institutions may help promote effective responses to that variability, but a commitment to enduring equity is necessary to foster long-term well-being for the Arctic and its peoples.

**Keywords:** arctic; sustainability; Indigenous peoples; history; economy

**Citation:** Huntington, H.P. What Does the Arctic's Unstable Past Say about a Sustainable Future? *Sustainability* **2021**, *13*, 8067. https://doi.org/10.3390/su13148067

Academic Editors: Stephanie Pfirman, Gail Fondahl, Grete K. Hovelsrud and Tero Mustonen

Received: 15 May 2021
Accepted: 17 July 2021
Published: 20 July 2021



## 1. Introduction

Sustainability is held up as a worthy and indeed necessary goal globally [1] and in the Arctic [2]. The basic idea is to live without squandering our geological, ecological, and societal reserves. The alternative is, of course, to continue to exceed the limits of what the planet can produce and replenish [3], leading eventually to gradual or catastrophic decline of human and ecosystem well-being. In this context, today's decisions about the Arctic will determine the range of tomorrow's possibilities, and the degree to which the Arctic can contribute to a sustainable global future.

A sustainable future is often envisioned as one in which poverty, inequity, and other forms of environmental and social insecurity are reduced or eliminated [1]. Implicit in this vision is a degree of stability that reduces the risk of upheavals and disruptions that can cause widespread economic hardship and spur mass migrations [4]. In considering what the Arctic's future could be, a review of history can help calibrate expectations in the range of the plausible rather than the utopian.

In this essay, I explore the history of human activity in the Arctic, focusing on livelihoods and economics while recognizing that cultural, spiritual, political, technological, environmental, and other factors matter as well. I write as an American scientist of European descent who has worked in the Arctic for all of my professional career. I cannot and do not claim to speak from an Arctic, much less an Arctic Indigenous, perspective.

The past is not a perfect analogue for the future, and I do not suggest that swings between prosperity and hardship are inevitable. Rather, an understanding of the past can help identify what contributes to, and what undermines, stability as seen from different perspectives [5]. Even in a time of rapid climate change [6], the Arctic will continue to exist in one form or another, though that form may be vastly different from the Arctic of today. But humans will still make decisions about the region, and those decisions will affect what happens to Arctic ecosystems and cultures. An ice-free Arctic could still be a relatively productive area with cultures that have adapted to new conditions, or it could be an industrial or military disaster area. What humans have done elsewhere in the world and in the Arctic to date provide insights into the consequences of the choices that may be made. The lessons of history may be useful in shaping the future of the Arctic, if Arctic and global societies are willing to heed them.

## 2. Early Human Presence in the Arctic

Humans have lived in the Arctic for tens of thousands of years [7]. Archeologists have identified several major periods and geographies defined by the lands and waters used, the tools developed, and the species hunted, fished, and gathered [8]. This record includes examples of remarkable continuity, such as the continuous occupancy of Point Hope, Alaska, for thousands of years [9]. The record also includes evidence of great disruption, such as the early settlement and then abandonment of Greenland prior to the Common Era [7], as well as many smaller-scale examples of hardship [5,10].

To varying degrees, the early peoples of the Arctic were self-reliant [11]. Trade provided the relatively few materials not available locally [12]. Their lives and livelihoods to some degree met modern definitions of sustainability, but far below today's expectations for material comfort and life expectancy. Environmental variability caused instability at small and large scales, creating a much greater risk of hardship and premature death than would be acceptable to most people now. Many of the early peoples doubtless found their lives satisfying and meaningful, and their deep-rooted respect for nature continues in Indigenous worldviews [13] and offers much in the way of values and experience to guide the future [14], but their material ways of life are not a particularly useful model for tomorrow's Arctic.

## 3. Contact and Colonization

Trade goods, traders, and explorers gradually made their way to the Arctic. Early interactions through the second millennium of the Common Era were likely an extension of existing trade routes and forms of contact. Territorial expansion also played a role, for example in the Norse colonization of Greenland and subsequent, if temporary, expansion into what is now Canada [15]. Arctic peoples themselves were hardly stationary, either. As one example, the Thule Culture of modern-day Inuit expanded rapidly from the Bering Strait region across Arctic North America to Greenland in the first half of the second millennium [16]. At smaller scales, too, migration, resettlement, and assimilation took place, for instance in the 19th century absorption of Koyukon Athabascan peoples in the upper Kobuk River Valley of Alaska into Iñupiaq ways of life and language [17].

Nonetheless, the influence and then visits and then settlements from southern lands brought about lasting change throughout the Arctic. A market for trade goods, including furs and ivory, led to the expansion of trade routes in both volume and geographical extent [18]. In Tsarist Russia, the expansion of empire also included a demand for tribute from remote peoples, typically in the form of furs and other local products [19]. Initially, such trade provided an outlet for surplus production of traditional goods. As demand and thus profits increased, however, traders became more directly involved, establishing annual trade fairs and permanent trading posts [18].

The quest for whale oil and baleen brought new waves of trade and newcomers to the Arctic [20,21]. Whaling crews typically did their own hunting and were also willing to trade with Indigenous communities. Their trade, however, was simply an opportunistic addition to their quest for whales, and the whalers thus had little stake in the long-term well-being of Arctic communities, in at least partial contrast to those who established more permanent trading posts [18].

Whaling at first was carried out from ships that came and left. Later, the whalers established shore-based stations in the Arctic, creating opportunities for Indigenous whalers to join in the commercial hunt, or at least the commercial sale of whale products [21]. The demand for whale products, however, was not sustainable, given the history of reducing and even eradicating whale populations around the world. The commercial bowhead whaling era in the Arctic was no different, with whale populations reduced to a fraction of their original size [22]. Whaling ended not because of concerns about sustainability, but because the market for whale products collapsed when alternatives were developed [21].

The trade goods reaching the Arctic through the fur and whale trades, including metal products, were highly sought after, and presumably helped increase material quality of life, even if at least some of the commercial harvests were unsustainable [21]. But beneficial goods were not the only introductions. Alcohol and diseases also came north [18], causing social disruption and death. Arctic peoples switched from living largely independently to many working as producers of raw materials for a global economy, disrupting traditional patterns and norms, and creating, among other things, economic inequality [23]. The Arctic fox boom in the early 20th century is one example, creating a network of trading posts, which exploited trappers to varying degrees [24], and led to families dispersing from communities across the landscape to find their own trapping grounds.

Trade in animal products produced locally was in theory sustainable and provided Arctic residents with opportunities to earn cash and trade goods. The introductions of alcohol and disease were perhaps not inevitable, but were still an all-too-predictable corollary to trade and settlement. In practice, whaling and trapping were not always carried out sustainably, to the detriment of many Arctic communities. St. Lawrence Island, in the northern Bering Sea, is one of the more extreme examples [21]. Whalers decimated the region's walrus population for ivory to supplement their profits, leaving the islanders with fewer sources of food. Alcohol and disease added to the disruption, leading to starvation in the late 1870s, after which only one of the island's communities remained.

In addition to the perils of unsustainable harvest and cultural disruption, the era of furs and whales also demonstrates the risks of dependence on global commodities markets. Bowhead whaling was profitable until baleen was no longer valuable. Arctic fox furs were highly desirable until fashions changed. Even if the take of animals had been ecologically sustainable, there was no guarantee of a lasting economic foundation. The Arctic economy was at the mercy of distant market forces, on which the Arctic had little or no influence. Colonization further eroded local control by imposing the rules and structures of distant governments on Arctic communities.

## 4. Post-World War II

By the mid-20th century, the whaling and trapping economies had disappeared or become too small to provide livelihoods for many people. Around the Arctic, national governments were taking a larger role, bringing cash, jobs, and social services into remote communities [25], which provided inroads for further social, economic, and educational assimilation [26] accompanied by cultural disruption. The Cold War led to the construction of radar and other remote installations, which provided employment [27] for those willing to adhere to the schedules of the wage economy. Even if much food was still produced locally, government support and subsidies were becoming ever more important.

The development of mineral and petroleum resources also increased rapidly. The Gold Rush in the Yukon and Alaska starting in the late 1800s was an unusual frenzy in some ways, and in other ways a precursor of what was to come. Oil had been produced starting in the 1920s in Norman Wells, along Canada's Mackenzie River in the Northwest Territories [28], and geologists had prospected oil seeps on Alaska's North Slope since at least the 1910s [29]. Svalbard's coal reserves were being mined by Norwegians and Soviets [30]. The Arctic was again seen as a land of extractive opportunity, drawing people north in various ways, though rarely with much thought about the well-being of, and often at great cost to, those already living there.

The growth of extractive industrial activities has inevitably led to conflicts with environmental sustainability, at least at local scales. Roads may interfere with migrations of caribou and reindeer [31]. Shipping and offshore oil and gas activities can disrupt marine mammal migrations and Indigenous hunting [32]. Transport of liquid natural gas from Russia's Yamal Peninsula and iron ore from Canada's Baffin Island now take place year-round [33], requiring ice-breaking vessels that create another form of disruption for hunters and marine mammals [34]. China's interest in the "Polar Silk Road" is but one indicator of expanding international interest in the Arctic [35]. None of these is necessarily unsustainable, but each activity threatens another incremental reduction in ecological health and productivity. Cumulatively, the course is all too well known from other parts of the world, as lowered expectations for what constitutes a healthy and productive ecosystem lead to altered local cultures and impoverished ecosystems [36].

Commercial fisheries, too, expanded northwards [37]. Salmon fisheries had played a major role in Alaska's bid to become a state, as it sought more control over management of fishes and fishing [38]. Demand for fish products and the technology to catch and process fish combined to make possible massive fisheries in the North Atlantic and North Pacific [39]. By the late 20th century, shrimp were a major source of Greenland's export earnings [37], and the Barents and Bering Seas fisheries were both lucrative and widely regarded as well managed [39].

While fisheries may seem a strong candidate for the foundation of a sustainable economy, they too depend on global markets and on continued willingness to accept management restrictions today to protect tomorrow's fish catches. Competition from salmon farming has greatly reduced the price for wild-caught salmon [40]. Climate change and ocean acidification may reduce the stocks of commercially valuable fish [41]. Conflicts about the allocation of fish catches or the role of by-catch may also disrupt support for current fisheries' management approaches [42]. If fisheries expand still farther north, new conflicts may arise domestically and internationally [43].

Politically, recent decades have seen expanded recognition of Indigenous rights, including land and resource ownership and access rights, creation of Indigenous corporations intended to create economic opportunity, and devolution of some powers to new and existing local and regional governments [44]. These developments are far from uniform across the Arctic, with most activity seen in North America and Greenland. Whether the creation of corporations and government structures is progress toward renewed levels of autonomy or a new form of assimilation is another question [45].

Today's Arctic may seem in some ways a reasonable balance of resource extraction, traditional cultures, fisheries, government support, and other economic pillars. On the other hand, today's Arctic is the product of happenstance, countless small decisions taken separately that together have created the various systems found in different countries. These systems each have their strengths and their weaknesses, but none provides a reliable basis for a stable and sustainable economy and society. Some major financial companies are reducing their investments in fossil fuels, which could reduce new Arctic oil and gas production in some regions such as Alaska [46]. Climate change is disrupting ecosystems, threatening fisheries [41] and causing more and more wildfires [6]. Arctic communities increasingly depend on government spending, including subsidies [47], but budgetary strain may undermine continued political willingness to support those expenditures. Today's decisions will shape tomorrow's Arctic, but where and how are those decisions being made?

## 5. The Future

Arctic regions have long aspired to create viable economies. They have also aspired to political control, up to and including independence [48]. Political independence is incompatible with economic dependence, and so economic development is often seen as a necessary ingredient in self-determination. Coupled with a desire to reduce the relative poverty common around the Arctic [47], there is thus a strong incentive to develop now and to worry about sustainability later.

In different regions at different times, the Arctic has enjoyed periods of economic prosperity. Not since before contact and colonization, however, has the Arctic enjoyed economic and political independence in the sense of having the ability to control its own destiny. To the contrary, the modern Arctic economy is primarily a reaction to global market demands and the willingness of southern governments to spend money in the North.

Then again, few if any countries anywhere have enjoyed a stable economy over the past century, if "stable" means an economy that has followed the same basic pathway without interruption. Economies change and evolve, and stability is achieved by diversity and innovation rather than by stasis. One can look at the history of the Arctic as one of flexibility, taking advantage of the opportunities of the moment rather than chasing an impossible dream of stable economic self-determination. Flexibility comes at a price, however, including periods of hardship as booms give way to busts. A long-term outlook could help counterbalance the appeal of the quick payoff and support investments that promote economic diversification and a greater degree of self-sufficiency.

If Arctic history suggests instability is the norm, then perhaps shaping tomorrow's Arctic should be reframed as creating the conditions that will allow tomorrow's Arctic the greatest scope for shaping itself. This idea is compatible with the concept of sustainability, insofar as today's choices should not limit the choices available tomorrow [49]. Resources used or damaged today are not available tomorrow, and thus society should be cautious in its choices and actions. At the same time, the more rigid the institutional and other arrangements that are made today, the scope of choices available tomorrow will be smaller, and adjustments will be harder to make.

Today there are competing visions for the Arctic [50], ranging from resource storehouse to Indigenous homeland to geostrategic theater and beyond. As long as the Arctic is sparsely populated and subject to modest levels of human activity, these visions can coexist to at least some degree. That possibility is unlikely to last. Sooner or later, those with an interest in the Arctic will have to determine who should make choices, and those making the choices will have to decide which pathways to follow and which to abandon. The current path is one of incremental and reactive compromise, which leads not to sustainability, but to an inevitable decline.

An alternative path is to focus on basic principles, specifically a commitment to living equitably within the social and ecological bounds of the Arctic, from which specific policies and actions can be developed and adjusted as needed. The Arctic today is still the home of intact ecosystems and vibrant cultures not through carefully planning, but because it lies at the margins of the human presence on Earth. Geographical chance will not suffice to protect the Arctic of tomorrow. An economy of reaction will not create a sustainable Arctic tomorrow. Decisions made elsewhere for other reasons will turn Arctic well-being into collateral damage as external ambitions take precedence. The only pathway to a sustainable Arctic future is to commit to the long-term well-being of Arctic cultures, landscapes, and waters as a priority, with other interests to follow only insofar as they are compatible with Arctic well-being.

**Funding:** This research received no external funding.

**Conflicts of Interest:** The author declares no conflict of interest.

## References

1. United Nations. *Sustainable Development Goals Report*; United Nations Publications: New York, NY, USA, 2019.
2. Petrov, A.N.; Vlasova, T. Towards an Arctic sustainability monitoring framework. *Sustainability* **2021**, *13*, 4800. [CrossRef]
3. Rockström, J.; Steffen, W.; Noone, K.; Persson, Å.; Chapin, F.S.; Lambin, E.F.; Lenton, T.M.; Scheffer, M.; Folke, C.; Schellnhuber, H.J.; et al. A safe operating space for humanity. *Nature* **2009**, *461*, 472–475. [CrossRef] [PubMed]
4. Oliver-Smith, A.; Shen, X. (Eds.) *Linking Environmental Change, Migration and Social Vulnerability*; UNU Institute for Environment and Human Security: Bonn, Germany, 2009.
5. Krupnik, I.; Crowell, A.L. (Eds.) *Arctic Crashes*; Smithsonian Scholarly Press: Washington, DC, USA, 2020.
6. Thoman, R.L.; Richter-Menge, J.; Druckenmiller, M.L. (Eds.) *Arctic Report Card 2020*; National Oceanic and Atmospheric Administration: Silver Spring, MD, USA, 2020.
7. Pitulko, V.V.; Tikhonov, A.N.; Pavlova, E.Y.; Nikolskiy, P.A.; Kuper, K.E.; Polozov, R.N. Early human presence in the Arctic: Evidence from 45,000-year-old mammoth remains. *Science* **2016**, *351*, 260–263. [CrossRef] [PubMed]
8. Friesen, T.M.; Mason, O.K. (Eds.) *The Oxford Handbook of the Prehistoric Arctic*; Oxford University Press: Oxford, UK, 2016.
9. Rainey, F.G. Whale hunters of Tigara. *Am. Mus. Nat. Hist.* **1947**, *41*, 231–283.

10. Mason, O.K.; Gerlach, S.C. Chukchi hot spots, paleo-polynyas, and caribou crashes: Climatic and ecological dimensions of North Alaska prehistory. *Arct. Anthropol.* **1995**, *32*, 101–130.
11. Krupnik, I. *Arctic Adaptations: Native Whalers and Reindeer Herders of Northern Eurasia*; University Press of New England: Hanover, NH, USA, 1993.
12. Fitzhugh, W.W.; Crowell, A. *Crossroads of Continents*; Smithsonian: Washington, DC, USA, 1988.
13. Petersen, H.C. *Kalaalit Ilisimasaat/Local Knowledge*; IPI Press: Hanover, NH, USA, 2010.
14. Zdor, E.; Huntington, H.P. Tradition connects the past, present, and future of the Arctic Ocean. In *Whither the Arctic Ocean?* Wassmann, P., Ed.; BBVA Foundation: Bilbao, Spain, 2021; pp. 195–203.
15. Thorsson, Ö. (Ed.) *Sagas of the Icelanders*; Penguin: New York, NY, USA, 2000.
16. McGhee, R. *The Last Imaginary Place: A Human History of the Arctic World*; Oxford University Press: Oxford, UK, 2005.
17. Burch, E.S., Jr. *The Iñupiaq Nations of Northwest Alaska*; University of Alaska Press: Fairbanks, AK, USA, 1998.
18. Bockstoce, J.R. *Fur and Frontiers in the Far North*; Yale University Press: New Haven, CT, USA, 2010.
19. Slezkine, Y. *Arctic Mirrors*; Cornell University Press: Ithaca, NY, USA, 1994.
20. Ross, W.G. *Arctic Whalers, Icy Seas: Narratives of the Davis Strait Whale Fishery*; Irwin: Toronto, ON, Canada, 1985.
21. Bockstoce, J.R. *Whales, Ice & Men*; University of Washington: Seattle, DC, USA, 1986.
22. Givens, G.H.; Heide-Jørgensen, M.P. Abundance. In *The Bowhead Whale*; George, J.C., Thewissen, J.G.M., Eds.; Academic Press: London, UK, 2020; pp. 77–86.
23. Blackman, M.B. *Sadie Brower Neakok: An Iñupiaq Woman*; University of Washington: Seattle, DC, USA, 1989.
24. Bockstoce, J.R. *White Fox and Icy Seas in the Western Arctic*; Yale University Press: New Haven, CT, USA, 2018.
25. Spencer, R.F. *The North Alaskan Eskimo: A Study in Ecology and Society*; U.S. Government Printing Office: Washington, DC, USA, 1959.
26. Okakok, L. Serving the purpose of education. *Harv. Educ. Rev.* **1989**, *59*, 405–422. [CrossRef]
27. Chasen, R.E. Distant early warning systems in the North American Arctic. *Polar Rec.* **1967**, *13*, 595–596. [CrossRef]
28. AMAP. *Assessment 2007: Oil and Gas Activities in the Arctic—Effects and Potential Effects*; Arctic Monitoring and Assessment Programme: Oslo, Norway, 2010.
29. de Leffingwell, E.K. *The Canning River Region, Alaska*; Government Printing Office: Washington, DC, USA, 1919.
30. Arlov, T. *Svalbards Historie*; Tapir: Trondheim, Norway, 2003. (In Norwegian)
31. Wilson, R.; Parrett, L.; Joly, K.; Dau, J. Effects of roads on individual caribou movements during migration. *Biol. Conserv.* **2016**, *195*, 2–8. [CrossRef]
32. Huntington, H.P.; Daniel, R.; Hartsig, A.; Harun, K.; Heiman, M.; Meehan, R.; Noongwook, G.; Pearson, L.; Prior-Parks, M.; Robards, M. Vessels, risks, and rules: Planning for safe shipping in Bering Strait. *Mar. Policy* **2015**, *51*, 119–127. [CrossRef]
33. Vetter, D. Russian Gas Tanker's Arctic Ice Voyage Branded "Irony of Our Time." *Forbes*, 24 February 2021. Available online: https://www.forbes.com/sites/davidrvetter/2021/02/24/russian-gas-tankers-arctic-ice-voyage-branded-irony-of-our-time (accessed on 19 July 2021).
34. Erbe, C.; Marley, S.; Schoeman, R.; Smith, J.; Trigg, L.; Embling, C. The effects of ship noise on marine mammals—A review. *Front. Mar. Sci.* **2019**, *6*, 606. [CrossRef]
35. Su, P.; Huntington, H.P. Using critical geopolitical discourse to examine China's engagement in Arctic affairs. *Territ. Politics Gov.* **2021**. [CrossRef]
36. Pauly, D. Anecdotes and the shifting baseline syndrome of fisheries. *Trends Ecol. Evol.* **1995**, *10*, 430. [CrossRef]
37. Hamilton, L.C.; Brown, B.C.; Rasmussen, R.O. West Greenland's cod-to-shrimp transition: Local dimensions of climatic change. *Arctic* **2003**, *56*, 271–282. [CrossRef]
38. Mackovjak, J. *Alaska's Salmon Traps*; Cross Sound Innovations: Gustavus, AK, USA, 2013.
39. Vilhjalmsson, H.; Hoel, A.H. Fisheries and aquaculture. In *Arctic Climate Impact Assessment*; Cambridge University Press: Cambridge, UK, 2005; pp. 691–780.
40. Knapp, G.; Roheim, C.; Anderson, J. *The Great Salmon Run: Competition between Wild and Farmed Salmon*; Traffic North America: Washington, DC, USA, 2007.
41. AMAP. *AMAP Assessment 2018: Arctic Ocean Acidification*; Arctic Monitoring and Assessment Programme: Tromsø, Norway, 2018.
42. Smiley, S. *Commercial and Subsistence Harvesters Speak out against Trawler Bycatch of Chinook Salmon*; KSTK: Wrangell, AK, USA, 2021.
43. Zou, L.; Huntington, H.P. Implications of the Convention on the Conservation and Management of Pollock Resources in the Central Bering Sea for the management of fisheries in the central Arctic Ocean. *Mar. Policy* **2017**, *88*, 132–138. [CrossRef]
44. Loukacheva, N. *The Arctic Promise: Legal and Political Autonomy of Greenland and Nunavut*; University of Toronto Press: Toronto, ON, Canada, 2007.
45. Mitchell, D.C. *Sold American: The Story of Alaska Natives and Their Land, 1867–1959*; University of Alaska Press: Fairbanks, AK, USA, 2003.
46. Herz, N. *Anxiety Creeps into Oil-Dependent Alaska as Banks Step back from Arctic Investment*; Alaska Public Media: Anchorage, AK, USA, 2020.
47. Glomsrød, S.; Duhaime, G.; Aslaksen, I. *The Economy of the North 2015*; Statistics Norway: Oslo, Norway, 2017.

48. Olsen, I.H. Greenland, the Arctic, and the issue of representation: What is the Arctic? Who has a Say? In *The Arctic and World Order*; Spohr, K., Hamilton, D.S., Eds.; Johns Hopkins School of Advanced International Studies: Washington, DC, USA, 2020; pp. 77–95.
49. World Commission on Environment and Development. *Our Common Future*; Oxford University Press: Oxford, UK, 1987.
50. Young, O. Shifting ground: Competing policy narratives and the future of the Arctic. In *The Arctic and World Order*; Spohr, K., Hamilton, D.S., Eds.; Johns Hopkins School of Advanced International Studies: Washington, DC, USA, 2020; pp. 47–62.

*sustainability*

MDPI

*Article*

# Gender Equality for a Thriving, Sustainable Arctic

**Embla Eir Oddsdóttir [1,*,†], Hjalti Ómar Ágústsson [2], Eva-Maria Svensson [3], Gunhild Hoogensen Gjørv [4], Sarah Seabrook Kendall [5], Malgorzata (Gosia) Smieszek [6,‡], Tahnee Prior [7,‡], Erika Hayfield [8], Karla Jessen Williamson [9], Marya Rozanova-Smith [10,11], Andrey Petrov [12] and Varvara Korkina Williams [13]**

1 Icelandic Arctic Cooperation Network, 600 Akureyri, Iceland
2 Gender Equality in the Arctic (GEA), Icelandic Arctic Cooperation Network, 600 Akureyri, Iceland; hjalti.o.agustsson@jafnretti.is
3 Department of Law, University of Gothenburg, 405 30 Göteborg, Sweden; eva-maria.svensson@law.gu.se
4 Peace Research Institute of Oslo (PRIO), NO-0186 Oslo, Norway; gunhild.hoogensen.gjorv@uit.no
5 Environment and Natural Resources, School of Engineering and Natural Sciences, University of Iceland, 102 Reykjavík, Iceland; ssk19@hi.is
6 UiT, The Arctic University of Norway, 9019 Tromsø, Norway; malgorzata.smieszek@uit.no
7 Schulich School of Law, Dalhousie University, Halifax, NS B3H 4R2, Canada; tahnee.prior@gmail.com
8 Department of History and Social Sciences, University of the Faroe Islands, 100 Tórshavn, Faroe Islands; erikah@setur.fo
9 Educational Foundations, University of Saskatchewan, Saskatoon, SK S7N 5A2, Canada; karla.jwilliamson@usask.ca
10 Elliott School of International Affairs, The George Washington University, Washington, DC 20052, USA; marya_rozanova@gwu.edu
11 Department of Social and Human sciences, Russian State Hydrometeorological University, 195196 St. Petersburg, Russia
12 ARCTICenter, University of Northern Iowa, Cedar Falls, IA 50614, USA; andrey.petrov@uni.edu
13 Indigenous Independent Scholar, Cedar Falls, IA 50613, USA; korkivaa@uni.edu
* Correspondence: embla@arcticiceland.is; Tel.: +354-864-5979
† Further information on co-contributing authors see Pan-Arctic Report—Gender Equality in the Arctic (arcticgenderequality.network).
‡ Further information on co-contributing authors see "Women of the Arctic", 96400 Rovaniemi, Finland; Plan A. (genderisnotplanb.com).

**Citation:** Oddsdóttir, E.E.; Ágústsson, H.Ó.; Svensson, E.-M.; Hoogensen Gjørv, G.; Kendall, S.S.; Smieszek, M.; Prior, T.; Hayfield, E.; Williamson, K.J.; Rozanova-Smith, M.; et al. Gender Equality for a Thriving, Sustainable Arctic. *Sustainability* **2021**, *13*, 10825. https://doi.org/10.3390/su131910825

Academic Editors: Stephanie Pfirman, Gail Fondahl, Grete K. Hovelsrud and Tero Mustone

Received: 20 July 2021
Accepted: 8 September 2021
Published: 29 September 2021



**Abstract:** On 21 May 2021, a milestone Pan-Arctic Report: Gender Equality in the Arctic was published in tandem with the Arctic Council's Ministerial Meeting held in Reykjavík, 19–20 May 2021. This article provides a brief review of the report and its major findings across six chapters that address key themes concerning gender equality in the Arctic: Law and Governance, Security, Gender and Environment, Migration and Mobility, Indigeneity, Gender, Violence, Reconciliation and Empowerment and Fate Control. A major conclusion of the report is that accessible, comparable, gender-disaggregated, and Arctic -specific data is severely lacking. Further, all chapters highlight the importance of gender-based analysis and gender mainstreaming in all decision-making processes at national and regional levels. The varying roles that gender—and its intersections with existing inequalities—plays in mediating the impacts of climate change and other socioeconomic transformations are also discussed throughout the report. The Arctic Council is identified as the main driver for implementing recommendations that were provided and discussed at the Council's Ministerial Meeting and in the Reykjavík Declaration 2021, where the eight ministers of Arctic states "Emphasize[s] the importance of gender equality and respect for diversity for sustainable development in the Arctic ... encourage[s] the mainstreaming of gender-based analysis in the work of the Arctic Council and call[s] for further action to advance gender equality in the Arctic". This report and its policy relevant highlights, address these priorities and serve as a knowledge base for promoting gender equality and non-discrimination in the Arctic.

**Keywords:** gender; equality; empowerment; engagement; Indigenous; data; mainstreaming; diversity and inclusion; migration; mobility; security; intersectionality; youth; Arctic Council

## 1. Introduction

There is an inherent gender imbalance in the on-going policy discussions and decision-making about the Arctic, as women are underrepresented in Arctic governing bodies, administration, business, and science. Although generalisations should be avoided, given the cultural and social diversity of the North, economic development throughout much of the region affects men and women differently. It is a cause for concern that future development in the North, for the most part, focuses on traditional male sectors such as oil and gas, mining, shipping, and tertiary industrial development. Moreover, disproportionate out-migration of adult females characterises many rural areas of the Arctic, primarily as a result of diminishing employment and due to a lack of educational opportunities for women. The resulting sex-ratio imbalance negatively affects the resilience and future sustainable development of Arctic communities, some of which are seeing very high death rates for males, especially from external causes.

The geopolitical and global economic significance of the Arctic region is growing, inter alia because of climate change. The Arctic Council (AC) and its Sustainable Development Working Group (SDWG) have emphasised gender equality in previous projects and initiatives and the importance of issues of gender and diversity is increasingly evident. Some examples of previous work and valuable input in this field with gender issues in the Arctic as their focal point, include the 2002 Conference in Inari, Finland, which focused on themes of women and work, gender, the self-determination of Indigenous peoples, and violence against women [1]. Furthermore, the first edition of the Arctic Human Development Report, published in 2004, featured a specific chapter on gender [2]. In the second edition, published in 2014, a different approach was taken, and gender issues were not addressed in a specific chapter but rather mainstreamed into individual chapters to various degrees [3].

Given the gender imbalance in Arctic policy discussions, and the pace of the changes transforming Arctic societies, further information is needed about the various impacts of gender. The milestone Pan-Arctic Report: Gender Equality in the Arctic [4], published by the Icelandic Ministry for Foreign Affairs and the Icelandic Arctic Cooperation Network as a product of the Icelandic Chairmanship in the AC from 2019–2021 [5] on 21 May 2021, addresses this gender imbalance by providing a comprehensive overview of issues related to gender equality in the Arctic. The report was published in tandem with the AC Ministerial Meeting held in Reykjavík 19–20 May 2021. Gender equality has been one of Iceland's priorities during its AC Chairmanship 2019–2021, under the theme of People and Communities. The report is a part of an international collaborative project under the AC's SDWG on Gender Equality in the Arctic [6] (GEA), dating back to 2013. Leads and co-leads include Iceland, Sweden, Finland, Canada, the United States, the Saami Council, the Aleut International Association, and a host of other additional partners. Initiated by the Ministry for Foreign Affairs in Iceland, in collaboration with the Directorate for Equality in Iceland and the Stefansson Arctic Institute the project has been led by the Icelandic Arctic Cooperation Network (IACN) under the leadership of Embla Eir Oddsdóttir, Director of IACN, and the GEA Team in Akureyri, Iceland. This article provides an overview of the GEA project phases to date as well as a review of the themes of the report, with major findings.

## 2. Background and Progression of the GEA Report Process

The purpose and objective of the GEA project has been to raise visibility and understanding of the importance of gender issues in the Arctic, to identify priorities and concrete strategies for increased diversity and gender balance in policy- and decision-making processes, and to provide information to facilitate sustainable policy-making for the future.

Phase I of the GEA project (GEA I) was an international conference Gender Equality in the Arctic—Current Realities, Future Challenges, which took place in Akureyri, Iceland in October 2014. It resulted in an eponymous conference report published in 2015 by the Icelandic Ministry for Foreign Affairs [7]. The main objective of the conference was to promote an extensive, policy relevant dialogue on gender equality putting current realities and future challenges into context, in light of climate and environmental changes as well as economic and social developments in the region. Another goal was to raise decision-makers' awareness of the situation of women and men in the Arctic and to strengthen cooperation among different people working with gender issues. The conference was organized into seven different themes, and in the conclusions of the conference, the participants identified issues that are relevant to the on-going discussions on gender equality in the Arctic. The conference in Akureyri brought together government representatives, policymakers, academics, and a wide range of stakeholders including members of the business community, resource managers and users, community leaders, and NGO representatives. Special emphasis was placed on Indigenous representation at the conference.

Following the success of GEA I, Phase II (GEA II) was launched in 2017. GEA II involved the building of a network of experts in the field and the creation of a website for the purpose of: promoting and expanding the dialogue on gender equality in the Arctic, providing a formal network of groups and experts interested in the topic, encouraging cooperation with and amongst existing networks, and providing an online platform for material and events relevant to Arctic Gender Equality.

Phase III of the Gender Equality in the Arctic project (GEA III) was launched in 2019 and includes a regular newsletter—the GEA Times [8]—in addition to various other networking and dissemination activities, through online media and events, expanding its database of gender related material. However, the focus of GEA III has primarily been a pan-Arctic report addressing the gendered dimensions of selected themes and gauging the current state of affairs to better understand how gender affects, and is affected by, policy- and decision-making processes within the Arctic. The report was developed by 10 lead authors and approximately 80 contributing authors from 15 states, including all eight Arctic States. The report's engagement process was a vital component in knowledge generation and development of the report and significant efforts were made to ensure inclusion and transparency during the process by actively soliciting feedback from peers and interested parties. A special emphasis was on the partnership with AC Permanent Participants and other Indigenous representatives, both through our Partners, the Editorial Committee, the Youth Advisory Group, the SDWG Social, Economic and Cultural Expert Group (SECEG), and through contributions to chapters from Indigenous experts, including from the Saami Council, the Aleut International Association, and the Arctic Athabascan Council, as well as the Paktuutit Inuit Women of Canada.

The 2021 report is intended to inform policy and provide the AC, policymakers, researchers, and stakeholders with a departure point from which to foster further dialogue and actions on gender issues in the Arctic. Each chapter provides a list of policy relevant highlights, almost 70 for all chapters, with suggestions for actions and/or opportunities for further research on the topics. In addition, the report provides recommendations for the Arctic Council based on the main conclusions of the report as a whole. The report's recommendations were discussed at the AC Ministerial Meeting in Reykjavík on 20 May 2021 and were included in the Ministerial Meeting declaration (the Reykjavík Declaration 2021) the AC "Emphasize[s] the importance of gender equality and respect for diversity for sustainable development in the Arctic and welcome[s] the Pan-Arctic Report, Gender Equality in the Arctic, Phase 3, encourage[s] the mainstreaming of gender-based analysis in the work of the Arctic Council and call[s] for further action to advance gender equality in the Arctic" [9] (p. 9). Moreover, chapter 4 on Sustainable Social Development in the Council's Strategic Plan 2021 to 2030, also accepted at the Ministerial Meeting in Reykjavík, the Council resolves to "promote gender equality and non-discrimination in the Arctic with the aim of contributing to sustainability and balanced participation in leadership and decision-making both in the public and private sectors" [10] (p. 5).

## 3. Methods and Approach

The GEA report process engages an international, multidisciplinary, multi-stakeholder and mixed method approach, including both ongoing state-of-the-art quantitative and qualitative research in addition to literature review and discourse analysis. Themes for the report were decided in a collaborative process that included diverse research communities, SDWG partners in the project, AC permanent participants and other Indigenous experts, youth representatives, and members of the SDWG's SECEG. The consultative process was instrumental in developing priority themes for the report as well as providing regular feedback on the process and development of chapters. Following the consultation process six themes emerged that became the focus of the report with chapters led by a total of 10 lead authors, with additional contributing authors ranging—depending on chapter themes—from six to 31 per chapter, all in all close to 80 contributing authors

Each lead author had significant freedom in their methodology for developing the chapters, thus some variations exist in their approaches. All chapters include contributions from multiple contributing authors which in some cases were weaved into the lead author's writing, whereas in other cases presented in a more clearly separated fashion, such as through case studies.

Once initial drafts were ready, feedback sessions for each chapter were held in October and November 2020. These were public online sessions in which lead authors—along with colleagues, contributors, and Indigenous and youth representatives—presented and discussed each chapter theme. All feedback sessions were recorded, transcribed, and sent to lead authors for review and integration into chapters.

Review and feedback on draft chapters was solicited from the project partners, the Editorial Committee, the Youth Advisory Group (YAG), all contributors, and other additional and relevant experts. Specific YAG Reviewers were asked to review each chapter. Finally, a formal peer review took place through external reviewers for each chapter. The Law and Governance chapter, as well as legal sections in the chapters on Gender and Environment (environmental law) and Empowerment and Fate Control, were reviewed by legal scholars and experts. Finally, the draft report went through the formal review process of the Arctic Council SDWG and its Heads of Delegations.

## 4. Results

On the basis of the multidisciplinary and multi-methods approach, six overarching themes were identified as central issue areas for gender equality in the Arctic. The first theme addressed law and governance, examining the formal obligations regarding gender equality in the public governance of the Arctic region, as expressed in political and legal documents, including special consideration of Indigenous Peoples. The second theme is security and, in particular, human security, focusing on the impacts of inequalities in the Arctic that are exacerbated by climate change, thereby identifying trends in insecurities from the individual and community levels to the state. Gender and the environment informs the third theme, providing an overview of the gendered dimensions of issues connected to a broadly understood Arctic environment, including the climate, oceans, land, biodiversity, natural resources, waste, and pollution. The fourth theme is migration and mobility, examining how migration and mobility issues in the Arctic are constructed through gender and why an understanding of migration and mobility requires a gendered approach. A fifth, crucial theme is the combination of Indigeneity, gender, violence, and reconciliation, demanding a mapping of the complex relations amongst violence; gender; social, economic, political, and legal systems; human health and well-being; culture; identities. Lastly, the overarching theme of empowerment and fate control identifies concrete strategies for political, economic, and civic gender empowerment in order to facilitate sustainable policy-making for the Arctic.

Each theme is represented as a chapter in the GEA III report. The report opens with the chapter on **Law and Governance**, led by Eva-Maria Svensson. It explores the political and legal commitments for which public governing bodies are accountable, how these bodies express their ambitions regarding gender equality in the Arctic, and how the commitments are fulfilled.

The political and legal obligations for the accountable subjects regarding equal rights between men and women, and gender equality, are extensive. The Arctic States are committed to following international, as well as corresponding regional, federal, national, and territorial, legal instruments and political agendas. The international legal instruments (treaties, conventions, or covenants) legally bind those states that choose to accept the obligations contained in them by becoming a party. States determine, for themselves, which instruments they will accept according to the principle of state sovereignty [11]. However, states are expected to ensure that their international obligations are upheld, and the degree to which they meet their international obligations is explored through comments made by monitoring bodies of international legal instruments.

Arctic States and Indigenous Peoples cooperate in several intergovernmental bodies, and the Arctic Council (the Council) is the leading intergovernmental forum for cooperation in the Arctic [12–14]. The Council has been criticised for not adequately prioritising gender equality, both internally and among Arctic States [15,16]. However, since 2013 the Council has been one of many supporters and cooperative partners in the Gender Equality in the Arctic Project, and it can be an important promoter of gender equality. Arctic States, Indigenous Peoples' Organisations, non-Arctic states, and a variety of international organisations issue Arctic policy documents, some of which include a focus on gender equality. The policies of Arctic States are of analytical relevance as they are representations of governments responsible for international, federal, regional, and national legal and political obligations, and included in the analysis are Arctic policies issued by Arctic States and two of the Indigenous Peoples' organisations (IPO) that are Permanent Participants in the Council.

Gender equality is a primary concern for the global community and all Arctic States, except for the U.S., have ratified the Convention on the Elimination of all forms of Discrimination Against Women (CEDAW, adopted 1979), which implies that they are obligated to ensure full equality of women before the law, protection against discrimination in the public and the private spheres, improve the de facto position of women, and address gender-based stereotypes that uphold unequal gender relations.

What emerges from the analysis in the chapter on Law and Governance is that governance in the Arctic does not prioritize gender equality and, more generally, that the goal of gender equality is not met within the region. Work regarding gender equality has tended to be reactive rather than proactive, and "gender equality, as well as equality between different ethnic groups, has not, so far, been prioritized despite far-reaching obligations for the concerned states" [14].

While most of the Arctic States have a gender equality policy in place, the Council's "rules of procedure contain no reference to gender and there is no gender policy for the Council as a whole. [Furthermore], while the secretariats are subject to gender regulations in accordance with the State in which they are located; there is no overall gender policy or guidelines which inform the Council's activities" (T. Barry, personal communication, 16 October 2020). Written policies rarely explicitly express, or take as their starting points, the political and legal obligations regarding gender equality and/or equal rights for men and women. With few and vague exceptions, the only genders addressed are men and women, and policies addressing gender equality and diversity are scant and vague.

The CEDAW Committee has expressed concerns about the lack of awareness of CEDAW in all Arctic States that have ratified the convention (all but the United States) and pointed out that some groups of women in the Arctic are vulnerable, especially Indigenous and rural women, and Arctic States do not adequately uphold their rights, for example when it comes to exposure to violence, equal participation in governing bodies, and economic self-support. The Special Rapporteur on the Rights of Indigenous Peoples has repeatedly raised serious concerns about the situation of Indigenous women and girls in Canada and the Native Women's Association of Canada and other institutions have reported that many Aboriginal women have been murdered or reported missing. Further, discriminatory and gender bias in policing is signaled, as is overrepresentation of Native women in the prison system. The rapporteur concluded that there appears to be a need for an Aboriginal program strategy for women sentenced at federal level.

Policy relevant highlights include a recommendation that the Council launch a gender equality policy, and a collaboration around gender equality in the development of new strategies is encouraged. Public governing bodies of the Arctic should acknowledge and apply a more far-reaching gender equality concept, including through an intersectional gender equality approach. Suggestions for research initiatives include identification and further analysis of controversial concepts imposed on the region and its population, such as individual rights, power, culture, and tradition.

The chapter on **Security**, led by Gunhild Hoogensen Gjørv and Sarah Seabrook Kendall, examines the links between inequalities in the Arctic and the experience of insecurity. Most Arctic States are characterized as examples of peace, security, and gender equality to the degree that the Arctic has been singled out as exceptional regarding peace and security. This characterization has, however, been contested, not least as it, all too often, ignores and even misrepresents insecurities experienced at individual or community levels [17]. Indeed, the "exceptional" peace, security, and gender equality image relies on a militarised understanding of security that is divorced from perceptions of security of Arctic peoples (Indigenous and non-Indigenous). A broader understanding of (human) security is based on the interaction of a combination of five factors: actors, practices, values, survival, and future [18], where it is further understood that perceptions of security are both subjective and context based [19,20]. Human security—which focuses particularly on the individual and community levels of security—has been broadly defined to include environmental, food, health, economic, political, personal, and community security [21]. Rather than arguing for a problematic, state-centric notion of Arctic exceptionalism, the chapter draws from global insights about insecurity and identifies important challenges and insecurities within the Arctic region itself. The chapter addresses gendered and human insecurities associated with climate change and provides brief examples of some of the gender/human insecurities experienced across the Arctic today.

Although Arctic governance has made significant strides, both with regards to priorities and representation of Arctic peoples, the developments have not been without critique (see section above; [22,23]). The applicability of the human security concept to the Arctic has likewise been debated and criticised [24,25]. A human security lens has, however, also been used to highlight the inequalities and injustices of governance systems. Inequalities often lie within the structures of formal institutions and informal social practices. Applying the concept of human security with an intersectional analysis—that is, examining how law and governance can contribute to inequalities depending on combinations of gender, race, ethnicity, sexual orientation, class, etc.—can be a useful framework for understanding the nature of security threats in the circumpolar Arctic, including the impacts of climate change [25–28]. For example, changes in the Arctic often result in insecurity and vulnerability of social and ecological systems, which are often rooted in marginalisation of northern populations through colonisation and continued oppression of Arctic Indigenous Peoples [29]. Women and girls, Indigenous Peoples, elders, and Two Spirit people are regarded as the most vulnerable Arctic populations [3].

The chapter on Security concludes that gender security perspectives are crucial to improving Arctic societal well-being and stability, and it emphasises the need for a broader, research-based understanding of security. It further highlights the tendency of inequalities and centre—periphery imbalances to lead to insecurities, as most Arctic regions are neglected or bypassed regarding services, support, and inclusion in broader political goals.

As such, security in the Arctic cannot be reduced to a militarised understanding and narrow, geopolitical considerations of states, and Arctic peoples generally remain more preoccupied with everyday security issues. While some scholars and policy makers have resisted the use of the human security concept for Arctic contexts based on the assumption of inclusion in welfare states. In reality, gaps remain, for example, in issues of health, housing, food, economy, environment, and personal and community violence. Further, male voices remain privileged in contemporary security dialogues, not least those relating to state-centred security interests. Recognition of the rights of northerners to participate in their own security dialogues, as well as identifying barriers for women's participation in these dialogues, is required as is understanding how security is perceived and experienced. Intersectional analysis including gender and other identity markers is integral to moving forward towards a more comprehensive understanding of security.

A broad and comprehensive approach to security is necessary to capture the nature and nuance of human insecurity in the Arctic. The most pressing human security threats in the region across the environmental, social, economic, and cultural dimensions can only be properly understood in collective terms. Consequently, far from being an inappropriate analytical framework, human security offers significant analytical traction through its capacity to capture physical and non-material security problems in the circumpolar Arctic that are scalable to smaller or larger communities, distinct peoples, or the region as a whole, and for its intersectional approach that understands the compounding and mitigating effects of distinct security issues and identities.

Policy relevant highlights, directed at Arctic States and the Arctic Council, include a call for the application of a comprehensive security approach with intersectional analysis to better address current and future insecurities. Further, reduced inequalities and consequent tensions, greater inclusion of local and regional bodies in broader political goals should be fostered. Moreover, a responsive climate change policy and mitigation should be based on an intersectional analysis and understanding of impacts, of climate change, on societies and inequalities. This requires a comprehensive, people-centric understanding of Arctic security.

The chapter on **Gender and Environment**, led by Malgorzata (Gosia) Smieszek and Tahnee Prior, claims that the Arctic cannot be fully understood without recognition of the relationship of Arctic Peoples and the environment, in which gender plays a central role. Gender and gender norms have implications for interactions with, activities in, and observations of the environment, as well as for access to, and participation in, management of natural resources. It influences conservation efforts and participation in decision-making bodies at all levels, as women and men are impacted differentially by environmental change and have important roles in environmental sustainability, only achieved through equal access to opportunity and shaping political agendas [30]. The chapter gives particular attention to variations in how people of different genders relate to their environment. It addresses the gendered impacts of development and environmental change, highlighting central dimensions of the gender–environment nexus in an Arctic context through illustrative cases in various localities and sectors, including mining, fisheries, and forestry.

A gender-specific analytical approach provides the basis for a comprehensive view of environmental and social issues, which can lead to more effective policies [31,32]. Gender equality is integral to effective and equitable sustainable development, and there has been a clear shift in the commitment to gender equality and recognition of gender in international environmental agreements over the past decades [32]. Still, gender remains marginal in the overall body of scholarship on climate change adaptation, resilience, and vulnerability [33–35]—both globally and, to an even greater extent, in the Arctic. A deeper comprehension of the vulnerability of Arctic Peoples and communities is required to strengthen necessary adaptation efforts [36,37], but approaches that are insensitive to gender and other indicators of social inequalities risk reinforcing existing vulnerabilities and can result in maladaptation [33,36,38]. Adaptation to climate change, in Arctic research and policy, should thus be reframed to systematically account for health, education, food security, and Arctic economies, all of which are simultaneously differentiated by gender.

Natural resources are vital to the livelihoods of all Arctic peoples and many Indigenous populations continue land-based lifestyles, central to communities' well-being and cultural survival [39]. Simultaneously, natural resource extraction and development are promoted as a pathway to creating better living conditions in the north [40–42]. Resource-based industries in the North are male-dominated, and the effects of resource development strongly gendered [43]. Gender equity has important multiplier effects in sustainable development through women's empowerment, and moving beyond statements on equity, diversity, and inclusion toward the implementation of policies that ensure principles of inclusion is crucial [44].

The chapter on Gender and Environment reflects how gender equality is a prerequisite and accelerator of progress towards sustainable development, and it reflects that centring gender equality in efforts to respond to changes can tap into underexplored potential that fosters people's ability to become agents of change in the face of future challenges.

As noted, the Arctic environment is central to its peoples' health, lifestyles, cultures, and livelihoods, and gender plays a central role in human–environment relations. There is an understanding that women and men are not only differently affected by the primary and secondary effects of climate change and other socioeconomic transformations, but they also play important and distinct roles in achieving environmental and social sustainability. However, research sensitive to gender is still fragmented and, until recently, remained on the margins of a rapidly growing body of Arctic scholarship and policy-relevant science [45–47]. The research agenda on gender and climate change, extractive industries, renewable energy, marine resources, and pollution in the North is far from complete; large gaps remain in knowledge which has predominantly been based on individual case studies, which do not provide a comprehensive gender-sensitive overview of developments in the Arctic. There is also a paramount lack of sex- and gender-disaggregated data, or reliance on patchy, outdated ones, across all environment-related issue areas. Furthermore, there is an overall gender blindness and lack of incorporation of gender-sensitive approaches or insights generated by gender analysis into most mainstream environmental, conservation, marine, and natural resource decision-making processes. To this extent, these processes, as noted in the first Global Gender and Environment Outlook, do not fully serve environmental or social interests [32] (p. 23).

The chapter concludes that gender equality is integral for effective, efficient, and equitable environmental protection. Further, all regions of the Arctic exhibit only sporadic engagement with gender and gender analysis, and there is a dearth of sex- and gender-disaggregated data across the Circumpolar North. Finally, there is a lack of systematic engagement with gender-based analysis and gendered perspectives within the Arctic Council and across its Working Groups.

Policy relevant highlights include a push for new data collections that are gender- and sex-disaggregated, as this will support policy- and decision-making and enhance adaptive capacity. Further, a call for the strategic application of a gender lens to the work of the Arctic Council, including through gender mainstreaming and intersectional approaches, as this supports policy development and decision-making, allows for more tailored actions, plans, policies, and programs. Finally, it is recommended that Arctic studies be expanded to incorporate a specific gender focus to account for the region's particular traits and characteristics.

The chapter on **Migration and Mobility**, led by Erika Anne Hayfield, discusses how migration and mobility in the Arctic are constructed through gender combining statistics with a qualitative context-based approach to understand space as gendered and the contextual nature of migration and mobility. The Arctic remains a place where people are constantly on the move and mobility across the region is complex with globalisation and technological developments further transforming mobility potential. Place-specific contexts are important for understanding Arctic mobilities and addressing migration and mobility requires a gendered approach [47].

Arctic places have diverse opportunity structures with "different conditions and barriers that directly and indirectly promote or hinder opportunities for individuals" [48] (p. 64). At the same time, local opportunity structures intersect with overarching macro structures, for example, national gender equality policy, the spatial patterning of economic development initiatives, or access to education. Therefore, migration decisions are complex and situated within local and national opportunity structures, but they are firmly woven into individual, social, and relational contexts.

Gendered migration is a major factor leading to an unequal balance of men and women in the Arctic [49], and in most regions, men outnumber women, especially in younger age groups. Women outnumber men in terms of out-migration, and there are higher levels of immigration, as well as domestic in-migration, of men. A skewed sex ratio may reinforce inequalities of women and men and is a driving force of female out-migration. The skewed sex ratio across the Arctic is a cause for concern for future social sustainability of the region.

Migration and mobility in the Arctic are as diverse as the peoples and the places they live. Low population density has implications for access to work, education, welfare, markets, and more. Distances and climate provide conditions for movement, and from these contexts local mobile cultures emerge. Mobility structures are complex but must be understood within local mobile cultures. The Arctic has a long history of Indigenous Peoples practicing mobilities, but these practices have been somewhat transformed through colonialisation. The field must move beyond mobilities associated with globalisation and urbanisation to better understand contemporary Indigenous mobilities.

The chapter on Migration and Mobility emphasises how studies on migration and gender need to employ an intersectional research approach and improve at involving other social categories. Gendered migration and mobility are still neglected areas in Arctic literature, and much of the extant literature is oriented to differences between women and men. Too few studies are grounded in feminism, masculinity studies, intersectionality, LGBTQIA2S+ and Indigenous gender perspectives.

An imbalance between women and men in the Arctic emerges, sometimes with highly skewed sex ratios where women are more educated than men, more inclined to seek higher education or work in larger urban areas, and thus more likely to out-migrate, whereas men are more likely to seek vocational education closer to home but travel further for work. The Arctic is a masculine space and women may perceive a lack of opportunities, not least in industries heavily dominated by men. There is evidence that masculinities are structured around work and being breadwinners, as opposed to attaining higher education and being primary carers. Colonialisation has transformed gender within Indigenous cultures, and, as a result, Indigenous women have become relatively marginalised within traditional economic and subsistence activities and are more likely to hold paid work.

Indigenous People are overrepresented amongst the homeless worldwide, as is the case for Alaska, Greenland, and Arctic Canada [50]. Histories of displacement, experiencing a loss of home, and being forced to move have resulted in intergenerational trauma, which, in some cases, is the root cause of homelessness. Such trauma is linked to racism towards Indigenous Peoples along with mental health issues, violence, incarcerations, and addictions [50]. Additionally, homelessness is a gendered phenomenon. For example, in Alaska, surveys indicate that women are overrepresented in figures for homelessness [50].

Young people, and especially young women, out-migrate from small communities in the Arctic, and there are indications that for those ascribing to LGBTQIA2S+ identities, cultures in small communities, and the Arctic in general, are not open enough.

Given that migration and mobility in the Arctic are highly gendered, the lack of knowledge on this topic, from a gender perspective, is both surprising and concerning. Young people, and especially women, out-migrate from the Arctic, yet most studies that address migration and mobility in the Arctic fail to include gender perspectives. What is more, studies on gender in the Arctic rarely include significant life issues such as migration. Thus gender, migration, and mobilities tend to be approached as standalone and isolated research topics.

Policy relevant highlights point to the need for further understanding of the complex processes involved in migration and mobility processes in the region, including those leading to out-migration of young people and women, and context sensitive integration strategies related to immigrant populations. Focus should be on developing, improving, and sustaining local opportunity structures—as well as material and welfare structures. Using a gendered intersectional approach, such a focus must also encompass industry development in the Arctic. Gender-sensitive support and recovery policies and services, including in terms of housing and homelessness, should be provided.

The chapter on **Indigeneity, Gender, Violence, and Reconciliation**, led by Karla Jessen Williamson, argues that discussing gender in the Arctic calls for awareness of the imposition of a foreign understanding of gender—binary and patriarchal—forced on Indigenous Peoples through colonisation. States tend to view gender and violence through a binary lens, and prevention of gendered violence is often organised through policies that do not adequately consider diversity or context.

The chapter addresses terminology related to gender, sexuality, and diversity as well as problems related to the imposition of Western binary perspectives on Arctic Indigenous communities. It further explores violence—not yet covered in a comparative fashion for the Arctic—including violence against Indigeneity and the consequent persistent inequalities between Indigenous and non-Indigenous populations. Worldviews and value systems of Western states have encroached upon Indigenous worldviews and value systems through processes of colonisation which impacted most aspects of Indigenous lives, reflected in persistent inequalities between Indigenous and non-Indigenous populations. Indigenous Peoples have a minority status in their respective states with direct implications for sense of belonging and quality of life in Indigenous communities; the inherent marginalisation of Indigenous Peoples means that Indigenous interests may not be aligned with the rest of the population.

Criminology studies suggest a high prevalence of violent crimes among Indigenous communities worldwide, partly due to the breakdown of Indigenous informal social controls because of dispossession and colonisation processes which involved traumatic "[i]mposition of foreign law, institutions, peoples, economies and beliefs" [51] (p. 33). Gendered violence continues to be a serious issue across the Arctic and the connection between socioeconomic inequalities and violent crimes is explored. However, Indigenous women and girls face disproportionate violent victimisation in the context of ongoing settler–colonial relations and a long history of targeted colonial violence against Indigenous Peoples. New governance structures are rapidly evolving as responsibilities are transferred from states to Indigenous Peoples amidst calls for decolonisation, self-determination, and devolution efforts. Although levels of capacity vary, different Arctic Indigenous Peoples address and develop their own responses to gender-based violence within Indigenous communities.

Truth and reconciliation commissions have been used in various contexts, although perhaps the most widely known is the South African Truth and Reconciliation Commission [52]. Broadly speaking, these commissions investigate human rights abuses by engaging with affected populations and attempting to "clarify the national narrative of affected populations", and "establish a set of facts as a basis of the truth about the history and evolution of a given conflict, to devise a new and more acceptable national narrative" [53] (p. 1). They have become an important mechanism in promoting accountability, reform, and fostering reconciliation [54]. While truth and reconciliation processes are subject to debate and their outcomes vary a great deal, reconciliation commissions have been established in the Arctic, including Canada, Greenland, and through the Sámi Truth and Reconciliation Commission.

The chapter on Indigeneity, Gender, Violence, and Reconciliation explored how data indicates socioeconomic disparities between Indigenous and non-Indigenous individuals in the Arctic while at the same time, Indigenous Peoples carry the burden of collective trauma from alienation and marginalisation brought on by processes of colonisation and assimilation policies. To some extent, such trauma may be addressed through Truth and Reconciliation processes to improve relationships, confront previous colonisation practices, and address social inequalities.

Women are overrepresented as victims of violent crimes, inclusive of acts such as sexual abuse, rape, and domestic violence. Indigenous women and girls face disproportionate violent victimisation in the context of ongoing settler–colonial relations and a long history of targeted colonial violence against Indigenous Peoples. States, self-governments, and communities strive to find ways of handling this serious and ongoing concern.

Gendered and intersectional data, including specific data on Indigenous and LGBTQIA2S+ populations, are severely lacking. To effectively analyse and understand the intersection of violence against Indigeneity, inequality, and social-economic contexts, as well as gendered violence in the Arctic, disaggregated and meaningful data is required for comparison.

Policy relevant highlights include a call for the Arctic Council and its working groups to promote the use of inclusive terminologies and apply gender mainstreaming. Further, a reminder that the United Nations Declaration on the Rights of Indigenous Peoples is relevant to all Council states, and a better understanding of inequalities faced by Indigenous populations is vital for effective policy-making. The Arctic Council is encouraged to create a mechanism—either through a task force or an expert group—to monitor the status of Indigenous Peoples within the Arctic states, provide status reports and recommendations on ways to address systemic inequalities. Finally, a call should be made to the SDWG to initiate a project on sharing knowledge of best practices to prevent and raise awareness of gendered violence in the Arctic.

The final theme addressed in the **Empowerment and Fate Control** chapter, led by Marya Rozanova-Smith, Andrey Petrov, and Varvara Korkina Williams, seeks to identify concrete strategies for political, economic, and civic gender empowerment in order to facilitate sustainable policy-making for the Arctic. Gender empowerment is defined as the capacity of all genders to exercise power in decision-making and the process by which they, individually and collectively, can help themselves and others maximise the quality of their lives. The term is closely linked to the concept of fate control, which is defined as the ability to guide one's own destiny and refers to a process that creates power in individuals over their own lives, society, and their communities [55].

Enabling gender equality by empowering all genders to effectively participate in modern society is one of the most important advances towards sustainable development, encompassing equal representation in the politics and public administration, labour market, and civil society [56] (SDG5). Recent studies demonstrate that, despite an increasing global trend towards gender equality in general, and women's empowerment in particular, it varies dramatically across countries, regions, and communities, as well as across spheres of engagement [57–60].

While the theme of gender empowerment in the Arctic regions has received limited attention, gender empowerment processes are particularly important in the Arctic, which is experiencing unprecedented climate-induced environmental change [61–64]. Simultaneously, divergent social, economic, and institutional changes are being observed in many Arctic regions [3,65]. These changes require novel approaches to understanding gender equality and empowerment in the Arctic that accounts for socioeconomic, political, cultural, and ethnic diversity.

The authors pursue the idea that all social, economic, ethnic, demographic, and gender groups must have an ability to thrive, in order to ensure the communities', regions', and nations' sustainable future. Gender empowerment is one of the most important elements of such thrivability, as it encapsulates the ability of all genders to possess fate control and pursue their individual and collective goals and aspirations as a part of a community.

The chapter on Empowerment and Fate Control suggest moving gender empowerment and fate control from the periphery to the centre of public discourse and decision-making, as well as making sure to incorporate Indigenous Peoples' traditions and perspectives on gender and gender equality in the theoretical and practical framework of gender knowledge building and policy.

Studies do not indicate a strong trend towards increasing female leadership and women's deeper involvement into regional economic and political affairs. However, local self-government institutions and civic initiatives in the Arctic are increasingly engaging women [66]. Despite the importance of the topic of gender empowerment and fate control, there is a significant gap in both public information sources and academic knowledge about the current state and emerging trends of political, economic, and civic gender empowerment in the Arctic. Gender indicators and indices are instrumental in capturing gender equality and empowerment processes across all sectors and at all levels of politics and government, economy, and civil society. The authors suggest a system of key variables to provide a basic framework for analysing gender empowerment in the Arctic (GEA indicators). This set of indicators will help monitor and compare the current state of gender empowerment across Arctic regions and communities and identify key patterns over time.

Gender empowerment is key to community sustainability, resilience, and thrivability. However, the Arctic is diverse, and there is no one-size-fits-all policy solution to gender empowerment gaps. Underrepresented genders' access to and participation in political, economic, and civic spheres needs to be improved. In some Arctic communities, a particular focus should be placed on men's empowerment and individual fate control. Gender mainstreaming in policy and research plays an important role in attaining gender empowerment at the circumpolar and national scales, and should be continued while placing more emphasis on regional to local (community) levels.

Policy relevant highlights include a call for the need to improve gender and sex- disaggregated data collection and access, to provide comprehensive, comparable and trackable data across the region; a suggestion to establish a system of monitoring based on gender empowerment indicators; to acknowledge and incorporate Indigenous Peoples' traditions and perspectives on gender and gender equality into legal, theoretical, and practical frameworks of gender knowledge; to mainstream gender equality and empowerment at all levels and in all spheres; to ensure an inclusive approach to gender equality.

## 5. Conclusions

What emerged during the process of writing the report was that accessible, comparable, gender-disaggregated, and Arctic specific data was severely lacking. Furthermore, where data is available for the Arctic, it is still lacking in specific data on Indigenous populations and LGBTQIA2s+. This makes any meaningful comparison between, and within, states near impossible in most cases and severely impedes efforts to adequately understand the dynamics of gender across the Arctic. One of the main recommendations of the report is, therefore, that the Arctic Council, as the leading political body within the Arctic with members from all Arctic states and the Indigenous Permanent Participants, as well as observer states, "should encourage and facilitate the development of guidelines for consistent and comparable data and definitions throughout the Arctic. This would entail, at a minimum, gendered and ethnically disaggregated data" [4] (p. 17).

All chapters also highlighted the importance of gender-based analysis and gender mainstreaming in all decision-making processes at national, regional, and local levels. This entails evaluating the effects of all actions, policies, and programmes on all genders to ensure that decisions do not perpetuate existing inequalities and create new ones. Moreover, temporary, special measures to reverse existing inequalities are recommended as necessary. Again, the Council is identified as the main driver for implementing the recommendations, both in its own work and by encouraging its Member States to set an example at national and regional levels. Indeed, the report's recommendations were discussed at the AC Ministerial Meeting in Reykjavík on 20 May 2021, and in the Ministerial Meeting declaration (the Reykjavík Declaration 2021) the AC "Emphasize[s] the importance of gender equality and respect for diversity for sustainable development in the Arctic and welcome[s] the Pan-Arctic Report, Gender Equality in the Arctic, Phase 3, encourage[s] the mainstreaming of gender-based analysis in the work of the Arctic Council and call[s] for further action to advance gender equality in the Arctic" [9] (p. 4). In Chapter 4 on Sustainable Social Development in the Council's Strategic Plan 2021 to 2030, also accepted at the Ministerial Meeting in Reykjavík, the Council resolves to "promote gender equality and non-discrimination in the Arctic with the aim of contributing to sustainability and balanced participation in leadership and decision making both in the public and private sectors" [10] (p. 5). The recommendations outlined in this report will help forward gender equality and non-discrimination in the Arctic, and it should, therefore, be seen as a key means of achieving the goals outlined by the Arctic Council at the Ministerial Meeting in Reykjavík.

**Author Contributions:** Conceptualization, E.E.O. and H.Ó.Á.; writing—original draft preparation, E.E.O. and H.Ó.Á.; writing—review and editing, E.E.O., H.Ó.Á., E.-M.S., G.H.G., S.S.K., M.S., T.P., E.H., K.J.W., M.R.-S., A.P. and V.K.W.; supervision, E.E.O.; project administration, E.E.O.; funding acquisition, E.E.O. All authors have read and agreed to the published version of the manuscript.

**Funding:** This Gender Equality in the Arctic Pan-Arctic Report was funded by: The Nordic Council of Ministers, the Icelandic Gender Equality Fund, the Ministry for Foreign Affairs of Iceland, Ministry for Foreign Affairs of Sweden, the Ministry for Foreign Affairs of Finland, the Ministry of Foreign Affairs of Norway, Government of the Faroe Islands, the Government of Canada, the United States Department of State, the U.S. National Science Foundation (PLR #2039884), the Stefansson Arctic Institute. Co-sponsored by: the University of Akureyri, the Institute of Arctic Studies at Dartmouth College, the Polar Institute and Wilson Center, the UArctic and its Institute for Arctic Policy.

**Institutional Review Board Statement:** Not applicable.

**Informed Consent Statement:** Not applicable.

**Data Availability Statement:** The Pan-Arctic Report Gender Equality in the Arctic is available in its entirety on the project website—Gender Equality in the Arctic (https://arcticgenderequality.network, accessed on 15 July 2021). Data supporting report results lies with the chapter lead-authors and in some cases in publicly available website or article, as indicated in chapter reference lists.

**Acknowledgments:** The GEA project and the Icelandic Arctic Cooperation Network (IACN) are especially grateful to representatives from the Arctic Council Permanent Participants, whose advice and consultation was invaluable at all stages. We would further like to thank the Project Partners; the Editorial Committee; the Arctic Council's Sustainable Development Working Group (SDWG), its Executive Secretary and its Social, Economic, and Cultural Expert Group (SECEG); the Youth Advisory Group; and all contributing authors of the report; Nordregio; Fínlína; and the Arctic Council Chairmanship team at the Icelandic Ministry for Foreign Affairs. Further, the report editors and lead authors are grateful to all contributing authors: Joanna Absalonsen, Rebecca Alty, Charlene Apok, Steven Arnfjord, Marit Aure, Stéphanie Barillé, Elizabeth (Sabet) Biscaye, Jessica Black, Courtney Carothers, Yolande Chapman, Julia Christensen, Brynhildur Davíðsdóttir, Stephan Dudeck, Torjus Solheim Eckhoff, Grid-Arendal; Chad R. Farrell, Dina Abdel Fattah, Olga Stefansdottir, Devlin Fernandes, Sara Fusco, Firouz Gaini, Siri Gerrard, Andrii Gladii, Anne Gotfredsen, Irmelin Gram-Hanssen, Will Greaves, J. Otto Habeck, Lawrence Hamilton, Erika A. Hayfield, Lara W. Hoffmann, Kamrul Hossain, Anna Maria Hubert, Runa Preeti Ísfeld, Gréta Bergrún Jóhannesdóttir, Tonje Margrete Winsnes Johansen, Rachael Lorna Johnstone, Tanja Joona, Sohvi Kangasluoma, Anna Karlsdóttir, Nordregio, Sarah Seabrook Kendall, Mara Kimmel, Siff Lund Kjærgaard, Jón Fannar Kolbeinsson, Fanny-Tsilla Koninckx, Jennifer Koshan, Natalia Kukarenko, Bridget Laroque, Kirsti Lempiainen, Liza Mack, Markus Meckl, Samantha Michaels, Suzanne Mills, Jana Mortensen, Päivi Naskali, David Natcher, Soren Stach Nielsen, Ernst Sumberg Olsen, Fiskaaling, Ivalo Olsvig, Silja Bára Ómarsdóttir, Maria Osipova, Per Jonas Partapuoli, Maria Pitukhina, Birger Poppel, Marie Kathrine Poppel, Alexandra Poturaeva, Olga Povoroznyuk, Magalie Quintal-Marineau, Marya Rozanova-Smith, Federica Scarpa, Tina H.P. Schoolmeester, Alexander Sergunin, Kyle Shaughnessy, Sigrún Sigurðardóttir, Sveinbjörg Smáradóttir, Kathryn Urban, Lilia Vinokurova, Vladislava Vladimirova, Pál Weihe, Nafisa Yeasmin, Nadezhda Zamyatina, Laura Zanotti, Karen Birna Þorvaldsdóttir and Bergljót Þrastardóttir.

**Conflicts of Interest:** The authors declare no conflict of interest. However, the Gender Equality in the Arctic Pan-Arctic Report is an Arctic Council Sustainable Development Working Group Project and received significant support from Arctic Council member states. The funders had no role in the design of the study; in the collection, analyses, or interpretation of data; in the writing of the manuscript, or in the decision to publish the results. E.M.S. received funding from the Ministry for Foreign Affairs of Sweden for the chapter on Law and Governance; G.H.G. received funding from the Ministry of Foreign Affairs of Norway for the chapter on Security; G.S. and T.P. received funding from the Ministry for Foreign Affairs of Finland for the chapter on Gender and Environment; A.A.H. received funding from the Government of the Faroe Islands for the chapter on Migration and

Mobility; K.J.W. received funding from Crown–Indigenous Relations and Northern Affairs Canada, Government of Canada for the chapter on Indigeneity, Gender, Violence and Reconciliation; M.R.S received funding from the National Science Foundation.

## References

1. Ministry of Social Affairs and Health. In Finland. Taking Wing. In Proceedings of the Conference on Gender Equality and Women in the Arctic, Saariselkä, Finland, 3–6 August 2002.
2. Einarsson, N.; Larsen, J.N.; Nilsson, A.; Young, O.R. (Eds.) *Arctic Human Development Report*; Stefansson Arctic Institute: Akureyri, Iceland, 2004.
3. Larsen, J.N.; Fondahl, G. *Arctic Human Development Report: Regional Processes and Global Linkages*; Nordic Council of Ministers: Copenhagen, Denmark, 2015. [CrossRef]
4. Agustsson, H.; Oddsdottir, E. *Pan-Arctic Report: Gender Equality in the Arctic*, 1st ed.; Iceland's Arctic Council Chairmanship 2019–2021 and the Arctic Council Sustainable Development Working Group, with the Icelandic Arctic Cooperation Network, the Icelandic Directorate for Equality, and the Stefansson Arctic Institute; Arctic Council, Sustainable Development Working Group: Tromsø, Norway, 2021; Available online: https://arcticgenderequality.network/phase-3/pan-arctic-report (accessed on 1 July 2021).
5. Arctic Council. Icelandic Chairmanship. Available online: https://arctic-council.org/en/about/previous-chairmanships/iceland-chair-2/ (accessed on 21 June 2021).
6. Gender Equality in the Arctic. Available online: https://arcticgenderequality.network/ (accessed on 21 June 2021).
7. Oddsdóttir, E.; Sigurdsson, A.M.; Svandal, S. *Gender Equality in the Arctic: Current Realities, Future Challenges Conference Report*; Ministry for Foreign Affairs: Reykjavik, Iceland, 2015. Available online: https://www.stjornarradid.is/media/utanrikisraduneyti-media/media/nordurslodir/gender-equality-in-the-arctic.pdf (accessed on 21 June 2021).
8. Gender Equality in the Arctic. The GEA Times. Available online: https://arcticgenderequality.network/gea-times (accessed on 21 June 2021).
9. Arctic Council. *Declaration of the Foreign Ministers of the Arctic States at the 12th Ministerial Meeting of the Arctic Council, Held in Reykjavik, Iceland, 20 May 2021 (Reykjavík Dedlaration)*; Arctic Council Secretariat: Tromsö, Norway, 2021; Available online: https://oaarchive.arctic-council.org/handle/11374/2600 (accessed on 1 July 2021).
10. Arctic Council. Arctic Council Strategic Plan 2021-203, 20 May 2021. Available online: https://oaarchive.arctic-council.org/handle/11374/2601 (accessed on 1 July 2021).
11. Besson, S. Oxford Public International Law. 2011. Available online: https://opil.ouplaw.com/view/10.1093/law:epil/9780199231690/law-9780199231690-e1472 (accessed on 15 November 2020).
12. Nord, D.C. *The Arctic Council: Governance within the Far North*; Routledge: London, UK, 2016.
13. Nord, D.C. *The Changing Arctic: Creating a Framework for Consensus Building and Governance within the Arctic Council*; Palgrave Macmillan: London, UK, 2016.
14. Svensson, E.-M. Gender equality in the governance of the Arctic region. *Nord. J. Law Soc.* **2017**, *1*, 16–64. [CrossRef]
15. Gunnarsson, Å.; Svensson, E.M. Gender equality in the Artic and North: Socio-legal and geopolitical challenges. *Nord. J. Law Soc.* **2017**, *1*, 6–15. [CrossRef]
16. Lahey, K.; Svensson, E.-M.; Gunnarsson, Å. Gender challenges & human capital in the Arctic. *Arct. Yearb.* **2014**, *1*, 183–201.
17. Hoogensen Gjørv, G. Human insecurities of marginalized peoples in the Arctic: The cost of Arctic and Nordic exceptionality. In *Breaking through: Understanding Sovereignty and Security in the Circumpolar Arctic*; Greaves, W., Lackenbauer, W., Eds.; University of Toronto Press: Toronto, ON, Canada, 2021.
18. Hoogensen Gjørv, G. Tensions between environmental, economic, and energy security in the Arctic. In *Northern Sustainabilities: Understanding and Addressing Change in a Circumpolar World*; Fondahl, G., Wilson, G., Eds.; Springer: Berlin/Heidelberg, Germany, 2017.
19. Roe, P. Actor, Audience(s) and Emergency Measures: Securitization and the UK's Decision to Invade Iraq. *Secur. Dialogue* **2008**, *39*, 615–635. [CrossRef]
20. Vaughn, J. The Unlikely Securitizer: Humanitarian Organizations and the Securitization of Indistinctiveness. *Secur. Dialogue* **2009**, *40*, 263–285. [CrossRef]
21. United Nations Development Programme. *Human Development Report 1994: New Dimensions of Human Security*; Oxford University Press: Oxford, UK, 1994.
22. Heininen, L.; Exner-Pirot, H.; Plouffe, J. Governance & governance in the Arctic: An introduction to Arctic yearbook 2015. In *Climate Change and the Arctic: Global Origins, Regional Responsibilities? Arctic Yearbook 2015*; Heininen, L., Exner-Pirot, H., Plouffe, J., Eds.; Northern Research Forum: Akureyri, Iceland, 2015; Available online: https://arcticyearbook.com/arctic-yearbook/2015/12-yearbook/2015-arctic-governance-and-governing/121-governance-governance-in-the-arctic-an-introduction-to-arctic-yearbook-2015 (accessed on 15 November 2020).
23. Hoogensen Gjørv, G.; Hodgson, K.K. 'Arctic' exceptionalism' or 'comprehensive security'? Understanding security in the Arctic. In *Redefining Arctic Security: Arctic Yearbook*; Heininen, L., Exner-Pirot, H., Barnes, J., Eds.; Arctic Portal: Akureyri, Iceland, 2019; pp. 218–230. Available online: https://arcticyearbook.com/images/yearbook/2019/Scholarly-Papers/11_AY2019_Hoogensen_Hodgson.pdf (accessed on 5 January 2021).

24. Griffiths, F. Not that good a fit? 'Human security' and the Arctic. In *Arctic Security in the 21st Century Conference: Report*; The Simons Foundation and the School for International Studies, Simon Fraser University: Burnaby, BC, Canada, 2008; pp. 55–61.
25. Hoogensen Gjørv, G. Virtuous imperialism or a shared global objective? The relevance of human security in the Global North. In *Environmental and human security in the Arctic*; Gjørv, G.H., Bazely, D.R., Goloviznina, M., Tanentzap, A.J., Eds.; Routledge: London, UK, 2014; pp. 58–80.
26. Dalby, S. *Security and Environmental Change*; Polity: Cambridge, UK, 2009.
27. Greaves, W. For whom, from what? Canada's Arctic policy and the narrowing of human security. *Int. J.* **2012**, *67*, 219–240. [CrossRef]
28. Greaves, W. Securing sustainability: The case for critical environmental security in the Arctic. *Polar Rec.* **2016**, *52*, 660. [CrossRef]
29. Hellstad, V. 'Nordlendinger Uønsket': En Studie av Nordnorsk Identitiet i Møte Med Oslo ['Northerners Unwanted': A Study of North Norwegian Identity in the Oslo Context]. Master's Thesis, University of Oslo, Oslo, Norway, 2010.
30. Nuttall, M. Global interdependence and Arctic voices: Capacity-building for sustainable livelihoods. *Polar Rec.* **2002**, *38*, 194–202. [CrossRef]
31. Alston, M.; Whittenbury, K. (Eds.) *Research, Action and Policy: Addressing the Gendered Impacts of Climate Change*; Springer: Berlin/Heidelberg, Germany, 2013; Available online: http://library1.nida.ac.th/termpaper6/sd/2554/19755.pdf (accessed on 10 January 2020).
32. United Nations Environment Programme. *Global Gender and Environment Outlook*; United Nations: New York, NY, USA, 2016.
33. Bunce, A.; Ford, J. How is adaptation, resilience, and vulnerability research engaging with gender? *Environ. Res. Lett.* **2015**, *10*, 123003. [CrossRef]
34. Pearse, R. Gender and climate change. *Wiley Interdiscip. Rev. Clim. Chang.* **2017**, *8*, 1–16. [CrossRef]
35. Santos, P.R.; Holm, N.; Olsen, J.; Hovelsrud, G.K. Gender and Climate Change Research: Moving Beyond Transformative Adaptation. In *Arctic Yearbook*; Heininen, L., Exner-Pirot, H., Barnes, J., Eds.; Arctic Portal: Akureyri, Iceland, 2020.
36. Vincent, K.E.; Tschakert, P.; Barnett, J.; Rivera-Ferre, M.G.; Woodward, A. Cross-chapter box on gender and climate change. In *Climate Change 2014: Impacts, Adaptation, and Vulnerability—Part A: Global and Sectoral Aspects—Contribution of Working Group II to the Fifth Assessment Report of the Intergovernmental Panel on Climate Change*; Field, C.B., Barros, V.R., Dokken, D.J., Mach, K.J., Mastrandrea, M.D., Bilir, T.E., Chatterjee, M., Ebi, K.L., Estrada, Y.O., Genova, R.C., et al., Eds.; Cambridge University Press: Cambridge, UK, 2014; pp. 105–107.
37. Kelly, P.M.; Adger, W.N. Theory and Practice in Assessing Vulnerability to Climate Change and Facilitating Adaptation. *Clim. Chang.* **2000**, *47*, 325–352. [CrossRef]
38. Holmgren, S.; Arora-Jonsson, S. The Forest Kingdom—With what values for the world? Climate change and gender equality in a contested forest policy context. *Scand. J. For. Res.* **2015**, *30*, 235–245. [CrossRef]
39. Cunsolo Willox, A.; Harper, S.L.; Edge, V.L.; Landman, K.; Houle, K.; Ford, J.D. The land enriches the soul: On climatic and environmental change, affect, and emotional health and well-being in Rigolet, Nunatsiavut, Canada. *Emot. Space Soc.* **2013**, *6*, 14–24. [CrossRef]
40. Glomsrød, S.; Duhaime, G.; Aslaksen, I. (Eds.) *The Economy of the North 2015*; Statistics Norway: Oslo, Norway, 2017.
41. Parlee, B. Resource development and well-being in northern Canada. In *Resources and Sustainable Development in the Arctic*; Southcott, C., Abele, F., Natcher, D., Eds.; Routledge: London, UK, 2019; pp. 132–155.
42. Southcott, C.; Abele, F.; Natcher, D.; Parlee, B. *Resources and Sustainable Development in the Arctic*; Routledge: London, UK, 2019.
43. Walker, H.; Reed, M.G.; Thiessen, B. *Gender and Diversity Analysis in Impact Assessment*; Canadian Environmental Assessment Agency: Toronto, ON, Canada, 2019.
44. Kuo, M. Consciously combating unconscious bias. *Science* **2017**. [CrossRef]
45. Hoogensen Gjørv, G. Finding gender in the Arctic: A call to intersectionality and diverse methods. In *The Interconnected Arctic: UArctic Congress 2016*; Latola, K., Savela, H., Eds.; Springer Polar Sciences: Berlin/Heidelberg, Germany, 2017; pp. 293–303.
46. Sinevaara-Niskanen, H. When gender matters: Equality as a source of Arctic sustainability. In *Resources, Social and Cultural Sustainabilities in the Arctic*; Tennberg, M., Lempinen, H., Pirnes, S., Eds.; Routledge: London, UK, 2019; pp. 131–143.
47. Vladimirova, V.; Habeck, J.O. Introduction: Feminist approaches and the study of gender in Arctic social sciences. *Polar Geogr.* **2018**, *41*, 145–163. [CrossRef]
48. Bæck, U.-D.K. Spatial maneuvering in education. Educational experiences and local opportunity structures among rural youth in Norway. *Nord. J. Comp. Int. Educ.* **2019**, *3*, 61–74.
49. Turunen, E. *Gender Balance of Total Population in the Arctic*; Nordregio: Stockholm, Sweden, 2019; Available online: https://nordregio.org/maps/gender-balance-of-total-population-in-the-arctic/ (accessed on 15 October 2020).
50. Christensen, J.; Arnfjord, S.; Carraher, S.; Hedwig, T. Homelessness across Alaska, the Canadian North and Greenland: A review of the literature on a developing social phenomenon in the Circumpolar North. *Arctic* **2017**, *70*, 349–364. [CrossRef]
51. Behrendt, L.; Porter, A.; Vivian, A. Factors affecting crime rates in six rural Indigenous communities. In *The Routledge International Handbook of Rural Criminology*; Donnermeyer, J.F., Ed.; Routledge: London, UK, 2014; pp. 33–44.
52. Parmentier, S.; Aciru, M. The whole truth and nothing but the truth? On the role of truth commissions in facing the past. In *Facing the Past: Amending Historical Injustices through Instruments of Transitional Justice*; Malcontent, P., Ed.; Intersentia: Cambridge, UK, 2016; pp. 225–246.

53. Sarkin, J. Towards an understanding of how truth commissions can use their amnesty powers to enhance their impact and legacy. In *The Global Impact and Legacy of Truth Commissions*; Sarkin, J., Ed.; Intersentia: Cambridge, UK, 2019; pp. 211–246. [CrossRef]
54. Ferrara, A. Assessing the long-term impact and legacy of truth commissions. In *The Global Impact and Legacy of Truth Commissions*; Sarkin, J., Ed.; Intersentia: Cambridge, UK, 2019; pp. 75–98. [CrossRef]
55. Dahl, J.; Fondahl, G.; Petrov, A.; Fejellheim, R. Fate control. In *Arctic Social Indicators: Follow-Up to the Arctic Human Development Report*; Larsen, J.N., Schweitzer, P., Fondahl, G., Eds.; Nordic Council of Ministers: Copenhagen, Denmark, 2010; pp. 129–146.
56. United Nations General Assembly. *Transforming Our World: The 2030 Agenda for Sustainable Development*; A/RES/70/1; United Nations: New York, NY, USA, 2015.
57. Alexander, A.C.; Bolzendahl, C.; Jalalzai, F. (Eds.) *Measuring Women's Political Empowerment Across the Globe: Strategies, Challenges and Future Research*; Palgrave Macmillan: Basingstoke, UK, 2018.
58. Sachs, J.; Schmidt-Traub, G.; Kroll, C.; Lafortune, G.; Fuller, G. *Sustainable Development Report 2019*; Bertelsmann Stiftung and Sustainable Development Solutions Network: Bonn, Germany, 2019.
59. United Nations Entity for Gender Equality and the Empowerment of Women. *Progress of the World's Women 2015–2016: Transforming Economies, Realizing Rights*; UNWOMEN: New York, NY, USA, 2016.
60. World Economic Forum. *Global Gender Gap 2018 Report*; World Economic Forum: Cologny, Switzerland, 2019.
61. Arctic Monitoring and Assessment Programme. *Adaptation Actions for a Changing Arctic: Perspectives from the Barents Area*; AMAP: Oslo, Norway, 2017.
62. Arctic Monitoring and Assessment Programme. *Adaptation Actions for a Changing Arctic: Perspectives from the Bering-Chukchi-Beaufort Region*; AMAP: Tromsö, Norway, 2017.
63. Arctic Monitoring and Assessment Programme. *Snow, Water, Ice and Permafrost in the Arctic (SWIPA)*; AMAP: Tromsö, Norway, 2017.
64. Larsen, J.N.; Anisimov, O.A.; Constable, A.; Hollowed, A.B.; Maynard, N.; Prestrud, P.; Prowse, T.D.; Stone, J.M.R. Polar regions. In *Climate Change 2014: Impacts, Adaptation, and Vulnerability—Part B: Regional Aspects. Contribution of Working Group II to the Fifth Assessment Report of the Intergovernmental Panel on Climate Change*; Barros, V.R., Field, C.B., Dokken, D.J., Mastrandrea, M.D., Mach, K.J., Bilir, T.E., Chatterjee, M., Ebi, K.L., Estrada, Y.O., Genova, R.C., et al., Eds.; Intergovernmental Panel on Climate Change: Geneva, Switzerland, 2014; pp. 1567–1612.
65. Rasmussen, R.O. (Ed.) *Megatrends*; Nordic Council of Ministers: Copenhagen, Denmark, 2011.
66. Rozanova, M.; Mikheev, V. Rethinking women's empowerment: Insights from the Russian Arctic. *Soc. Sci.* **2020**, *9*, 14. [CrossRef]

*sustainability*

*Article*

# Stay or Leave? Arctic Youth Prospects and Sustainable Futures of the Russian Arctic Communities

**Marya Rozanova-Smith [1,2,3]**

1 Department of Geography, Elliott School of International Affairs, The George Washington University, Washington, DC 20009, USA; marya_rozanova@gwu.edu
2 ARCTICenter, University of Northern Iowa, Cedar Falls, IA 50614, USA
3 Department of Social and Human Sciences, Russian State Hydrometeorological University, Saint Petersburg 192007, Russia

**Abstract:** Based on quantitative and qualitative analysis, this paper attempts to answer a research question that is critical for many Arctic communities: "What makes local youth want to leave?" Using the Russian Arctic cities of Naryan-Mar, Salekhard, and Novy Urengoy (Nenets and Yamalo-Nenets regions) as case studies, this article explores how local youth contribute to social sustainability and define the futures of their Arctic cities. The study identifies new variables relevant to the youth cohort built on the Urban Sustainability Index and social sustainability model. Based on 400+ questionnaires and interviews with Indigenous and non-Indigenous youth, education professionals, and public officials, this study looks at the youth's educational and professional strategies, social activities and cultural consumption, migration patterns, and civic engagement in a broader context. This article also discusses how local youth feel disempowered in building their futures and highlights the importance of access to educational opportunities and wider career choices in the Arctic.

**Keywords:** Arctic; youth; Indigenous youth; migration; social sustainability; Russia

**Citation:** Rozanova-Smith, M. Stay or Leave? Arctic Youth Prospects and Sustainable Futures of the Russian Arctic Communities. *Sustainability* **2021**, *13*, 12058. https://doi.org/10.3390/su132112058

Academic Editors: Stephanie Pfirman, Gail Fondahl, Grete K. Hovelsrud and Tero Mustonen

Received: 20 July 2021
Accepted: 25 October 2021
Published: 1 November 2021



## 1. Introduction

Fostering urban sustainability in the Arctic is one of the most pressing and challenging tasks in the rapidly changing Circumpolar North, and it will be for many years to come. In the Russian Arctic, since the beginning of its active exploration in the 20th century, rapid resource-based industrialization has resulted in an unprecedented rate and scale of urbanization, which has turned the remote Arctic regions into 'hot spots' of human and social mobility. Today, the highly industrialized areas of the Russian Arctic are reaching rates of urbanization comparable to the Russian average (74.66%), and the Yamalo-Nenets (YaNAO) and Nenets (NAO) regions (83.95% and 73.76%, respectively) are no exception.

Developed in the Soviet times as the regional urban centers of NAO and YaNAO, Naryan-Mar (1935), Salekhard (1938), and Novy Urengoy (1975) (Figure 1) were primarily designated to drive the exploitation of natural resources and soon became symbols of Soviet pride through heroic Arctic conquest. In the process, these cities became magnets for young professionals both dreaming of new feats and searching for upward social mobility and economic benefits. Cultivated over time, a diverse range of administrative functions failed to make these cities' economies diversified enough to sustain themselves in the situation of natural resource depletion or lower demand on the global market in times of substantial transition to renewable energy [1]. Today, both study regions are showcases of Arctic economies that are still dominated by natural resource industries, the government sector, and traditional subsistence activities [2].

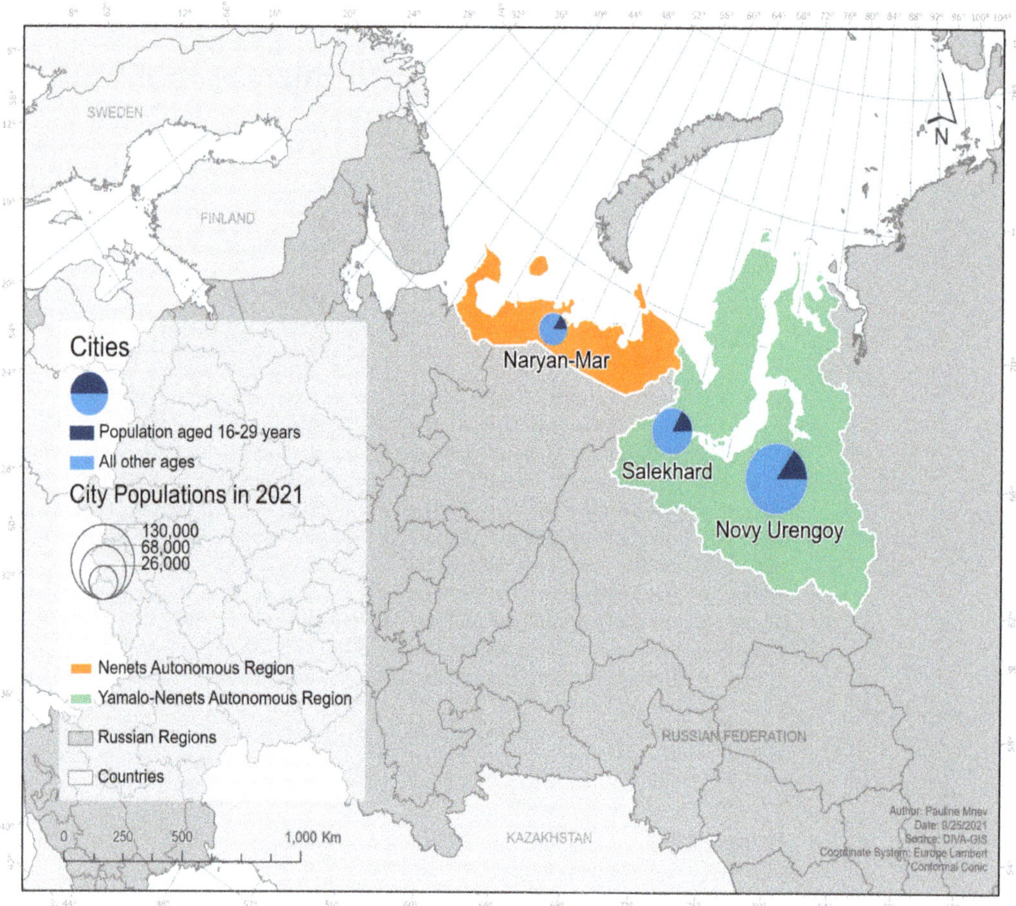


**Figure 1.** Map of the study sites showing the share of the city population aged 16–29 years. Author's calculations based on [3].

Rich in reserves of hydrocarbon raw materials, the local economies of NAO and YaNAO are heavily reliant on oil or gas extraction. NAO's estimated natural resources are 1 billion tons of oil and 500 billion cubic meters of gas. In recent years, 80–85% of the regional budget revenue came from taxes levied on oil companies. However, 84% of the tax revenue goes to the federal budget, and Naryan-Mar has limited resources for socioeconomic development and innovative practices [4,5]. YaNAO houses approximately 60% of Russia's gas reserves and 14% of its oil reserves, making it a highly attractive location for investments and development. The region is experiencing intensive economic growth. It is introducing mega-projects in the gas industry (with over 80% of all of Russia's natural gas production and over 60% of all-Russian LNG production), experiencing increases in urban infrastructure, and aggressively expanding transportation systems. Today, YaNAO is running one of the world's largest LNG projects. Related to oil and gas transportation, Yamal has one of the largest gas pipeline systems in the Circumpolar North. It is also one of the federal budget "donor regions", providing more than 10% of Russia's federal budget revenues [6]. Both the NAO and YaNAO regions are home to Indigenous Peoples pursuing a traditional way of life, and ambitious development projects are often confronted with intensified competition for land use with an Indigenous subsistence economy [7].

The three Arctic cities explored in this paper are different yet indicative cases of the overreliance on natural resource-based regional economies in the Arctic. Naryan-Mar and Salekhard—regional capitals (administrative centers) with regional legislative institutions—are located on Indigenous ancestral lands and are home to the Arctic Indigenous Peoples. Naryan-Mar's population of 25,536 (2021) is heavily dominated by ethnic Russians, with the second-largest group being Indigenous Peoples (Nenets and Komi (14.01%)) [8]. In Salekhard, with a population of 51,186 (2021), there are strong trends towards greater diversity. Among the numerous top ethnic groups are Russians, Tatars, Ukrainians, and Indigenous Peoples of the North (8.47%), including Nenets, Khanty, and Komi-Zyrayane [9]. The most populated city Novy Urengoy (118,115 residents, 2021), unofficially called the "Gas Capital of Russia", is a truly ethnically and culturally diverse YaNAO industrial capital with new strong trends towards even greater ethnic, cultural, and religious complexity [9,10]. The exception here is the absence of Arctic Indigenous Peoples: Only a few Indigenous families have settled in Novy Urengoy as for centuries Nenets People have viewed this location as "The Land of Fire", a deathtrap.

Today, NAO and YaNAO hold the top two positions on the list of Russian regional GDP per capita [11], which is approximately ten (!) times higher than the national average [6]. At the same time, both NAO and YaNAO are the only regions in Russia that do not have universities, and NAO has no scientific research centers. Their absence not only presents a clear barrier for the emergence of a modern knowledge-based economy in these resource bases, but can also greatly affect the local youth and their education, career, and life choices.

Based on quantitative and qualitative analysis, this paper attempts to answer the critical research questions: "What makes local urban youth in the two wealthiest regions of Russia—the Nenets and Yamal Nenets regions—want to leave permanently?" and "How does this influence broader-scale patterns of social sustainable development in the Arctic?"

Assessment of multiple risks associated with regional economies, environment, and local communities within the theoretical framework of sustainability is gaining more attention in science and recognition in policymaking across Arctic regions. Despite the initial predominant focus on economic sustainability aspects (this vision of sustainability as economic sustainability is still prevalent in official documents in the public administrative sphere in Russia [12] (Decree of Administration of Naryan-Mar N 422 (31 March 2015) "Ob utverzhdenii plana pervoocherednykh meropriyatiy po obespecheniyu ustoychivogo razvitiya ekonomiki i sotsial'noy stabil'nosti v MO "Gorodskoy okrug" Gorod Naryan-Mar ["On approval of the Plan of Priority Measures to Ensure Sustainable Economic Development and Social Stability in the City of Naryan-Mar"]; Strategy of Socioeconomic Development of Naryan-Mar till 2030; Strategy of Socioeconomic Development of Salekhard till 2030; Strategy of Socioeconomic Development of Novy Urengoy till 2030); see also: [13]) and environmental sustainability aspects, studies are now shifting towards a more comprehensive approach including a social component [12–16]. The concept of 'social sustainability,' a relatively new conceptual analysis area in wide-ranging sustainability, is firmly placed at the forefront of emerging Arctic urban sustainability studies [12,13,17–19].

The conceptualization of social sustainability is problematic [20] due to its multifaceted complexity and dynamism. Out of the myriad of employed definitions, in this research, 'social sustainability' is presented through the prism of a future-oriented construct of "sustainable urban communities", which are broadly defined as "places where people want to live and work, now and in the future" [21] (p. 6).

In Russia, academic and public discussions conceptualize the future of the North and its urban areas generally through the prisms of applied economic theory [22] with elements of instrumental rationality, economic geography [23], or allusions to a Soviet historical legacy of Arctic exploration and post-Soviet narratives about the Arctic and its role in national identity and pride as, e.g., described in [24]. To a lesser extent, these discussions apply concepts of social psychology such as belonging and emotional attachment to place and community [13,23].

Often overlooked in sustainability studies [25], youth need a special focus of attention in the future-focused approach as drivers of economic change and contributors to local communities' development. This study identifies the following key components of social sustainability [20,26,27] as especially relevant to urban youth of various age cohorts between 14 and 35 years old in the three Polar cities of Naryan-Mar, Salekhard, and Novy Urengoy:

- Opportunities for education and training (including well-performing higher education institutions).
- The scale of migration and its patterns.
- A wide range of high-quality jobs available for local youth.
- Opportunities for cultural, sports, and leisure activities.
- Social integration and social contribution (community and voluntary sector).

To measure these components, a system of youth-relevant indicators has been developed. Primarily based on ISO 37120 (Sustainable cities and communities—Indicators for city services and quality of life) (ISO, 2018) and the findings of the Program for International Research and Education project "Promoting Urban Sustainability in the Arctic" (PIRE Project) (PIRE), this paper also introduces new variables (e.g., entertainment and civic activities) (see Appendix A, Table A1). Due to limitations on data availability, other components of social sustainability, such as "Feeling of belonging (emotional attachment) to a place and community", "Affordable housing", and "Political engagement and empowerment" are not a part of analysis in this paper.

Among the selected social sustainability components, local youth migration is "one of the main drivers of changes in the urban landscape" [28] (p. 108). In many ways, other components can be described as underlying determinants of migration trends that affect migration flows, particularly youth out-migration.

In Arctic social studies, youth have been a special focus for decades [29–36]. Despite an increasing research interest in Russian Arctic youth's portrait and migration motivations [37–41], our knowledge in this sphere is still scarce. This limits our understanding of the factors contributing to social sustainability in the Arctic regions for the years ahead. In an attempt to fill some knowledge gaps, this paper presents Arctic youth's voices on their educational and career strategies, migration patterns, and future prospects.

## 2. Materials and Methods

### 2.1. Rationale for the Selection of Study Sites

This study uses the three Arctic cities of Naryan-Mar, Salekhard, and Novy Urengoy in NAO and YaNAO as case studies that are indicative of many Circumpolar regions. Based on functional classification, the focal cities represent two major models in the Arctic: administrative centers (Naryan-Mar and Salekhard) with the dominance of the government sector in the economy, and an industrial center (Novy Urengoy) with an economy based on natural resource extraction. In addition, NAO and YaNAO are the two wealthiest major economic centers among the Russian regions (GDP per capita), which helps to shed light on challenges and opportunities that Arctic youth can experience in other affluent Circumpolar regions.

### 2.2. Design of Selected Social Sustainability Indicators

As part of a contribution to studies on Arctic urban sustainability, this research applies metrics created by ISO 37120 [42], which have been modified to study youth. Aimed at finding some gaps in the system of ISO indicators related to the North's specific features, this novel approach includes analysis of variables that are likely to affect the future social sustainability of the Arctic urban communities. As presented in Appendix A (Table A1), the selected variables include relevant values covered by ISO and complementary variables explicitly designed for urban youth. Among them are demographics, education, economics, culture, sports, entertainment, civil society, and political empowerment indicators.

### 2.3. Methods

Data for indicators were collected from primary and secondary sources of information and included regional and municipal statistical datasets (Rosstat), ISO 37120 developed by the PIRE project's experts, reports from government bodies, municipal development plans, and regional strategies, and municipal reports.

The quantitative analysis of indicators complements qualitative findings based on informal surveys (questionnaires) that included young people, education and labor market professionals, interviews with local officials and Indigenous leaders, and participant observations undertaken during the author's qualitative research in the focal study areas.

In this study, questionnaires were chosen as the most efficient survey method as they allow for information to be gathered from large audiences and for results to be compared. They can also be used in future research to measure change, as well as preserving the anonymity of survey participants. Overall, 406 participants contributed to this study in the form of informal surveys (questionnaires) and interviews. The youth survey enrolled local young people—258 high school and vocational college students, including predominantly Indigenous high school students from the boarding school in NAO (Naryan-Mar) (50 respondents); high school students in NAO (Naryan-Mar) (36 respondents); high school students in YaNAO (Salekhard and Novy Urengoy) (84 respondents); and vocational college students in NAO (Naryan-Mar) (40 respondents) and YaNAO (Salekhard) (48 respondents). Questionnaires addressed the following topics: youth's life strategies (where they see their future); education and career strategies (who they want to be, where they are planning to proceed with their education, where they want to work); advantages and barriers to professional/personal growth in their home regions; students' networks outside their cities (friends and relatives); and students' leisure time and hobbies. To ensure that the study posed no risks to the participants, the questionnaires did not include the sensitive topic of political engagement and empowerment.

To understand the broader context, provide additional insights, and better understand local labor market conditions and challenges for local youth, 132 education and employment services professionals (52 in YaNAO and 80 in NAO) were surveyed. The questionnaires were focused on the most pressing issues that young city-dwellers face in the NAO and YaNAO regions, including education and career opportunities as well as local labor market conditions. All surveys were carried out face to face. They contained both open-ended and closed questions; the latter also included a continuous rating scale to measure the strength of attitudes. In addition, sixteen in-depth, semi-structured, in-person interviews about local labor market conditions and opportunities for the local youth were conducted with regional officials and Indigenous leaders who live and work in the focal cities of NAO and YaNAO. This method was preferred as it is the most effective method for qualitative research for understanding the societal context and exploring respondents' opinions and experiences. Also, this method was chosen because of its practicality, as it enables researchers to reach this focus group. Initial interview respondents were selected through personal contacts, and the interviewee pool was developed using the snowball sampling method. All the interview participants were notified about the research and its objectives, and their consent for participation was received. Interviews were analyzed in detail primarily through thematic analysis.

## 3. Results

This section is organized in the following way. Each thematic part first presents available statistical information and analytical data in the broader context to examine key components of social (socioeconomic) sustainability primarily based on ISO 37120 in the three focal Polar cities, and then complements and links them to results of surveys during the author's qualitative fieldwork. The presented complementary qualitative findings voicing Arctic youth's vision of their education, career, life strategies, leisure time activities, and social connections support the quantitative analysis of social sustainability components and indicators of urban sustainability relevant to the Arctic youth.

*3.1. Educational and Labor Market Opportunities for Local Youth: General Overview*

Among the important common characteristics of the Northern cities of Naryan-Mar, Salekhard, and Novy Urengoy that to a certain extent determine both the local youth's education and career paths and life strategizing are: (1) no higher education institutions, and (2) a dominant natural resource-reliant economy that creates a highly competitive labor market for high-, semi-, and low- skilled workers.

3.1.1. Paradox: No universities vs. highest rates of university degrees among residents

Nowadays, NAO and YaNAO are the only two regions in the Russian Federation that do not have higher education institutions. By not investing in human capital in these "donor regions" through the higher education system, the federal government de facto forces employers to bring in well-trained crème de la crème professionals from elsewhere. Correlating with ISO indicator 6.6 "Number of higher education degrees per 100,000 population" (Appendix A, Table A1), available data on the share of the employed population aged 25–64 with higher education show that NAO and YaNAO are experiencing an increasing influx of highly skilled labor (Figure 2).

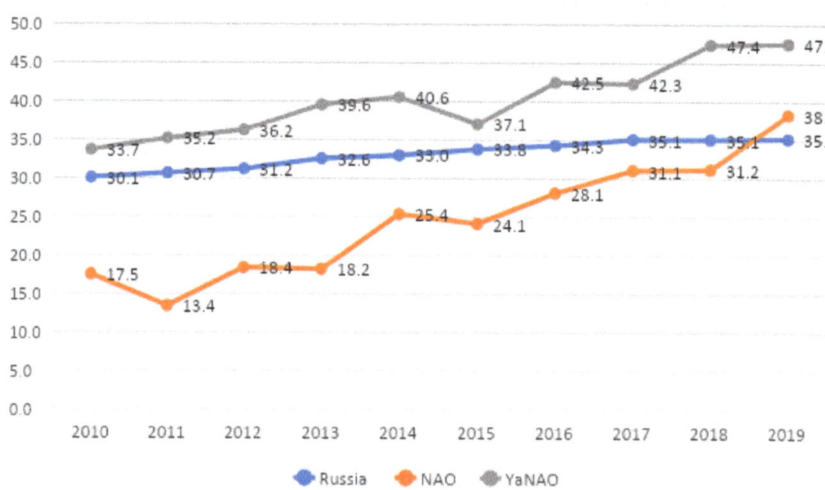


**Figure 2.** Dynamics of changing share of the employed population aged 25–64 with higher education (in %). Author's calculations based on Federal State Statistics Service of the Russian Federation (Rosstat), 2020 (annual sample surveys of the labor force).

This situation illustrates one of many Arctic paradoxes: two regions with no university education services have the highest university degree rates among residents in the country. As a result of this paradox, local high school and vocational college students are most likely to consider educational migration as the only way out (see survey results, Section 3.2 below). Often this decision to leave is a point of no return. Similarly to many other Circumpolar regions, return migration remains low: "leaving a region in pursuit of higher education increases the chance of a student starting a family [and/or new career path—author's] close to their place of education, which can decrease the likelihood of eventually returning to their region of origin" [43] (p. 183).

Although the Russian official statistics do not provide data on "city-to-city" and return migration [44] and accurate numbers are unavailable, webometric research findings by Moscow State University confirm the general trend: young people do not return to their Arctic home cities with university diplomas [45] and are likely to settle elsewhere in better climatic conditions and with a wider range of professional development opportunities. These results also concur with findings from surveys among education and employment professionals (72% described youth's permanent out-migration as a strong ongoing trend) and the author's interviews with local officials who recognize these processes as a serious concern.

From a historical perspective, youth exodus, primarily non-Indigenous, from the Russian Arctic is not a new phenomenon [46–49] in the dynamic and continuously changing Arctic, but rather a "continuation of a family cycle, migration to the North–migration from the North, which takes between one and three generations to complete" [50] (p. 61). However, the scale of today's ongoing youth educational out-migration and the exacerbating trend of youth reduction in the working-age cohort signal an ongoing demographic shift (Figure 3) and future labor supply challenges. In all focal cities' official Strategies of Socioeconomic Development until 2030, youth exodus is described as a clear threat to local economy and a challenge for city youth-oriented policies [51–53].

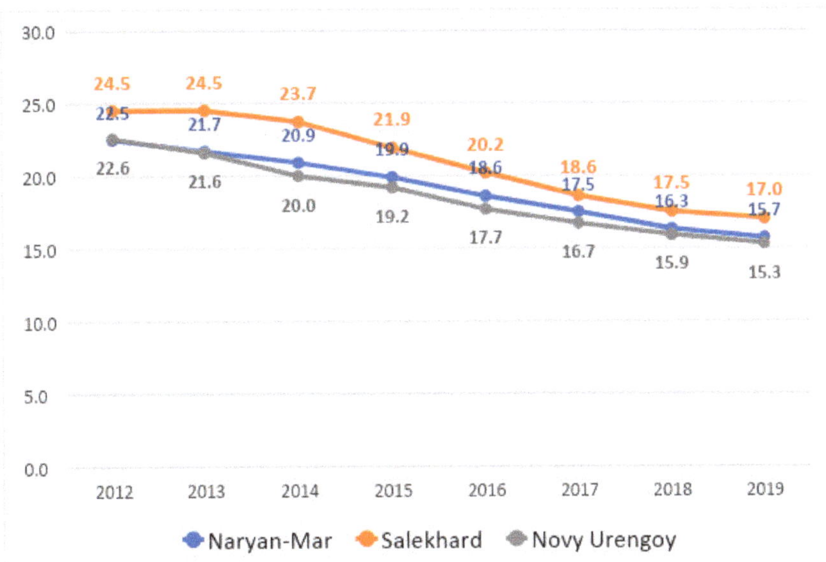


**Figure 3.** Dynamics of changing share of young adults in the 16–29-year-old cohort in Naryan-Mar, Salekhard, and Novy Urengoy, 2012–2019 (in %). Author's calculations based on [3].

Despite ongoing negative demographic trends, there is still no indication of a strong deficit of young people of working age for local economies. Out-migration of young adults meets a counter-trend—both Northern regions (urban centers) turned into magnets receiving a surplus of young people of working age (predominantly between 20 and 30), both high- and low-skilled professionals [28,54] (3).

3.1.2. Arctic Labor Market and Left behind Locals

The economies of NAO and YaNAO are heavily dominated by natural resource industries, the government sector, and traditional subsistence activities [2]. Depending on the city's central role as an industrial or administrative center, the most lucrative sphere of employment with higher wage rates in focal urban areas remains oil and gas companies (Novy Urengoy) and the government sector (Naryan-Mar and Salekhard) [5,6]. Overall, in NAO, the oil sector generates 8700 jobs (25% of the total NAO labor resources) [5], and the government of the regional capital of Naryan-Mar is the top employer (17% of total city employees). In YaNAO, the oil and gas sector generates 27% of the total jobs [6]; in Novy Urengoy, the gas capital of YaNAO, the fuel and energy complex employs 40.7% of the total labor resources [6], while in Salekhard, the government is one of the top sectors of employment with 27% of total city employees [52]. The question remains as to who can obtain high-quality jobs in those key economic sectors.

As a long-established and globally widespread practice in extraction industries [55–59], primarily oil and gas in these regions, corporations heavily rely on an "imported" temporary and fly-in, fly-out (FIFO) workforce and are reluctant to hire locals outside of a narrow range of qualified professionals specially trained for their needs or unskilled workers for miscellaneous services (e.g., cleaning, catering, etc.). A significant gap is also diagnosed between the level of remuneration in the extractive industries and other sectors of the economy. For instance, in NAO, the average monthly salary in extractive industries is more than 100 thousand rubles, while in other sectors of the economy, it is about 50–70 thousand rubles [53].

In addition, the regional labor markets are also saturated with newcomers from Russia's northwestern and central regions and the North Caucasus (predominantly from Dagestan and Chechen Republic) [60], who intend to stay for prolonged periods of time, as well as seasonal labor migrants from 'near-abroad' (primarily from the Central Asian countries of Kyrgyzstan, Uzbekistan, and Tajikistan), and diverse labor migrants from Azerbaijan [10,61,62]. It is plausible that competition for jobs between semi- and low-skilled young locals and newcomers is so intense that the former ones who have not left their hometowns for better education and training perceive themselves as "left behind" vulnerable Northerners. For instance, in Salekhard, the survey presented in the Strategy of Socioeconomic Development of Salekhard until 2030 indicates a high level of vulnerability among low/semi-skilled workers: 100% of respondents without higher degrees expressed serious concerns about poverty and unemployment; in comparison, only 56% of respondents with higher degrees were concerned about poverty, and 49% about unemployment.

An effort to address the lack of available high-quality jobs for young residents was initiated in NAO in 2016. The authorities introduced quotas for all companies with more than 50 employees working in the region to hire local residents [63]. However, the implementation of this initiative is problematic: quotas, even when "formally" filled, do not guarantee the locals prospective and well-paid positions.

Top oil and gas extraction companies—Rosneft (NAO) and Lukoil, Novatek, and Gazprom (YaNAO)—have invested in youth human capital by supporting educational initiatives, yet the outcomes are very modest and do not make any difference in the local labor market in the broader context of Arctic social sustainability. For instance, under an NAO administration initiative, in 2012, Rosneft launched an educational project of small-numbered high school classes with intense specialized training for the best high school students (10th–11th grade) in the region. Since 2012, out of almost 100 graduates from Rosneft classes, only one has been employed in this oil company, and five students have got scholarships at the Ukhta State Technical University in the neighboring Komi region [64]. In Salekhard and Novy Urengoy, similar educational initiatives are being implemented by Gazprom (since 2010), Novatek (since 2018), and Lukoil (since 2020). In addition to these specially organized high school classes, Gazprom also runs a company-owned Vocational School in Novy Urengoy and provides employment opportunities to some graduates for positions that do not require university diplomas. Although neither the Ministry of Education of YaNAO nor Gazprom publish information about graduates' employment results and career paths, based on the results of interviews and general observations, one can suggest they are also not impressive.

Indigenous youth have been especially affected by highly limited state-funded scholarship opportunities and the de facto abolition of the Soviet system of targeted enrollment of Indigenous students. To address this issue, new initiatives for new enrollment programs have been presented by the Russian Association of Indigenous Peoples of the North (RAIPON) [65,66], which may have a positive effect on educational and career opportunities for Indigenous youth in the near future. Also, in 2020, the YaNAO government initiated a regional limited support program for Indigenous students to compensate for their education costs [67].

Typical of many Arctic resource-based economies, gender segregation in labor markets [68–70] sharply appears in both Arctic regions and primarily affects women by narrowing their chances for economic empowerment to traditionally female occupations in the social and NGO sphere, education, medicine, and public (municipal) administration. Nevertheless, there are strong signs of young women's growing interest and involvement in entrepreneurship and social entrepreneurship, particularly. Since 2018, both regional governments subsidize small business initiatives, organize educational training programs, and provide opportunities for mothers with small children to get back to work (for instance, the Yamal educational project "Mom Is an Entrepreneur" [71]). YaNAO also has one of the highest ratings of governmental support out of all Russian regions for socially oriented NGOs [72] to promote and encourage predominantly female social entrepreneurship. Although the YaNAO and NAO programs are not aimed at overcoming traditional gender-based labor division, they firmly intend to create more opportunities for women in the local labor markets.

### 3.2. Educational Opportunities and Career Prospects through the Eyes of the Local Youth

The results of the youth survey revealed that to be competitive in the Arctic labor market, young people approaching high school or vocational school graduation who seek higher education opportunities and higher social status have to leave.

In Naryan-Mar, 9% of male and 8% of female high school students stated their intention to live in their home region (Table 1), while in Salekhard and Novy Urengoy, it is only 4 (!) % for both genders, respectively (Table 2).

**Table 1.** Settlement preferences among high school students (14–17 y.o.) in Naryan-Mar (%, participating respondents).

| Planning to Live in | No | | Not Sure | | Yes, Most Probably | |
|---|---|---|---|---|---|---|
| | Male | Female | Male | Female | Male | Female |
| NAO | 72 | 64 | 18 | 28 | 9 | 8 |
| Another region in Russia | 9 | 8 | 36.5 | 20 | 54.5 | 72 |
| Another country | 54.5 | 76 | 27.5 | 4 | 18 | 20 |

Source: Author's survey, conducted with Dr. Andrey Gretsov (high school students from Naryan-Mar).

**Table 2.** Settlement preferences among high school students (14–17 y.o.) in YaNAO cities of Salekhard and Novy Urengoy (%, participating respondents).

| Planning to Live in | No | | Not Sure | | Yes, Most Probably | |
|---|---|---|---|---|---|---|
| | Male | Female | Male | Female | Male | Female |
| YaNAO | 77 | 64 | 19 | 32 | 4 | 4 |
| Another region in Russia | 8 | 5 | 11 | 19 | 81 | 76 |
| Another country | 54 | 45 | 31 | 40 | 15 | 15 |

Source: Author's survey, conducted with Dr. Andrey Gretsov (high school students from Salekhard and Novy Urengoy).

In the boarding school in Naryan-Mar, among predominantly Indigenous students, 17% of male and 26% of female students see their future in NAO (Table 3). On the one hand, these numbers reflect that Indigenous people in the Arctic keep strong bonds with their lands. On the other hand, they may also reflect that students' parents who live in rural areas and are involved in subsistence economy might have less financial aid to pay tuition and living costs to send their children to obtain better degrees [73].

**Table 3.** Settlement preferences among predominantly Indigenous high school students (14–17 y.o.) in Naryan-Mar (%, participating respondents).

| Planning to Live/Study in | No | | Not Sure | | Yes, Most Probably | |
|---|---|---|---|---|---|---|
| | **Male** | **Female** | **Male** | **Female** | **Male** | **Female** |
| NAO | 57 | 36 | 26 | 38 | 17 | 26 |
| Another region in Russia | 24 | 24 | 10 | 38 | 66 | 38 |
| Another country | 76 | 83 | 14 | 10 | 10 | 7 |

Source: Author's survey, conducted with Dr. Andrey Gretsov (high school students of the Pyrerka Boarding school in Naryan-Mar).

Nearly one third of vocational students from our focus groups expressed their intent to stay in their regions; others, to a greater or lesser degree, are considering relocation as a part of their life strategy (Tables 4 and 5).

**Table 4.** Settlement preferences among vocational college students (17–21 y.o.) in Naryan-Mar (%, participating respondents).

| Planning to Live in | No | | Not Sure | | Yes, Most Probably | |
|---|---|---|---|---|---|---|
| | **Male** | **Female** | **Male** | **Female** | **Male** | **Female** |
| NAO | 23.5 | 26 | 43 | 42 | 33.5 | 32 |
| Another region in Russia | 33 | 48 | 40 | 20 | 27 | 32 |
| Another country | 80 | 72 | 20 | 24 | 0 | 4 |

Source: Author's survey, conducted with Dr. Andrey Gretsov (vocational students in the colleges of Naryan-Mar).

**Table 5.** Settlement preferences among vocational college students (17–21 y.o.) in YaNAO cities of Salekhard and Novy Urengoy (%, participating respondents).

| Planning to Live in | No | | Not Sure | | Yes, Most Probably | |
|---|---|---|---|---|---|---|
| | **Male** | **Female** | **Male** | **Female** | **Male** | **Female** |
| YaNAO | 18.5 | 50 | 46.5 | 16.5 | 35 | 33.5 |
| Another region in Russia | 33 | 44 | 37 | 17 | 30 | 39 |
| Another country | 54 | 78 | 33 | 11 | 13 | 11 |

Source: Author's informal survey, conducted with Dr. Andrey Gretsov (vocational students in the colleges of Salekhard).

Among the vocational students who indicated a willingness to relocate to other cities/regions/countries, 86% planned to obtain higher degrees and get better skills to compete in the job market.

### 3.3. *Employment Opportunities through the Eyes of the Local Youth*

Despite the relative economic prosperity of these two Arctic regions and low unemployment rates among the young (ISO Indicator 5.4: 3.7% for Salekhard and Novy Urengoy, and 4.7% for Naryan-Mar [74] (Appendix A, Table A1)), local Arctic youth—both Indigenous and non-Indigenous—anticipate difficulties in finding suitable jobs and are mostly pessimistic about their prospects in hometowns.

For instance, college students found themselves in a challenging position: In YaNAO, 50% of females and 30% of males expressed strong concerns about their employment and job possibilities, pointing out a "lack of jobs"—mainly due to competition in the local labor market; in NAO, those percentages were 60% and 50%, respectively.

High school students also indicated significant obstacles to getting a job in the local labor market. In NAO, 36% of female and 18% of male students pointed out "lack of vacant jobs/lack of jobs in the certain specialty"; 27% of male students also mentioned "low educational level", and 18% "high competition" and "low paid jobs". In boarding school, students (predominantly Indigenous) indicated even deeper concerns: 48% of females and 24% of males pointed out "lack of vacant jobs/lack of jobs in the certain specialty"; 21% of females also indicated "low educational level", and 14% of males—"high competition".

In YaNAO, 26% of females and 27% of males indicated "low educational level"; 24% of female and 15% of male students pointed out "lack of vacant jobs/lack of jobs in the certain specialty"; 26% of female and 15% of male students pointed out "lack of professional experience". Also, 27% of males indicated "high competition".

In all groups, students of both genders often mentioned, among other factors, "lack of professional experience" (often required by employers), "low paid jobs", "family issues", including "family's disapproval of professional choice", "laziness", and "low confidence", "lack of connections", "financial issues" (no possibilities to relocate or to start a small business), "limited job vacancies available", "a highly narrow, single-industry economy and the [economic— author's] underdevelopment in the region", etc.

Continuing professional development and career building in the Arctic is also viewed as problematic for respondents. Among the main reasons, they mentioned "narrow range of professions", "underdevelopment of the region", and "not enough opportunities to carry out certain activities" (female students, NAO); "few centers with diverse areas of professional activity", "it is hard to get the initial capital for professional growth and get a job with a good salary and [with the prospect — author's] of growth on the career ladder; it is challenging, you need to be a first-class professional" (male students, YaNAO). At the same time, as was illustrated by the survey participant, good education combined with professional competence is the key to success in their hometowns: "With a good education and professional skills, competition is reduced at times" (male college student, NAO).

The study also revealed gender differences in students' career preferences and that gender-related imposed choices of professions are more prevalent among females. Female students tend to think big and outside the box. Overall, they may have great potential for professional growth and future empowerment. However, even though they are inspired to be geologists, policewomen, ecologists, customs officers, chemists, movie producers, pharmacists, architects, and prosecutors, many emphasized the gap between personal aspirations and realities. They admitted that they were more likely to become (pre)schoolteachers, nurses, etc. For instance, in Naryan-Mar, predominantly Indigenous female students (14–17 y.o.), who came from remote Indigenous communities or originated from families pursuing traditional, including nomadic, lifestyles in remote rural areas of the NAO, demonstrate significant shifts in career strategizing while admitting the insurmountable obstacles to achieving their goals:

> "I really want to work in law enforcement agencies, in the police. However, I'm not sure whether I can enter and learn this profession with my academic performance. In this case, I am thinking of applying to be a preschool teacher" (female student, NAO); "I would love to become a choreographer, but I am thinking about the profession of a physical education teacher in a school" (female student, NAO).

Among other barriers for youth's employment in their Arctic regions that cannot be seen in official statistics and reports, the respondents pointed out various forms of nepotism and cronyism in the labor market, with elements of the closed culture of corporatocracy:

> "If you do not have good connections and money, then you will not have any professional growth" (female student, NAO); "If you do not have friends, it is hard to get a job" (male student, YaNAO); "You can get a job (good, well paid) ONLY through CONNECTIONS" (male vocational student, YaNAO); "A large number of private entrepreneurs hire relatives" (male vocational student, YaNAO).

Out of many advantages of working in the North, the young respondents primarily acknowledged the government economic and social support system built in the Soviet era to stimulate Arctic regional development [75–78]. For instance, among the advantages of working in YaNAO, 50% of female and 53% of male college students mentioned higher salaries and Northern allowances, and 17% of females and 20% of males mentioned long vacations. Also, 27% of male students highlighted early retirement with additional allowances vs. just 6% of females, respectively ("Working in the YaNAO, your length of service is calculated one year for two, or, in some cases, for three years. In other words, you can earn a pension faster here and receive a Northern allowance on top of your pension" (female college student, YaNAO)). In NAO, 36% of female and 53% of male college students mentioned higher salaries and Northern allowances.

On the one hand, high school respondents believed that the underdeveloped service sphere and undiversified Arctic economy potentially open vast opportunities for starting their own businesses (17% of female students in YaNAO and 20% of female students in NAO vs. 0% of males, respectively): "There are advantages for the development of small businesses, since there are not enough here" (female student, YaNAO); " ... to open a production line of goods that are not available in the region" (female student, NAO). On the other hand, high school students, especially females, less reliant on the paternalistic model of governance, see very few or no advantages and prospects in the Arctic regions (22% of females in YaNAO and 28% in NAO).

In YaNAO, among the five top-rated future prospective labor sectors in their regions, students, especially females, identified those related to the government and government-funded sectors (schools, hospitals) or big businesses with a high level of government involvement (oil and gas industries). Female students also view jobs in the traditional female domain as the most prospective—44% in medicine and 33% in education—while male students chose the male-dominated IT sphere (see Figure 4).

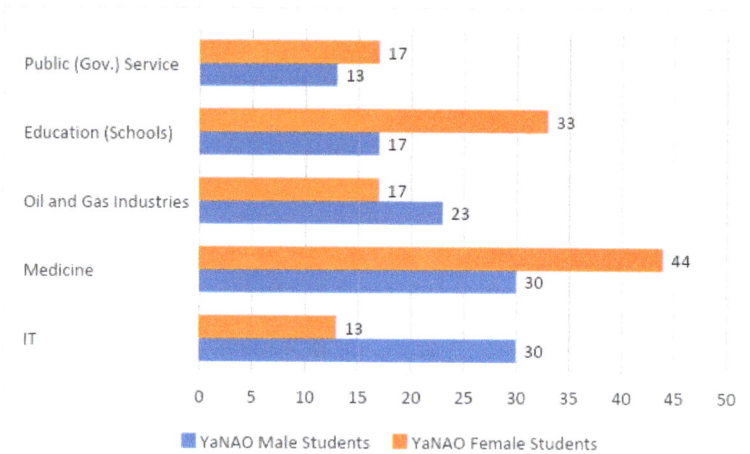


**Figure 4.** YaNAO vocational students' choices of future prospective labor sectors in their regions. Top 5 identified (%).

In NAO, with minor exceptions similar to Yamal, both male and female students are also in a path dependency situation. While females consider education (44%) and medicine (36%) among their top spheres, males view the oil and gas industries and reindeer husbandry as the core of Nenets's traditional economy, as well as transportation and construction as the most attractive and promising professions in NAO (Figure 5).

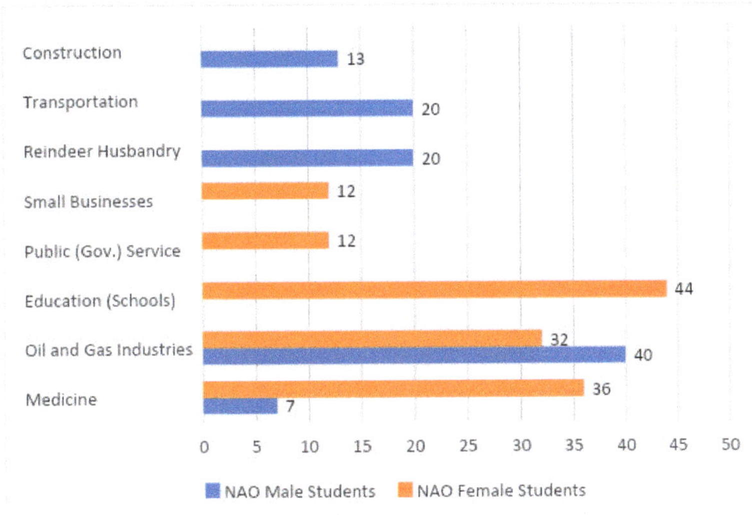


**Figure 5.** NAO vocational students' choices of future prospective labor sectors in their regions. Top 5 identified (%).

### *3.4. Opportunities for Cultural Activities, Sports, and Entertainment: General Overview of Cultural Infrastructure*

Regarding cultural infrastructure, "the number of cultural institutions, such as theaters, movie houses, and libraries, is only a starting point for understanding the cultural vitality of the city" [13] (p. 284) (ISO indicator 17.1, "Number of cultural institutions and facilities per 100,000 population"). The cultural indicator values (28 for Naryan-Mar, 22 for Salekhard, and 17.2 for Novy Urengoy (Appendix A, Table A1)) do not explain the real spectrum of opportunities for diverse young urbanites in the cultural sphere and therefore require a broader context.

In NAO and YaNAO, revenues coming into the regional budgets from extraction businesses allow local authorities to invest in cultural infrastructure, yet the policy implications differ in each of the three cities. Rooted in the Soviet tradition of formal cultural recognition [79], cultural policy priorities are largely determined by cities' ethnic composition and share of Indigenous population (Table 6).

Given the increasing trend of Indigenous urbanization, governments in both regional capitals—Salekhard and Naryan-Mar—acknowledge Indigenous heritage and emphasize symbolic recognition policies by "Indigenizing the urban landscape", supporting and promoting refined forms of Indigenous cultures [79]. While meaningful to Indigenous youth, these cultural initiatives do not equally benefit all Arctic youth groups.

The private sector in the sphere of culture is mostly underdeveloped, especially in Naryan-Mar and Salekhard, because of the low number of their population that does not generate great consumer demand. For instance, in YaNAO, the share of private businesses in the total number of organizations providing cultural services is 25% (for comparison, in Russia it is 53%, 2018) [6] (p. 56). To a lesser or greater extent, all three cities are experiencing a deficit of services of high importance for urban youth social life—cinemas, commercial concert halls hosting diverse and popular artists, diverse meet-up places for amateur interest groups, non-government youth creativity centers, as well as professional theatres s [80] (Figure 6).

**Table 6.** Ethnic composition of the cities: largest ethnic groups (%).

| Major Ethnic Groups | Naryan-Mar | Salekhard | Novy Urengoy |
|---|---|---|---|
| | **Non-Indigenous** | | |
| Russians | 79.41 | 61.27 | 64.14 |
| Ukrainians | 2.35 | 5.80 | 10.76 |
| Tatars | 0.56 | 8.50 | 4.99 |
| Belarusians | 0.70 | 0.74 | |
| Azeris | 0.63 | - | 1.95 |
| Bashkirs | - | 0.58 | 1.69 |
| Nogais | - | - | 2.61 |
| Kumyks | - | - | 2.06 |
| Chechens | - | - | 1.12 |
| Kyrgyz | - | 0.96 | - |
| | **Indigenous** | | |
| Nenets | 6.71 | 2.83 | - |
| Komi/Komi-Zyrayane | 7.30 | 2.56 | - |
| Khanty | - | 3.08 | - |

Source: The Russian Census of 2010 (latest available data).

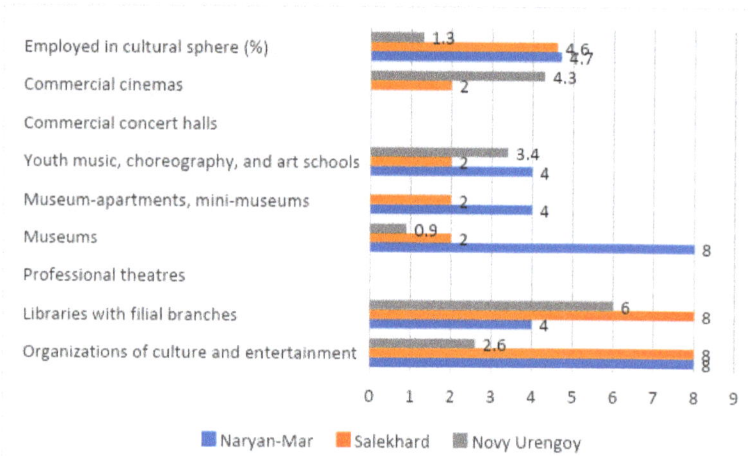


**Figure 6.** Number of cultural institutions (facilities) per 100,000 people, and employment in the cultural sphere (as a percentage of total employed). Sources: Author's calculations based on Municipal Statistics, Federal State Statistics Service of the Russian Federation (Rosstat), 2019 (latest available data); Regiony Rossii. Sotsial'no-ekonomicheskiye pokazateli – 2019. Rosstat:Moscow, 2019; 2gis.ru.

### 3.4.1. Cultural Facilities in Naryan-Mar: Diversity in the Monotony

Despite a high value for the 17.1 ISO indicator, not all cities' cultural facilities meet the expectations of different local youth groups. Naryan-Mar has a rather ethnically homogenous society with a relatively significant share of Indigenous population and limited cultural diversity policies: The cultural institutions are mostly folklore-oriented and work as fabrics for refining primarily Arctic Indigenous art and culture. For instance, under the umbrella of the budget-funded Culture Center "Arktika" and Ethnocultural Center, out of thirteen dance, choir, and theatrical studios with young people, only four are relevant to youth modern mainstream culture. The rest are folk-specific and, although culturally diverse, limited to Northern (including Russian) cultures only. Growing ethnic groups (both new and well-established but small-numbered) are not very well represented in the cultural sphere. Despite officially registered Dagestani, Tatar-Bashkir, and Chuvash associations [81], and Azeri people's visible presence, they are not perceived as an integral "permanent" part of the cultural and social fabric and thus not equally displayed in the public cultural domain.

Such a practice of selective cultural inclusion might also be influenced by the general public's views in this region on diversity policies: A public survey conducted in 2019 revealed that only 53% of respondents believed that all residents of ethnically defined autonomies, such as NAO, have equal rights regardless of their ethnic background; 37.5% expressed a strong belief that Indigenous Peoples are entitled to have more rights in their ancestral homeland than other ethnicities; 9% did not have an opinion about this matter [81].

Functionalities of other city cultural institutions are somewhat narrow and old-fashioned and play a minimal role in young city dwellers' everyday lives. The Central Library of NAO in Naryan-Mar offers special events and programs mainly oriented to children and older adults. Nenets Regional Museum of Local Lore has rich permanent collections and professional tours led by historians and Indigenous researchers. However, they are rather tourist-oriented and not a constant point of attraction for young locals.

#### 3.4.2. Cultural Facilities in YaNAO: Move towards a Greater Diversity

In YaNAO, both focal cities with their more significant mobility patterns present "an ongoing progressive move from a Eurocentric culture toward greater hybridization" [82] (p. 3) and more versatile cultural policies that are acclimating to better meet youth's expectations. The Department of Culture of the YaNAO focuses on preserving and developing cultural heritage and the inclusion of trendy, modern styles.

Novy Urengoy, as a young single-industry city founded in 1975 without much of a historical legacy beyond gas exploration, does not have a variety of museums except the Gazprom Museum and Art Museum, the only one in Yamal. Despite that, the city's support for popular youth cultures and diverse activities makes it more attractive for the young generation. It also contributes to and defines the Arctic shifting identity with emerging forms of Arctic ethnic and cultural diversity [83] and provides more opportunities for young newcomers to integrate into the Arctic mosaic social fabric.

In Salekhard, government-supported ensembles and studios that welcome young people reflect growing cultural diversity. Among them are the folklore ensembles of the peoples of the North Caucasus "Siyanie gor", Tatars and Bashkirs "Duslyk", and Mari people "Mari Kundem". Overall, out of eleven government-funded cultural projects, only three are ethnically neutral, and out of eight ethnically oriented projects, five promote the Indigenous cultures of the North.

The Department of Culture of YaNAO made a successful attempt to revitalize old-fashioned "monofunctional" cultural institutions to make them constant points of attraction for diverse cultural, intellectual, and social activities (for instance, the project "New Library of Yamal" [84]; a good example here is the remodeling and reformatting of the Yamal National Library in 2017). Importantly, it has also supported inclusivity in the cultural sphere by establishing a studio of creativity for people with disabilities.

#### 3.4.3. Cultural Economy for Indigenous Youth

In Salekhard and Naryan-Mar, the Soviet cultural recognition policies' legacies are still vital and set the agenda in the cultural domain. As demonstrated in some Indigenous studies, the developed cultural economy is of crucial importance to many Indigenous urbanites: Although Indigenous communities' sustainability was initially based on "an appreciation of intimate relationships between humans and the local environment [...], in the urban setting, these relationships have been mediated by infrastructural development" [80], particularly cultural infrastructure. The cultural sphere is of crucial importance as Indigenous urbanites often play the "social roles of professionals in organizing ethnic processes in the urban space" [85]. The indicator "Employment in the cultural sphere (as a percentage of total employed)" (Appendix A, Table A1) demonstrates high rates of government-funded employment in the cultural sphere for Naryan-Mar and Salekhard (4.6% and 4.7% respectively of total employed vs. 1.3% in Novy Urengoy). Although additional research is needed, it is possible that these employment opportunities also serve to strengthen the social and cultural capital of local Indigenous Peoples and may contribute to Arctic social substantiality.

### 3.5. Sports Infrastructure

All three Arctic cities have a high number of sporting facilities (Figure 7 presents Indicator 17.1 (Appendix A, Table A1)). Generously (co-)sponsored by the oil and gas corporations, they are an example of successful promoting and popularizing of sport among youth.

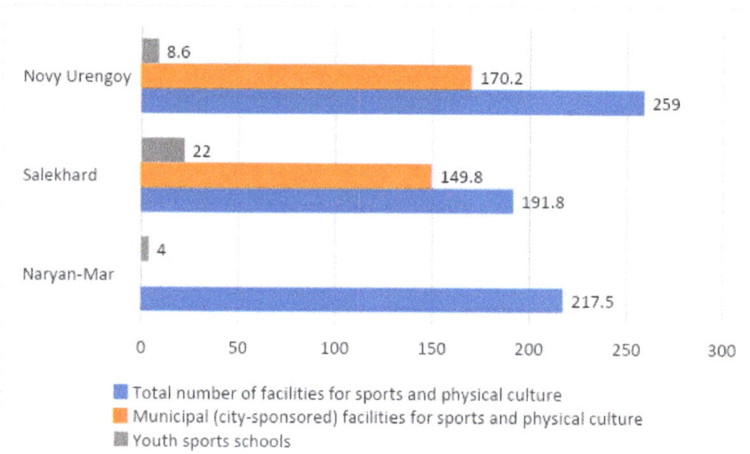


**Figure 7.** Number of sporting facilities per 100,000 people. Sources: Municipal Statistics, Federal State Statistics Service of the Russian Federation (Rosstat), 2019 (latest available data).

Similar to the cultural sphere, which acknowledges Indigenous heritage, popular sports in Salekhard and Naryan-Mar are complemented by sports sections facilitating the development of Indigenous traditional sports and supporting professional competitions. In the case of sports, a higher value of the sports variable indicates an improvement in the quality of life of Arctic youth.

### 3.6. Arctic Youth's Leisure Time

Youth involvement in cultural, sporting, and other social activities and the degree of social integration and levels of social contributions to their Arctic communities among youth can be measured through the component of leisure time and its structure (based on [86]).

Similar to the general population of Russian youth [87], the structure of Arctic youth's leisure time is not very diverse. It varies depending on gender, specific age group, status (high school students or vocational students), ethnicity (Indigenous/non-Indigenous), and time of year. Leisure time structure also varies based on the availability and affordability of cultural, sporting, and entertainment facilities in hometowns.

In wintertime, both cohorts of the young generation—high school and vocational students—mostly prefer to stay at home but also spend some of their leisure time outside, in the fresh air (see Figures 8 and 9). While spending time indoors (at home or in youth-oriented centers), female students at school age, in particular, are increasing their cultural capital. They are inclined to do something creative (e.g., writing, painting, playing music, drawing, cooking, vocal singing, caring for botanic flowers, photo sessions, stucco figures modeling, etc.) and read books.

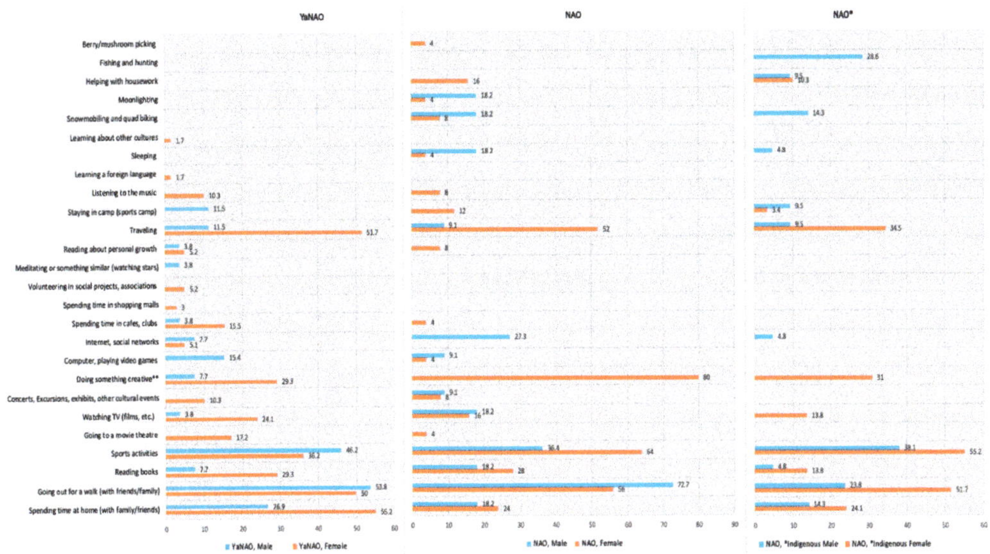

**Figure 8.** High school students from NAO and YaNAO, open question: "What activities are you engaged in during your leisure time? Where do you spend your leisure time?" (% of all collected responses in each group). * Students at the boarding school, predominantly Indigenous students. ** Includes activities such as writing, painting, playing music, drawing, cooking, vocal singing, caring for botanic flowers, photography sessions, modeling stucco figures, etc., and time spent in youth centers for creativity.

Based on survey results, indoor and outdoor sports activities play a vital role in the respondents' lives, and youth appreciate existing facilities in all three focal cities. With the exception of YaNAO male school students, sports are more prevalent among females. Both female and male respondents have a comprehensive range of sporting activities. Females are engaged in skiing, ice skating, cycling, rhythmic gymnastics, dancing, boxing, swimming, and volleyball, while male respondents prefer snowboarding, football, swimming, martial arts, volleyball, basketball, boxing, skiing, and biking. Almost one third of high school and vocational male students in NAO (28.5% and 33.3% respectively) and vocational male students in YaNAO are engaged in hunting and fishing, and Indigenous male students also practice reindeer herding.

In contrast to vibrant sports facilities, cultural infrastructure and public entertainment facilities are not satisfactory for local youth. Based on their responses, the share of school students going to cafes (in YaNAO, 15.5% of females and 3.8% of males; in NAO, 4% of females), movies (in YaNAO, 17% of females; in NAO, 4% of females), concerts, exhibits, and other cultural events (in YaNAO, 10.3% of females; in NAO, 8% of females and 9.1% of males) is relatively low, while vocational students and Indigenous school students did not mention engagement in any cultural and social activities.

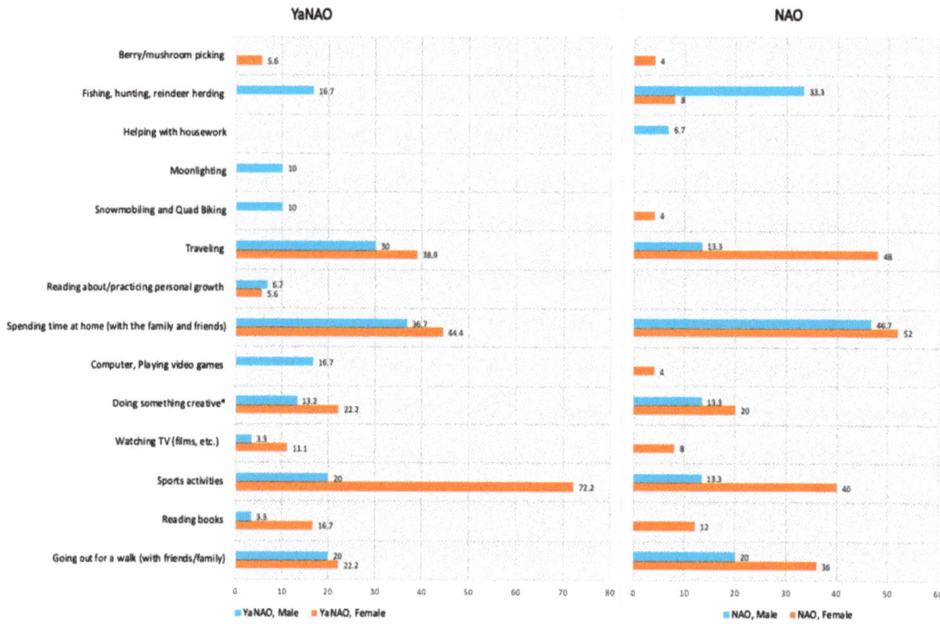


**Figure 9.** Vocational students from YaNAO and NAO, open question: "What activities are you engaged in during your leisure time? Where do you spend your leisure time?" (% of all collected responses in each group). * Includes activities such as writing, painting, playing music, drawing, cooking, vocal singing, caring for botanic flowers, photography sessions, modeling stucco figures, etc., and time spent in youth centers for creativity.

Two of the most common comments among youth are that "there are very few places to go" (female student, NAO) and "these places are not cheap" (female student, YaNAO).

Local recreational activities such as tourism are still minimal in these Arctic regions due to transport isolation and undeveloped tourism clusters. Based on students' responses, when it comes to traveling, they go to the "South" of Russia for a vacation to stay with their relatives, youth camps, the warm sea, or abroad, or they visit their relatives in the Northern villages in the Arctic tundra. Those staying in their hometowns mentioned active outdoor recreational activities with friends and spending time in the suburbs.

> "In summer, I travel with my parents to the sea; in winter, because of the severe cold, I stay at home and read" (female student, YaNAO); "I usually spend summer in the village and do nothing interesting, because there is nothing to do there" (female student, boarding school, NAO); "I stay at home in winter, and go out in summer" (male vocational student, YaNAO).

As the surveys show, youth spend plenty of time socializing with immediate relatives and friends. Following a general trend for Russia [87] (p. 113), the respondents did not indicate that they allocate time and effort to public activities or community service: Out of all collected questionnaires, only two respondents mentioned their social contribution through involvement in volunteering activities. The harsh Arctic climatic conditions could be one of the reasons for such a low level of participation as they may naturally limit the youth's engagement in these types of work. Long Polar winters and extreme cold also push young urbanites to stay at home, while in the summertime, Northerners tend to leave their cities.

## 4. Discussion

The study results show that youth migration decisions are complex and multifaceted. The present analysis is based on the dualist pull and push factors of migration. Although decisions to stay or leave are influenced by a broad range of social, professional, and economic factors and often shaped within a family context, our study identified that like in many other remote regions of the Circumpolar North, the lack of good-quality (higher) education opportunities is a key push factor and a driver of local youth out-migration [29–36,88].

In the sphere of education, the survey respondents from high schools (predominantly female students) in all three cities pointed out an acute problem common to most Arctic communities—the shortage or limited range of available educational services [88]. In particular, they identified the lack of higher education institutions in their regions as a substantive personal problem. Without higher education and relevant professional experience, local young people have few chances to build their careers in the extraction industries and other businesses or legislative and executive branches of government that would allow them to raise their social status, improve their material well-being, and fulfill their dreams.

The students' professional considerations are complex. Based on the school students' choices of specific professions, it is possible to presume their future educational and migration strategies. With a slight exception of Indigenous male students, all focal groups in Naryan-Mar, Salekhard, and Novy Urengoy expressed a strong interest in professions requiring higher education. Moreover, many of those preferred professions (e.g., orientalist, diplomat, movie producer, forensic expert, ecologist, architect, chemist, linguist, scientist-physicist/astrophysicist, etc.) are unavailable or have limited availability in these remote and economically less dynamic and developed Arctic regions. Indigenous male students were generally focused on occupations that do not require much modern (Western) professional training. Although more research is needed, it is possible to assume that they probably would be more interested in "Indigenized" education programs that have been established in other Arctic states with an emphasis on traditional knowledge and practices [88], should these programs become available in Russia. The surveys also indicate that local young people, especially vocational students, feel tremendous pressure to succeed in the Arctic labor market. Competition in the local labor markets can be defined as another strong push factor. Local youth without higher education face direct competition with low- and semi-skilled newcomers while not being able to apply for high-skill jobs that require higher education. This situation is also common for most remote Arctic regions where companies are reluctant to hire locals [55–59]. Among other push factors, the respondents of youth surveys and education/employment professionals pointed out corruption (mainly in the form of nepotism and cronyism) in the public sector of employment and broader labor market, with the dominance of the closed culture of corporatocracy.

Although locals do not have access to many high-wage jobs in their regions and are de facto limited with a minimal list of available occupations, the Northern labor market, with its social benefits system, seems attractive for many local young people. Established by law, so-called "Northern allowances" compensating workers for the higher cost of living and offering long paid vacation with travel compensations, early retirement, and a shorter working week (particularly for women) [75–78] are perceived as the key pull factor to retain youth. Also, the dominant non-diversified and single-industry economy, which is usually defined as a push factor, can be viewed as a pull factor. For instance, female students believe in other advantages of working in the North, besides benefits and social protection measures. They see many vacant niches (e.g., service industries and small businesses) and envision opportunities to develop small businesses to diversify the Arctic economy.

In the sphere of social integration, urban Arctic youth demonstrate low engagement in community services and the voluntary sector. Despite such "social atomization" (revealed in fragmentation of social networks and internally divided communities) of the young in the public domain, they are not experiencing an "individual atomization". Just the opposite, based on questionnaires, young people are attempting a high level of individual integration by maintaining strong social networks within their smaller inner circles of close friends and relatives or people who share their interests. The consequences for communities include a decreasing level of mutual trust outside residents' inner circle of friends and relatives, as well as a reduced level of public participation and sense of commitment toward improving their places [89], solidarity, and joint efforts to invest in the "common good" to fill a social vacuum between different "inner circles". As the value of community as a whole and intergenerational equity and trust are decreasing, nepotism is growing, along with the sharp divides between different groups of settlers (old-timers and newcomers, and FIFO workers) and between settlers and Indigenous communities. The underlying causes of Arctic youth's "social atomization" are in many ways rooted in their future (forced) life choices to relocate, thus contributing to a reduced feeling of attachment to their communities [90]. A critical imperative that affects Arctic youth's attachment to place is a well-rooted tradition of upward migration to the North inherent in many settler families and limited to one to three generations [50]. Young people are often encouraged by their parents to leave their hometowns for a "better life in the mainland": "We've been through a lot here, built our careers and everything, let our children to have a better life in good climate and better career opportunities" (Public official, Salekhard); "We came here [Naryan-Mar – author's] for career opportunities, [big—author's] "Northern money" and privileges [with intention – author's] to work for 10–15 years and then to go back to mainland. Actually, this is a good place to raise kids. The city is safe and compact with many opportunities for sports and creativity classes. But we are not planning to stay here after our public service is done" (Public official, Naryan-Mar); "Of course, I want my daughter to go to the University and to find a job somewhere else. What has she seen here? Look, there are even no real trees here, just permafrost all around and cold" (Public official, Novy Urengoy). This also often prevents young people from establishing stronger ties with their hometowns and reduces their willingness to contribute to their communities.

The study findings point to existing datasets that may be implicated in an intersectional approach to provide a more nuanced understanding of what determines individual young people's life strategies and defines their decisions to stay or leave. Among social categories that can be used for the intersectional analysis of youth migration, the study identified gender, age, Indigeneity, and socioeconomic status.

Gender is an important factor informing young people's career and life strategies. For instance, in the sphere of career preferences, the study revealed gender differences that reflect the structural peculiarities of the Arctic economies across the Circumpolar North [2,69,70,91–93]. Despite a general persistent trend over recent decades of the feminization of human capital (the gap between women with higher attainment in educational credentials and less educated men is increasing) in most Arctic regions [70,88,94], survey findings also show that in NAO and YaNAO, female students may find themselves in a path dependency situation. For instance, females often consider education, medicine, and the government sector among their top spheres. The surveys also identified that gender-related imposed choices of occupations are more prevalent among females. Male students' choices were also limited to male-dominated economic sectors, as they view the oil and gas industries, IT, transportation and construction, and reindeer husbandry (among Indigenous students) as the most attractive professions. On the one hand, young people find themselves in a situation of the Arctic's prevalent gender-based occupational clustering [69,70,91–93]. On the other hand, their professional choices make this system even more entrenched.

Age can be a determining factor for out-migration. In this research, the number of those willing to stay in the region grew with increasing age of the respondents: representatives of the youngest age group of high school students (14 to 17 y.o.) are mostly inclined to relocate, while vocational college students' (17–21 y.o.) life strategies vary.

Indigeneity identifies different approaches in life strategizing among Indigenous and non-Indigenous youth. In the Arctic countries, the ability of Indigenous people to access institutions of higher learning, while improving, is still problematic [88] (pp. 381–383; 387–389) [95–97]. In the education market, Indigenous young people face a lack of affordable, diverse educational opportunities, outside the scope of "traditional" professions available to them in local vocational schools and colleges. Limited in their career choices, female respondents focus on career building in a more traditional female domain (i.e., primary education, medicine, culinary arts, culture). Their choices of occupations in the cultural sphere may also be related to the legacy of the Soviet times, with government support of the Indigenous cultural economy and traditional activities in the Arctic, which provides some employment opportunities in specially created niches and, to some degree, reduces the competition for Indigenous youth [79]. At the same time, male students are often trapped in the male domain or follow their Indigenous path by choosing professions such as auto mechanics, drivers, reindeer breeders, etc.

The socioeconomic status category is related to the socioeconomic disparity in the focal regions and exposes the gap between students from relatively wealthy families with enough funds to send out their children to the universities and those families that lack financial means and whose children (have to or choose to) stay in their Arctic communities with few chances for career development, economic independence, and empowerment [73]. In the case of the Indigenous youth, although their stay improves the social sustainability indicator related to the component of migration and residential stability (vs. turnover) and makes them true stewards of Arctic lands, this existing gap in life opportunities and strategies of Indigenous vs. non-Indigenous students reveals a significant inequality in Arctic communities.

Based on these research findings, the Arctic communities face tremendous risks associated with youth "flight", making their future social sustainability uncertain. To retain young people or to attract them (back) to the Arctic [94,98], heavy investments in human capital and structural changes in the economy are inevitable. The necessity of Arctic university and research centers as a point of local youth attraction and development of a knowledge-based economy [99] in the Arctic has been recognized and implemented by many governments in the Circumpolar region (Canada, Iceland, Greenland and Faroe Islands, Finland, Norway, USA, and in some Arctic regions of Russia). In our two study regions, it is still in the distant future: While the NAO government included the establishment of a university branch in the NAO Strategy 2030 [5], in the YaNAO Strategy 2035 [6], a university is not even mentioned, leaving fewer chances for well-being and prosperity for local young people.

## 5. Conclusions

Using the example of the three Northern indicative cities of Naryan-Mar, Salekhard, and Novy Urengoy, the study examined key components of social sustainability primarily based on ISO 37120, complementing statistical information and analytical data with survey results. The study identified preponderant factors that to a large extent determine the local youth's life strategizing and define their decisions to stay or leave, which ultimately may affect social (socioeconomic) sustainability in these regions. Among the key factors are: (1) limited educational opportunities and lack of higher education institutions; (2) a limited range of high-quality jobs available for local youth in a highly competitive labor market for high-, semi-, and low-skilled workers; (3) limited opportunities for cultural and leisure activities; and (4) a low level of youth engagement in community services and the voluntary sector, revealing young people's low attachment to place [89,98,100].

From a broader perspective, the life strategies of the young generation of Northerners in Russia and their individual choices to stay in their Arctic communities or leave are a part of significant migration trends and patterns in the Circumpolar North [29–36]. In many Arctic countries, the prevalence of a psychological mood for out-migration among the local young people [29,101] puts them in a position where they are "stuck between their dreams and what they feel is realizable" [29] (p. 46) or move away seeking a way out.

The three Russian Arctic cities of Naryan-Mar, Salekhard, and Novy Urengoy showcase how insufficient investment in human and social capital, particularly relevant to the cohort of young people (e.g., through good educational and community facilities and wider employment opportunities for local youth), creates communities where local youth feel disempowered and pessimistic about their futures in the Arctic. The youth survey's findings on education, employment opportunities, and leisure time structure demonstrate that a majority of high school and vocational students view educational out-migration as a necessary condition for them to fulfill their dreams and realize their ambitions.

By analyzing survey results in the broader socioeconomic contexts of NAO and YaNAO, this article argues that Arctic regional economic prosperity, even in times of high and long-lasting demand for natural resources on the global market, does not necessarily benefit the locals, particularly the youth, nor lead to the social sustainability of Arctic communities. The combination of factors such as industrialization boom and economic 'bonanza' can serve to depict one of many Arctic paradoxes: Growing industries create new jobs and career opportunities that mostly fit and benefit not locals but rather newcomers and FIFO workers and, in turn, trigger young residents' out-migration and increase vulnerabilities in local communities. One can observe here a dilemma that is common for many remote Arctic areas where young individuals' self-interests often conflict with the overall common good for society and communities' social sustainability: "while a community may suffer from out-migration, individuals relocating elsewhere may experience an improvement in their quality of life" [102] (p. 62).

To improve the situation of the out-migration of young people, it is necessary to move Arctic youth from the periphery to the center of public policy discourse and decision making. This may include political actions to be taken in terms of prioritizing the provision of high-quality professional training programs and higher educational opportunities, providing greater investments in diverse social and cultural infrastructure, and implementing prioritization of youth-oriented affirmative action policies (e.g., quotas) for employing local youth in the labor market. Last but not least, the engagement of young people in defining problems and drawing up policies is vital to allow younger generations to have control over their own futures in the Arctic and responsibility for the future and social sustainability of their communities.

## 6. Limitations and Future Directions

The research was limited to three focal Arctic cities and did not include other Russian Arctic regions with university centers experiencing a youth flight. The lack of comprehensive statistical data on "city-to-city" and return migration limited the scope of analysis. A lack of relevant socioeconomic data did not allow the author to connect social sustainability indicators, governmental programs, laws and regulations, and industry and non-governmental sector initiatives with youth development trends. A non-probability sampling method was used for the youth survey due to limited access to students in educational institutions. The research conclusions may also be limited as not all dimensions of diversity (e.g., gender, ethnicity, Indigeneity) were addressed in the youth survey, which did not allow the study to utilize an intersectional approach. To ensure that the study posed no risks for the student participants, the questionnaires did not include the central topic of political engagement of the youth and structural barriers to empowerment. Future research will close some of these gaps.

**Funding:** This research was funded by NSF (Program for International Research and Education project "Promoting Urban Sustainability in the Arctic" (PIRE)), award number 1545913.

**Institutional Review Board Statement:** Approved by the RSHU.

**Informed Consent Statement:** Informed consent was obtained from all subjects involved in this study.

**Data Availability Statement:** Not applicable.

**Acknowledgments:** I would like to thank Marlene Laruelle (GWU), Robert Orttung (GWU), and Andrey N. Petrov (ARCTICenter, UNI) for conceptual advice on this article and support of the field research in the Nenets and Yamalo-Nenets Autonomous Regions as part of the Program for International Research and Education (PIRE) project "Promoting Urban Sustainability in the Arctic" (NSF Award #1545913). My deep appreciation goes to Nadezda Zmyatina (MSU) for her valuable consultation on the focal Arctic cities. Special thanks go to the Russian State Hydrometeorological University (RSHU) for its valuable organizational support of this fieldwork, and especially to its Rector, Valeriy L. Mikheev. I would further like to express my deep gratitude to all study participants in Naryan-Mar, Salekhard, and Novy Urengoy who shared their career plans, hopes, and concerns. Finally, my deep appreciation goes to Zoe Garbis (GWU) for copy editing, Pauline Mnev (GWU) for designing the map of the study site, and the three anonymous reviewers for their highly valuable comments and great ideas for future research directions.

**Conflicts of Interest:** The author declares no conflict of interest.

## Appendix A

**Table A1.** Urban sustainability indicators relevant to Arctic youth.

| № | Indicator | Naryan-Mar | Salekhard | Novy Urengoy |
|---|---|---|---|---|
| | **Demographics** | | | |
| 13.4.3 | Percentage of population who are youths (15–24 y.o.) | 9.5 | 9.5 | 10.4 |
| | **Economics** | | | |
| 5.4 * | Youth unemployment rate | 4.7 | 3.7 | 3.7 |
| 5.5 * | Number of businesses per 100,000 population | 3590 | 2610 | 2410 |
| ** | Employment in cultural sphere (as a percentage of total employed) | 4.6 | 4.7 | 1.3 |
| | **Education** | | | |
| 6.6 * | Number of higher education degrees per 100,000 population | 46,812 | 37,529 | 37,360 |
| & | Number of universities in the city | 0 | 0 | 0 |
| | **Cultural, Sporting, and EntertainmentInfrastructure** | | | |
| 17.1 *** | Number of cultural institutions and facilities per 100,000 population (modified indicator) | 28 | 22 | 17.2 |
| 17.1 *** | Number of sporting facilities per 100,000 population (modified indicator) | 217.5 | 191.8 | 259 |
| & | Number of restaurants, bars, cafes, and fast food restaurants per 100,000 population | 123.2 | 94.2 | 91.5 |
| & | Number of seats available at restaurants, bars, cafes, and fast food restaurants per 100,000 population | 6528 | 4161 | 5756 |
| & | Number of public caterings per 100,000 population | 31.8 | 15.7 | 49.13 |
| & | Number of seats available at public caterings per 100,000 population | 4898 | 2356 | 3829 |
| | **Contribution to Civil Society** | | | |
| & | Number of youths participating in volunteer activities per 100,000 population | n/a | n/a | n/a |
| | **Political Empowerment** | | | |
| & | Number of young adult deputies (18–35 y.o.) elected in the City Council per 100,000 population | 0 | 1.96 | 0 |

* PIRE's indicator; ** Based on methodology by Vera Kuklina and Natalia Shishigina [80]; *** Modified indicator. In the framework of the ISO 37120 methodology, cultural institutions and sporting facilities are not separated. & The author's indicator.

## References

1. Hill, F.; Gaddy, C. *The Siberian Curse: How Communist Planners Left Russia Out in the Cold*; Brookings Institution Press: Washington, DC, USA, 2003.
2. Larsen, J.N.; Petrov, A.N. The Economy of the Arctic. In *The Palgrave Handbook of Arctic Policy and Politics*; Coates, K., Holroyd, C., Eds.; Palgrave Macmillan: Cham, Switzerland, 2020. [CrossRef]
3. Federal State Statistics Service. Available online: https://eng.rosstat.gov.ru/ (accessed on 16 July 2021).
4. Ministry of Economic Development of the Russian Federation. *Program on Energy Efficiency and Energy Development in Nenets Autonomous Okrug (No. 229-p, 12 July 2017)*; Administration of the Nenets Autonomous Okrug: Naryan-Mar, Russia, 2017.
5. Ministry of Economic Development of the Russian Federation. *The Strategy of Socio-Economic Development of the Nenets Autonomous Okrug until 2030 (No. 256-sd, 7 November 2019)*; Administration of the Nenets Autonomous Okrug: Naryan-Mar, Russia, 2019.
6. Ministry of Economic Development of the Russian Federation. *The Strategy of Socio-Economic Development of the Yamalo-Nenets Autonomous Okrug until 2035 (No. 478, 24 June 2021)*; Administration of the Yamalo-Nenets Autonomous Okrug: Salekhard, Russia, 2021.
7. Petrov, A.; Vlasova, T. Migration and socio-economic well-being in the Russian North: Interrelations, regional differentiation, recent trends and emerging issues. In *Migration in the Circumpolar North: New Concepts and Patterns*; Huskey, L., Southcott, C., Eds.; CCI Press: Edmonton, AB, Canada, 2010; pp. 163–192.
8. Rosstat, Arkhangelsk. Available online: https://arhangelskstat.gks.ru/ (accessed on 12 January 2021).
9. Rosstat. The Russian Census of 2010. Available online: https://eng.rosstat.gov.ru/folder/76215 (accessed on 8 September 2020).
10. Yarlykapov, A. Divisions and Unity of the Novy Urengoy Muslim Community. *Probl. Post-Communism* **2020**, *67*, 338–347. [CrossRef]
11. Rosstat. *Regiony Rossii. Sotsial'no-Ekonomicheskiye Pokazateli*; Rosstat: Moscow, Russia, 2019. Available online: https://rosstat.gov.ru/accounts (accessed on 4 June 2021).
12. Sergunin, A. Planning for Sustainability: The Russian Case. In *Urban Sustainability in the Arctic: Measuring Progress in Circumpolar Cities*; Orttung, R., Ed.; Berghan Books: New York, NY, USA; Oxford, UK, 2020; pp. 228–252.
13. Orttung, R. *Conclusion. Next Steps for Measuring Arctic Urban Sustainability In Urban Sustainability in the Arctic: Measuring Progress in Circumpolar Cities*; Orttung, R., Ed.; Berghan Books: New York, NY, USA; Oxford, UK, 2020; pp. 278–288.
14. Petrov, A.N. Human Capital and Sustainable Development in the Arctic: Towards Intellectual and Empirical Framing. In *Northern Sustainabilities: Understanding and Addressing Change in the Circumpolar World*; Springer International Publishing: Cham, Switzerland, 2017; pp. 203–220. [CrossRef]
15. Petrov, A.; BurnSilver, S.; Chapin, T.; Fondahl, G.; Graybill, J.; Keil, K.; Nilsson, A.; Riedlspieger, R.; Schweitzer, P. Arctic Sustainability Research: A White Paper for the International Conference on Arctic Research Planning III (ICARP III). Summary and Findings. 2016. Available online: http://icarp.iasc.info/images/articles/Themes/WP_Summary_Sustainability_science_ICARP3_draft1.pdf (accessed on 2 July 2021).
16. Fondahl, G.; Wilson, G.N. Exploring Sustainabilities in the Circumpolar North. In *Northern Sustainabilities: Understanding and Addressing Change in the Circumpolar World*; Springer International Publishing: Cham, Switzerland, 2017; pp. 1–9. [CrossRef]
17. Partridge, E. Social Sustainability. In *Encyclopedia of Quality of Life and Well-Being Research*; Michalos, A.C., Ed.; Springer: Berlin/Heidelberg, Germany, 2014; p. 6178.
18. Colantonio, A. *Traditional and Emerging Prospects in Social Sustainability. Measuring Social Sustainability: Best Practice from Urban Renewal in the EU*; 2008/02: EIBURS Working Paper Series; November 2008; Oxford Institute for Sustainable Development (OISD): Oxford, UK, 2008.
19. Omann, I.; Spangenberg, J.H. Assessing social sustainability. The social dimension of sustainability in a socio-economic scenario. In Proceedings of the 7th Biennial Conference of the International Society for Ecological Economics, Sousse, Tunisia, 6–9 March 2002.
20. Dempsey, N.; Bramley, G.; Power, S.; Brown, C. The social dimension of sustainable development: Defining urban social sustainability. *Sustain. Dev.* **2009**, *19*, 289–300. [CrossRef]
21. Office of the Deputy Prime Minister (ODPM). *UK Presidency: EU Ministerial Informal on Sustainable Communities Policy Papers*; ODPM: London, UK, 2005.
22. Fujita, M.; Krugman, P.; Venables, A.J. *The Spatial Economy: Cities, Regions, and International Trade*; MIT Press: Cambridge, MA, USA, 1999.
23. Zamyatina, N.; Goncharov, R. Arctic Urbanization: Resilience in a Condition of Permanent Instability: The Case of Russian Arctic Cities. In *Resilience and Urban Disasters: Surviving Cities*; Borsekova, K., Nijkamp, P., Eds.; Edward Elgar: Cheltenham, UK, 2018; pp. 136–154.
24. Laruelle, M. *Russia's Arctic Strategies and the Future of the Far North*; M. E. Sharpe: Armonk, NY, USA; London, UK, 2014.
25. Kravale-Pauliņa, M.; Olehnoviča, E.; Ostrovska, I.; Ivanova, A.; Šipilova, V. Youth Policy Monitoring as a Tool for Developing Social Sustainability in Local Municipality. *Probl. Educ. 21st Century* **2008**, *76*, 350–363. [CrossRef]
26. Boström, M. A missing pillar? Challenges in theorizing and practicing social sustainability: Introduction to the special issue. *Sustain. Sci. Pract. Policy* **2012**, *8*, 3–14. [CrossRef]
27. European Investment Bank. The Charter of European Sustainable Cities and Towns Towards Sustainability. 2005. Available online: https://www.eib.org/attachments/jessica_bristol_accord_sustainable_communities.pdf (accessed on 13 February 2021).

28. Laruelle, M. Assessing Social Sustainability. Immigration to Russia's Arctic cities. In *Sustaining Russia's Arctic Cities. Resource Politics, Migration, and Climate Change*; Orttung, R., Ed.; Berghan Books: New York, NY, USA; Oxford, UK, 2017; pp. 88–111.
29. Karlsdóttir, A.; Jungsberg, L. (Eds.) *Nordic Arctic Youth Future Perspectives*; Nordregio Working Paper 2015:2; Nordregio: Stockholm, Sweden, 2015.
30. Pedersen, B.K. Young Greenlanders in the urban space of Nuuk. *Études Inuit Studies* **2008**, *32*, 91–105. [CrossRef]
31. Rygaard, J. The city life of youths in Greenland. *Études Inuit Studies* **2008**, *32*, 33–54. [CrossRef]
32. Hamilton, L.; Rasmussen, R.O. Population, sex ratios and development in Greenland. *Arctic* **2010**, *63*, 43–52. [CrossRef]
33. Carson, D.; Rasmussen, R.O.; Ensign, P.; Huskey, L.; Taylor, A. (Eds.) *Demography at the Edge: Remote Human Populations in Developed Nations*; Ashgate: Burlington, VT, USA, 2011.
34. Rasmussen, R.O. Polar Women Go South. *J. Nordregio* **2007**, *7*, 20–22.
35. Taylor, A.J. Current evidence of 'female flight' from remote Northern Territory Aboriginal communities—Demographic and policy implications. *Migr. Lett.* **2011**, *8*, 77–89. [CrossRef]
36. Hamilton, L. Footprints: Demographic Effects of Out-Migration. In *Migration in the Circumpolar North: Issues and Contexts*; Huskey, L., Southcott, C., Eds.; Canadian Circumpolar Institute: Edmonton, AB, Canada, 2010; pp. 1–14.
37. Bolotova, A.; Karaseva, A.; Vasilyeva, V. Mobility and Sense of Place among Youth in the Russian Arctic. *Sibirica* **2017**, *16*, 77–123. [CrossRef]
38. Osipova, O.V.; Maklashova, E.G. Migration intentions of the Arctic youth in the context of subjective evaluations of the social wellbeing. *Arct. North* **2016**, *24*, 13–24. [CrossRef]
39. Osipova, O.; Maklashova, E. Molodezh' Arktiki: Migratsionnyye namereniya i sotsial'nyye ozhidaniya. *J. Sib. Far East. Stud.* **2014**, *10*, 6–44.
40. Petrov, A.N.; Korkina, V.V.; Vlasova, T.K.; Rozanova, M.S.; Ulturgasheva, A. Arctic Youth from Russia. In *Arctic Youth and Sustainable Futures*; Larsen, J.N., Ingimundarson, J.H., Eds.; Nordic Council of Ministers: Copenhagen, Denmark, 2022; Forthcoming.
41. Bolotova, A. Living or Leaving? Youth and the Liveability of Single-industry Towns in the Russian Arctic. Available online: https://www.helsinki.fi/sites/default/files/atoms/files/ai-2-20-netti.pdf (accessed on 2 February 2021).
42. International Organization for Standardization. Sustainable Cities and Communities—Indicators for City Services and Quality of Life. 37101:2016. Available online: https://www.iso.org/standard/61885.html (accessed on 2 October 2020).
43. Rasmussen, R.O.; Roto, J.; Hamilton, L.C. West-Nordic Region. In *Arctic Social Indicators: ASI II: Implementation*; Larsen, J.N., Schweitzer, P., Petrov, A., Eds.; Nordic Council of Ministers: Copenhagen, Denmark, 2004; pp. 155–209.
44. Mkrtchyan, N.V. Problemy v statistike vnutrirossiyskoy migratsii, porozhdennyye izmeneniyem metodiki ucheta v 2011 g. *Demogr. Obozr.* **2020**, *7*, 83–99.
45. Zamyatina, N.Y.; Yashunsky, A.D. Virtual geography of virtual population. *Monit. Public Opin. Econ. Soc. Chang.* **2018**, *143*, 117–137. [CrossRef]
46. Heleniak, T. Arctic Populations and Migration. In *Arctic Human Development Report: Regional Processes and Global Linkages*; Larsen, J.N., Fondahl, G., Eds.; Nordic Council of Ministers: Copenhagen, Denmark, 2004; pp. 53–104.
47. Heleniak, T. Where Did All the Men Go? The Changing Sex Composition of the Russian North in the Post-Soviet Period, 1989–2010. *Polar Rec.* **2019**. [CrossRef]
48. Kashnitsky, I.; Mkrtchyan, N.; Leshukov, O. Interregional Migration of Youths in Russia: A Comprehensive Analysis of Demographic Statistics. *Vopr. Obraz./Educ. Stud. Mosc.* **2016**, 169–203. [CrossRef]
49. Konyshev, V.N.; Sergunin, A.A.; Subbotin, S.V. Sotsial'naya mobil'nost' v rossiyskoy Arktike [Social mobility in the Russian Arctic]. *Natsional'nye interesy prioritety i bezopasnost' [Natl. Interests Priorities Secur.]* **2017**, *13*, 1189–1196. (In Russian) [CrossRef]
50. Zamyatina, N.; Yashunsky, A. Migration cycles, social capital, and networks: A new way to look at Arctic mobility. In *New Mobilities and Social Changes in Russia's Arctic Regions*; Laruelle, M., Ed.; Routledge: London, UK; New York, NY, USA, 2017; pp. 59–84.
51. Appendix "Strategy of Socio-Economic Development of the Municipal Formation of the City of Novy Urengoy until 2030". Available online: http://www.newurengoy.ru/docs/15603-strategiya-socialno-ekonomicheskogo-razvitiya-municipalnogo-obrazovaniya-gorod-novyy-urengoy-do-2030-goda.html (accessed on 14 September 2020).
52. The Strategy of Social and Economic Development of the City of Salekhard until 2030s. Available online: https://www.salekhard.org/city/socs/strategiya-razvitiya-mo/ (accessed on 14 September 2020).
53. On Approval of the Strategy of Socio-Economic Development of the Nenets Autonomous Okrug until 2030. Available online: http://docs.cntd.ru/document/561620008 (accessed on 14 September 2020).
54. Laruelle, M. Introduction. In *New Mobilities and Social Changes in Russia's Arctic Regions*; Laruelle, M., Ed.; Routledge: London, UK; New York, NY, USA, 2017.
55. Ivo, J.H.B.; Stephen, J.D.H.; Svein, G. What's next for Big Oil. In *McKinsey Quarterly*; What Global Executives Think about Growth & Risk; McKinsey: New York, NY, USA, 2005.
56. Tysiachniouk, M.S. Disentangling Benefit-Sharing Complexities of Oil Extraction on the North Slope of Alaska. *Sustainability* **2020**, *12*, 5432. [CrossRef]
57. Nuykina, Y.V. Vliyaniye vakhtovogo metoda raboty na prinimayushchiye goroda Rossiyskogo Severa (na primere goroda Vorkuty). *Izvestiya Komi nauchnogo tsentra URO RAN* **2013**, *2*, 107–116.

58. Berman, M.; Loeffler, R.; Schmidt, J.I. Long-term benefits to Indigenous communities of extractive industry partnerships: Evaluating the Red Dog Mine. *Resour. Policy* **2020**, *66*, 101609. [CrossRef]
59. Tysiachniouk, M.; Andrey, N.P.; Vera, K.; Natalia, K. Between Soviet Legacy and Corporate Social Responsibility: Emerging Benefit Sharing Frameworks in the Irkutsk Oil Region, Russia. *Sustainability* **2018**, *10*, 3334. [CrossRef]
60. Sokolov, D. Ugra, the Dagestani North: Anthropology of mobility between the North Caucasus and western Siberia. In *New Mobilities and Social Changes in Russia's Arctic Regions*; Laruelle, M., Ed.; Routledge: London, UK; New York, NY, USA, 2017; pp. 176–193.
61. Kulik, S.; Eidemiller, K. K voprosu ob islamskom regionalizme v Arkticheskoi zone RF. In *Arktika: Istoriya i Sovremennost': Trudy Mezhdunarodnoi Nauchnoi Konferentsii, St. Peterburg, Russia, 20–21 April 2016*; Kulik, S.V., Ed.; Izdatel'skii Dom "Nauka": Moscow, Russia, 2016.
62. Oparin, D. Locals and immigrants on the Yamal Peninsula. Social boundaries and variations in migratory experience. *Asian Ethn.* **2018**, *19*, 251–269. [CrossRef]
63. *Law of the Nenets Autonomous Okrug, 8 November 2016, № 246-OZ «O kvotirovanii Rabochikh mest dlya otdel'nykh Kategoriy Grazhdan v Nenetskom Avtonomnom Okruge»*; Assembly of Deputies of the Nenets Autonomous Okrug: Naryan-Mar, Russia, 2016.
64. NAO Official Website. Traditsionnaya Tseremoniya Posvyashcheniya v Neftyaniki Uchenikov «Rosneft'-Klassa» Sostoyalas' v Posolke Iskateley. 14 September 2019. Available online: http://adm-nao.ru/press/governor/22256/ (accessed on 14 September 2020).
65. Interfax. Putin podderzhal ideyu Tselevogo Obucheniya Predstaviteley Malochislennykh Narodov Kraynego Severa. 29 November 2019. Available online: https://academia.interfax.ru/ru/news/articles/3902 (accessed on 19 September 2020).
66. Interfax. V Rossii Budet Sozdana Organizatsiya po Realizatsii Gosudarstvennoy Natsional'noy Politiki. 20 January 2020. Available online: https://www.interfax-russia.ru/moscow/news/v-rossii-budet-sozdana-organizaciya-po-realizacii-gosudarstvennoy-nacionalnoy-politiki (accessed on 4 January 2021).
67. Decree of the Government of the Yamalo-Nenets Autonomous Okrug of 12 December 2018, N 1661-R On approval of the comprehensive program "Sustainable development of indigenous peoples of the North in the Yamalo-Nenets Autonomous Okrug for 2018–2022". Available online: https://docs.cntd.ru/document/550269204 (accessed on 17 December 2020).
68. Ross, M.L. *The Oil Curse: How Petroleum Wealth Shapes the Development of Nations*; Princeton University Press: Princeton, NJ, USA, 2012.
69. Malgorzata, S.; Prior, T. Gender and Environment. In *Gender Equality in the Arctic*; Oddsdóttir, E., Ágústsson, H.O., Eds.; Icelandic Ministry for Foreign Affairs: Reykjavik, Iceland, 2021; pp. 98–143.
70. Rozanova-Smith, M.; Petrov, A.; Williams, V.K. Empowerment and Fate Control. In *Gender Equality in the Arctic*; Oddsdóttir, E., Ágústsson, H.O., Eds.; Icelandic Ministry for Foreign Affairs: Reykjavik, Iceland, 2021; pp. 224–267.
71. Website "Moi Biznes YaNAO". Available online: http://mb89.ru/news/ofitsialno/na-yamale-vpervye-sostoitsya-proekt-mama-predprinimatel-/ (accessed on 2 January 2021).
72. Website "Ministerstvo Ekonomicheskogo Razvitiya Rossiyskoy Federatsii". Available online: http://nko.economy.gov.ru/PortalNews/Read/5109 (accessed on 2 January 2021).
73. Decree of the Government of the Russian Federation of 21 November 2007, N 1661-R <On the Concept of the Federal Target Program "Economic and Social Development of the Indigenous Peoples of the North, Siberia, and the Far East until 2015".> (Chapter 3). Available online: http://www.consultant.ru/document/cons_doc_LAW_136920/ (accessed on 29 August 2020).
74. Program for International Research and Education Project "Promoting Urban Sustainability in the Arctic (PIRE), ISO 5.4 Indicator. Available online: http://blogs.gwu.edu/arcticpire/project-publications/data/ (accessed on 23 October 2020).
75. Reisser, C. Russia's Arctic Cities: Recent Evolution and Drivers of Change. In *Sustaining Russia's Arctic Cities: Resource Politics, Migration, and Climate Change*; Orttung, R., Ed.; Berghahn: New York, NY, USA, 2016; pp. 1–22.
76. Lishchuk, Y.N.; Kapelyuk, S.D. Problemy regulirovaniya oplaty truda rabotnikov Kraynego Severa i priravnennykh k nemu mestnostey. *Ekonomika* **2020**, *16*, 97–113. [CrossRef]
77. Volgin, N.A.; Shirokova, L.N.; Mosina, L.L. Aktual'nyye voprosy razvitiya rossiyskogo severa: Kompensatsionnyye i stimuliruyushchiye sistemy. *Uroven' zhizni Naseleniya Regionov Rossii* **2018**, *2*, 34–46. [CrossRef]
78. Novikova, I.V. Perspektivy ispol'zovaniya dal'nevostochnykh garantiy gosudarstva kak instrumenta sodeystviya zanyatosti naseleniya v regione. *Uroven' zhizni Naseleniya Regionov Rossii* **2016**, *1*, 206–219.
79. Rozanova, M. Indigenous Urbanization in Russia's Arctic: The Case of Nenets Autonomous Region. Sibirica. *J. Sib. Stud.* **2019**, *18*, 54–91.
80. Kuklina, V.; Shishigina, N. Yakutsk: Culture for Sustainability. In *Urban Sustainability in the Arctic: Measuring Progress in Circumpolar Cities*; Orttung, R., Ed.; Berghan Books: New York, NY, USA; Oxford, UK, 2020; pp. 101–128.
81. Government of NAO. Available online: http://smi.adm-nao.ru/otnosheniya-v-nao/sociologicheskoe-issledovanie-obshestvennogo-mneniya-po-voprosam-toler/ (accessed on 2 August 2020).
82. Laruelle, M. Postcolonial Polar Cities? New Indigenous and Cosmopolitan Urbanness in the Arctic. *Acta Borealia* **2019**, *36*, 149–165. [CrossRef]
83. Laruelle, M. Urban Regimes in Russia's Arctic Cities: Testing a Concept in a New Environment. *Arctic* **2020**, *73*, 53–66. [CrossRef]
84. YaNAO Library. Available online: http://nb.yanao.ru/#top (accessed on 2 August 2020).
85. Khakhovskaya, L. Aborigeny v gorode: Etnokul'turnyi oblik zhitelei Magadana. *Sibirskie Istoricheskie Issledovaniia* **2004**, *2*, 39–59.

86. Keyes, C.L.M. Social well-being. *Soc. Psychol. Q.* **1998**, *61*, 121–140. [CrossRef]
87. Gudkov, L.; Zorkaya, N.; Kochergina, E.; Pipiya, K.; Ryseva, A. *Russia's 'Generation Z': Attitudes and Values 2019/2020*; Chekhov Printing House: Moscow, Russia, 2020; Available online: http://library.fes.de/pdf-files/bueros/moskau/16134.pdf (accessed on 11 November 2020).
88. Hirshberg, D.; Petrov, A. Education and Human Capital. In *Arctic Human Development Report: Regional Processes and Global Linkages*; Nordisk Ministerråd: Copenhagen, Denmark, 2015.
89. De Dominicis, S.; Fornara, F.; Ganucci Cancellieri, U.; Twigger-Ross, C.; Bonaiuto, M. We are at risk, and so what? Place attachment, environmental risk perceptions and preventive coping behaviours. *J. Environ. Psychol.* **2015**, *43*, 66–78. [CrossRef]
90. Bramley, G.; Morgan, J. Building competitiveness and cohesion: The role of new housebuilding in central Scotland's cities. *Hous. Stud.* **2003**, *18*, 447–471. [CrossRef]
91. Ashwin, S. *Adapting to Russia's New Labour Market: Gender and Employment Behaviour*; Routledge: London, UK, 2006.
92. Novikova, S.S. Social state and gender equality in modern Russia. *Contemp. Probl. Soc. Work.* **2016**, *2*, 72–78. [CrossRef]
93. Oddsdóttir, E.; Sigurðsson, A.M.; Svandal, S. (Eds.) *Gender Equality in the Arctic: Current Realities, Future Challenges*; Icelandic Ministry for Foreign Affairs: Reykjavik, Iceland, 2015.
94. Rasmussen, R.O. (Ed.) *Megatrend*; Nordic Council of Ministers: Copenhagen, Denmark, 2011.
95. Stonechild, B. *The New Buffalo: The struggle for Aboriginal post-secondary education in Canada*; University of Manitoba Press: Winnipeg, MB, USA, 2006.
96. Preston, J.P. The urgency of postsecondary education for Aboriginal peoples. *Can. J. Educ. Adm. Policy* **2008**, *86*, 1–22.
97. Rasmussen, R.O.; Barnhardt, R.; Keskitalo, J.H. Education. In *ASI I, 2010. Arctic Social Indicators—a follow-up to the Arctic Human Development Report*; Larsen, J.N., Schweitzer, P., Fondahl, G., Eds.; TemaNord. Nordic Council of Ministers: Copenhagen, Denmarks, 2010; pp. 67–90.
98. Heleniak, T. The role of attachment to place in migration decisions of the population of the Russian North. *Polar Geogr.* **2009**, *32*, 31–60. [CrossRef]
99. Temple, P. (Ed.) *Universities in the Knowledge Economy: Higher Education Organisation and Global Change*; Routledge: London, UK, 2012.
100. Tuhkunen, A. *Between Location and a Sense of Place Observations Regarding Young People's Migra-tion Alacrity in Northern Europe*; University of Tampere Publ.: Tampere, Finland, 2007.
101. Hayfield, E.A.; Aure, M.; Barillé, S.; Farrell, C.R.; Gaini, F.; Gerrard, S.; Gotfredsen, A.; Gram-Hanssen, I.; Hamilton, L.; Hoffmann, L.W.; et al. Migration and Mobility. In *Gender Equality in the Arctic*; Oddsdóttir, E., Ágústsson, H.O., Eds.; Icelandic Ministry for Foreign Affairs: Reykjavik, Iceland, 2021; pp. 144–191.
102. Fondahl, G.; Crate, S.; Filippova, V.V. Sakha Repubic (Yakutia), Russian Federation. In *Arctic Social Indicators: ASI II: Implementation*; Larsen, J.N., Schweitzer, P., Petrov, A., Eds.; Nordic Council of Ministers: Copenhagen, Denmarks, 2014; pp. 57–91.

sustainability MDPI

Article

# The Contribution of Natural Resource Producing Sectors to the Economic Development of the Sakha Republic

**Shinichiro Tabata**

Slavic-Eurasian Research Center, Hokkaido University, Sapporo 060-0809, Japan; shin@slav.hokudai.ac.jp

**Abstract:** This paper provides basic materials for considering the sustainability of natural resource development in the Arctic, taking the Sakha Republic as a case study of the Russian Arctic regions. The author clarifies the contribution of the mining industry to the economic development of Sakha with special attention paid to the contribution to government budgets by numerical and statistical analysis of regional and municipal data. The paper demonstrates that the mining industry has been a driving force of the economic growth of Sakha and that the oil sector has sharply increased its presence while the diamond sector has decreased its presence. Simultaneously, it reveals that the mining industry is unevenly developed in Sakha, which has caused significant inequality in per capita Gross Municipal Product (GMP). Then, the analysis of the paper shows that Sakha's contribution to the federal budget has increased significantly in recent years due to growing oil production and that the diamond sector is still more influential than the oil sector in the contribution to the republican and local budgets.

**Keywords:** Sakha; diamond; oil; government budget; economic growth; mining industry

**Citation:** Tabata, S. The Contribution of Natural Resource Producing Sectors to the Economic Development of the Sakha Republic. *Sustainability* **2021**, *13*, 10142. https://doi.org/10.3390/su131810142

Academic Editors: Stephanie Pfirman, Gail Fondahl, Grete K. Hovelsrud and Tero Mustonen

Received: 19 July 2021
Accepted: 7 September 2021
Published: 10 September 2021



## 1. Introduction

The purpose of this paper is to provide basic materials for considering the sustainability of natural resource development in the Arctic, taking the Sakha Republic as a case study of the Russian Arctic regions. This paper is a part of the Arctic Challenge for Sustainability II (ArCS II in short), which is a national flagship project of Arctic research in Japan in the period 2020–2025. In this project, I am responsible for one of the sub-programs, entitled "Energy resource development and regional economy," which aims to study the impact of energy resource development on regional economies in the Russian Arctic, particularly in the Sakha Republic and Yamalo-Nenets Autonomous Okrug.

I think that before deciding to launch or continue a natural resource development project, we must evaluate the meaning or merits of this project. For this purpose, it is indispensable to fully understand the present contribution of natural resource production to the economic development in these regions. Fiona Hill and Clifford Gaddy argued in their famous book, The Siberian Curse [1], that it might be a mistake to make a decision to develop Siberia if we consider the full burdens or costs of developing these areas.

Sakha is famous for its diamond production. Until recently, Sakha had produced more than 90% of diamonds in Russia (see Section 2.1), and Russia's share in the world was 32.8% in 2019 [2] (p. 372). Recently, the production of crude oil has increased tremendously. In this paper, I attempted to statistically clarify the contribution of the mining industry to the economic development of Sakha with special attention paid to the contribution to government budgets. There is a big difference in taxation between the oil and diamond sectors. Taxes on oil play an important role in the federal budget, while those on diamond do so in the regional budget. I illustrate this difference in this article. Note that in this article, a subject of the federation (republic, krai, oblast, etc.) is called a region.

To the best of my knowledge, there are few literatures in English and in Russian that deal with these issues in Russia in general, and in Sakha in particular [3–5]. One of the reasons is the lack of relevant statistical data on the regional level. Gross Regional Production (GRP) data and industrial production data in a region are poor in Russia, while there is a certain improvement on the federal level. Concerning budget execution (performance) data and tax payment data, however, relatively detailed data are available. This paper is a preliminary attempt to take advantage of these data.

In the next section, the contribution to economic growth is analyzed. Section 3 deals with the contribution to government budgets. Website information of data used in this paper is provided in Appendix A.

## 2. Contribution of the Mining Industry to Economic and Industrial Growth in Sakha

### *2.1. Contribution to Republican Economy*

First, I analyzed the contribution to the GRP growth of Sakha by industry. The largest contribution by the mining industry is evident from Figure 1. In this figure, the blue line shows the GRP growth rate of Sakha, which was higher than Russia's GDP growth rate in recent years. In the period 2014–2019, while the average growth rate of Russia's GDP was only 0.9%, that of Sakha's GRP was 2.9% (calculated from Rosstat's website). The stacked bars in Figure 1 show the contribution of each industry. The large contribution of the mining industry is followed by the construction and transportation sectors. (Note that there was a change in the classification of economic activities from OKVED to OKVED2 in Russia's GDP and GRP statistics in 2016. Consequently, data until 2015 and those from 2016 are not compatible in Figure 1. For example, the transportation sector was included in the sector called "Transport and communications" until 2015, while it was included in "Transportation and storage" since 2016. In the legend of Figure 1, names of the sector (economic activities) are those of the old classification.) These two sectors are closely related to the mining industry due to the construction of oil and gas pipelines and the transportation of oil and gas through pipelines. In fact, the mining industry accounted for 50.6% of Sakha's GRP in 2019. The share of the construction sector (9.6%) and transportation and storage sector (6.2%) was also significant.

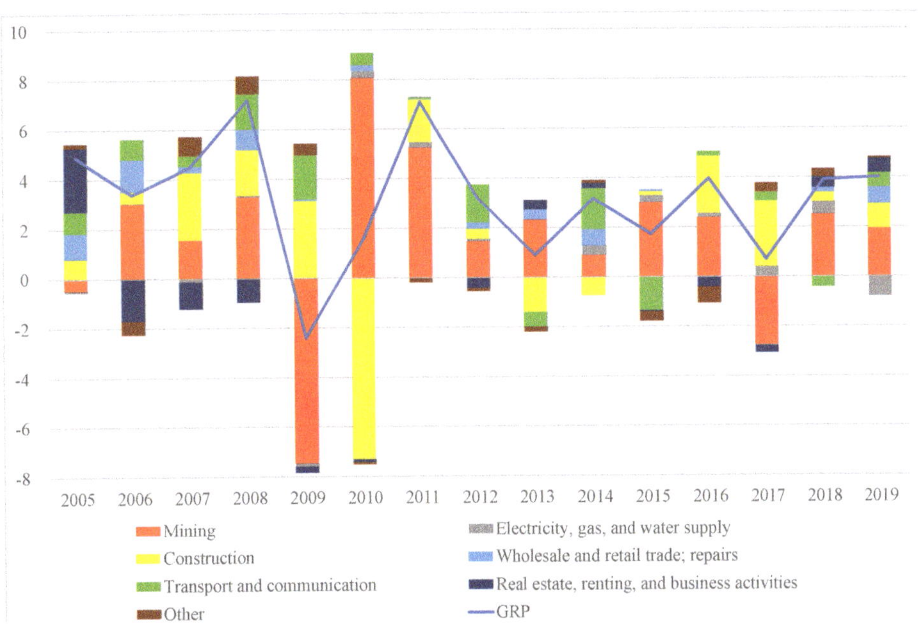


**Figure 1.** Contribution to Sakha's GRP growth by industry, in percentages, 2005–2019. Sources: Compiled by the author from Sakhastat's website.

Concerning the breakdown of the mining industry in the GRP statistics of Sakha, available data are limited: I only have data shown in Table 1. (Note that in Table 1, Figures 2 and 3, I regard the production of other mining as diamond production since there seem to be no major minerals other than diamond in the mining industry of Sakha, excluding fuel-energy and metals. In fact, Pulyaevskaya writes that the share of the diamond sector in 2010 was 20.5% [4] (p. 164), which is almost the same as the percentage shown in Table 1 (20.4%). In Table 1, the sum of coal, oil and gas, metal, and diamond is equal to the mining industry, except for a small discrepancy (42 million rubles) in 2010. In Sakha, major products of the metal sector are gold, silver, and antimony [6].) According to this table, the share of the mining industry increased together with the oil and gas sector, while the share of the diamond sector decreased to a certain degree. This tendency is confirmed by the analysis of industrial statistics. Figure 2 shows the contribution by subsectors of the mining industry to the growth of industrial production. The blue line shows the growth of mining production, the average growth rate of which was 9.7% in the period 2010–2019, and the stacked bars show the contribution by each subsector (I calculated these contributions with the volume of goods shipped as weight). We see the largest contribution by the oil sector. The contribution of the diamond and coal sectors was not significant in the past decade. As shown in Figure 3, the share of the diamond sector in the mining industry was larger than 50% until 2009, but since then, it has decreased considerably. (Note that Figure 3 shows the volume of goods shipped (Ob″em otgruzhennykh tovarov), not the volume of goods produced. As is the case with GRP, there was a change in the classification of economic activities in industrial statistics. Data after 2017 in Figure 3 were derived from the new classification format (OKVED2). The service in this figure is a new sector in the new classification. It seems that such activities were included in some of the other sectors until 2016.) On the other hand, the share of the oil and gas sector has grown rapidly since 2010.

**Table 1.** Structure of the mining industry in Sakha's GRP, 2006, 2010, 2015.

| | 2006 | 2010 | 2015 |
|---|---|---|---|
| | | In million rubles | |
| Total GRP | 206,845 | 386,825 | 749,987 |
| Mining industry | 80,571 | 154,548 | 361,253 |
| including | | | |
| Coal | 9228 | 18,289 | 17,890 |
| Oil and gas | 2988 | 43,424 | 127,718 |
| Metal | 6138 | 14,045 | 33,933 |
| Diamond | 62,217 | 78,748 | 181,712 |
| | | In percentages | |
| Total GRP | 100.0 | 100.0 | 100.0 |
| Mining industry | 39.0 | 40.0 | 48.2 |
| including | | | |
| Coal | 4.5 | 4.7 | 2.4 |
| Oil and gas | 1.4 | 11.2 | 17.0 |
| Metal | 3.0 | 3.6 | 4.5 |
| Diamond | 30.1 | 20.4 | 24.2 |

Sources: Compiled by the author from [3] (pp. 82–83).

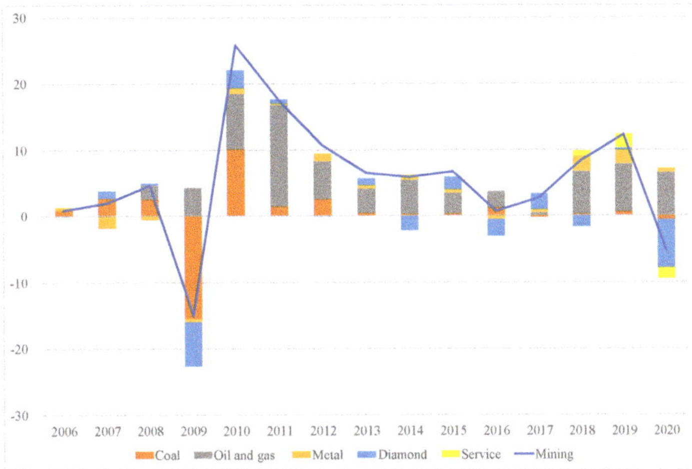


**Figure 2.** Contribution to the growth of the mining industry by subsector in Sakha, in percentages, 2006–2020. Sources: Compiled by the author from Sakhastat's website.

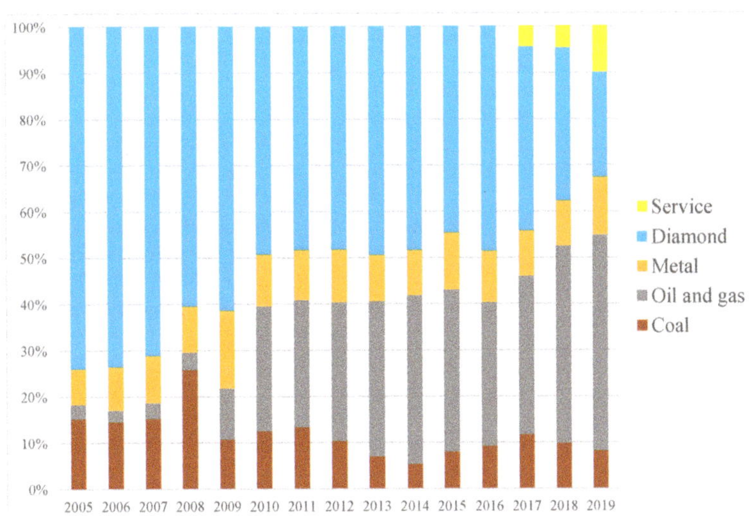


**Figure 3.** Structure of the mining industry of Sakha, in percentages, 2005–2019. Note: Structure of the volume of goods shipped. Sources: Compiled by the author from Sakhastat's website.

In fact, crude oil production in Sakha increased quite rapidly thanks to the development of the Talakan oil field (Figure 4). In 2019, Sakha accounted for 2.6% of oil production in Russia (calculated from [7] (p. 726) and Sakhastat's website). It should be noted that more than 90% of crude oil produced in Sakha is exported through the East Siberia—Pacific Ocean (ESPO) pipeline [8] (p. 134). In fact, the percentage of export to production was 94.7% in 2010 and 95.6% in 2013, if we calculate it using export data reported on the same page. It is obvious that the oil sector has become a driving force for the economic development of Sakha.

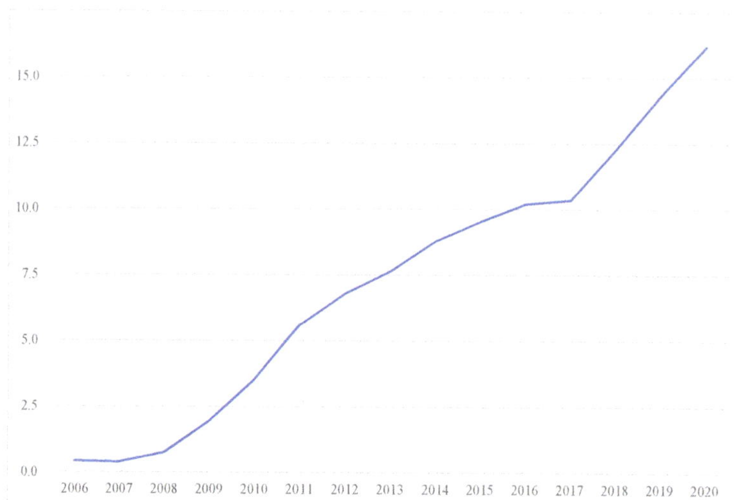


**Figure 4.** Crude oil production in Sakha, in million tons, 2006–2020. Sources: Compiled by the author from the website and other sources of Sakhastat.

The production of natural gas remained in the range of 1.9–2.0 billion $m^3$ during the years 2009–2018. It jumped to 2.9 billion $m^3$ in 2019 and 6.8 billion $m^3$ in 2020 thanks to the production in the Chayanda gas field, which is one of the main sources of the "Power of Siberia" pipeline going to China (these production data were obtained from the website and other sources of Sakhastat). The share of Sakha in the natural gas production of Russia was 1.0% in 2020, and it is expected to grow rapidly in the near future [9] (calculated from the websites of the Ministry of Energy and Sakhastat).

On the contrary, diamond production is forecasted to decrease in Sakha as well as in the world [10]. In fact, its production in Sakha peaked in 2017 (36.9 million carats) and decreased to 34.3 million carats in 2019 (calculated by the author from the production data of five mining and processing divisions of Alrosa (diamond-producing monopoly, see below) published in its Annual Reports). The share of Sakha in the diamond production of Russia dropped from 97% in 2010 and 87% in 2015 to 76% in 2019 (Data of diamond production in Russia are available from the website of Russia's Ministry of Finance). There was an attempt to create a Territory of advanced development (TOR in Russian), called "Diamond valley" in Sakha, but it was not realized [6,11]. The rest was produced in Arkhangelsk.

The impact of diamond and crude oil production on Sakha's economy is difficult to compare. I can roughly estimate the export value of diamond and crude oil in 2018 as USD 3.9 billion and USD 5.8 billion, respectively. Concerning diamond, Sakha's export value is obtained from the website of the Federal Customs Service. With respect to the crude oil exports of Sakha, we cannot obtain these data from the same source since most of the export data of crude oil are registered in Moscow, where the company headquarters are located. Therefore, I estimated them from Russia's export value of oil and the share of Sakha in the export quantity of Russia, assuming that 95% of Sakha's production was exported, as indicated above.

It seems to follow from this that the impact of the oil sector is larger. The problem, however, is that most of the rents or value-added of the oil sector are not realized in Sakha but transferred to Moscow, where the company headquarters are located. Most of the rents of the oil sector are not realized in the mining industry but realized as trade and transportation margins and taxes due to the low producers' prices [12]. On the other hand, in the case of diamond, the headquarters are located in Sakha. Therefore, it seems that most of the rents are realized in Sakha. In addition, as we see in Section 3, most of the taxes paid by the diamond sector are used in the Republic, while most of them from the oil sector are transferred to Moscow.

*2.2. Contribution to Local Economy*

In Sakha, we can use the indicator of Gross Municipal Product (GMP) that shows value-added production in a municipality (as for municipalities in Sakha, see Appendix B). GMP is calculated only in several regions of Russia [5] (p. 136). In this subsection, I examine how the mining industry contributes to the economy of a municipality, taking advantage of this indicator.

Although the mining industry is well developed in Sakha, its distribution is quite uneven. Table 2 shows the share of major producing districts in the production of main mineral resources in Sakha in 2019. For example, Lensk district accounted for 66.4% of crude oil production in Sakha. The production of the other mineral resources is also concentrated in a few districts. As a result, there are considerable differences in the share of the mining industry in GMP (Figure 5). Note that this share in the GRP of Sakha was 51.1% in 2016. This share exceeds 70% in five districts, including Nyurba, Anabar, Mirny, Oymyakon, and Lensk. On the other hand, in the majority of other municipalities, this share is less than 30%.

**Table 2.** Major districts of mining production in Sakha; share in percentages, 2019.

| | | | | | | |
|---|---|---|---|---|---|---|
| **Crude Oil** | Lensk | 66.4 | Mirny | 33.6 | | |
| **Natural Gas** | Vilyuysk | 63.0 | Lensk | 30.2 | Mirny | 6.8 |
| **Coal** | Neryungri | 95.5 | | | | |
| **Diamond** | Mirny | 54.8 | Nyurba | 30.0 | Anabar | 15.2 |
| **Gold (2015)** | Oymyakon | 47.1 | Aldan | 35.1 | Olyokminsk | 6.6 |

Note: The share of Anabar in Diamond includes the Bulun and Olenek districts (see below). Sources: Compiled by the author from the website and other sources of Sakhastat and [13] (2020).

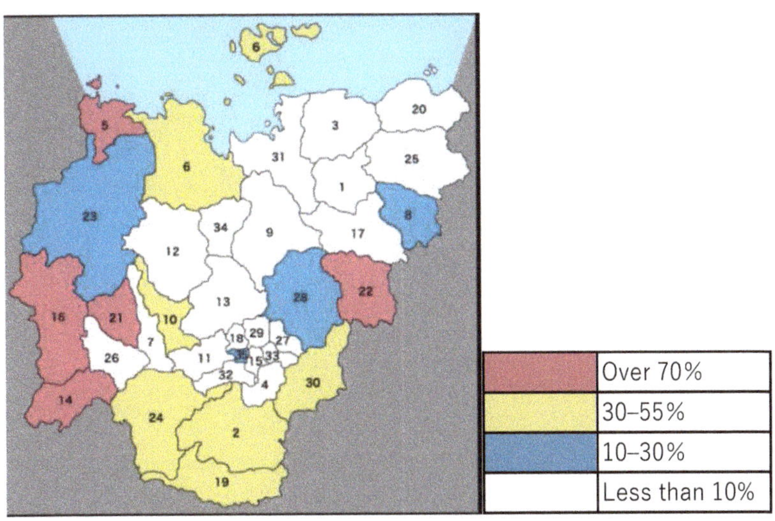


**Figure 5.** Share of the mining industry in GMP in Sakha, 2016. Note: Refer to Appendix B for the names of municipalities. Sources: Compiled by the author from data obtained from Sakhastat.

There are also big differences in per capita GMP (Figure 6). Note that the average per capita GMP in Sakha was 904 thousand rubles in 2016 (Calculated by dividing total

GMP by total population of Sakha. The arithmetic mean of per capita GMP is 754 thousand rubles and its median is 423 thousand rubles). In only six districts, per capita GMP exceeds one million rubles. This implies that in the other 30 municipalities, per capita GMP is less than the average of Sakha. Inequality seems to be large. The Gini coefficient is 0.50.

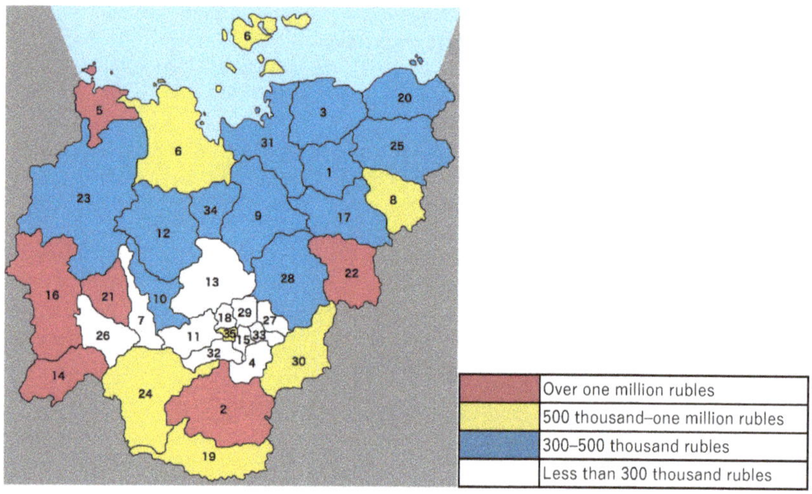


**Figure 6.** Per capita GMP of Sakha, 2016. Sources: Calculated by the author from the website and other sources of Sakhastat.

The six districts with high per capita GMP include Lensk (4.1 million rubles), Anabar (3.2 billion rubles), Mirny (3.0 million rubles), Nyurba (2.0 million rubles), Oymyakon (1.9 million rubles), and Aldan (1.0 million rubles). In the top five districts in terms of per capita GMP, the share of the mining industry exceeds 70% (Figure 5). In the Aldan district, the share of the mining industry is 44.0 percent. The correlation between the share of the mining industry and per capita GMP is 0.86, which demonstrates the significant contribution of the mining industry to the economy of municipalities.

I conducted a multiple linear regression analysis to compare the impact of oil, diamond, and gold production on per capita GMP in 2016. The result is as follows (R2 of 0.75).

$$\text{Per capita GMP} = 461.4 + 40.8\ (\text{share of oil}) + 38.9\ (\text{share of diamond}) + 25.8\ (\text{share of gold})$$

All independent variables are statistically significant ($p < 0.01$). This seems to suggest that the impact of oil production on the economy of a municipality is slightly stronger than that of diamond production. (Note that data of diamond production in carats were obtained from the production data of five mining and processing divisions of Alrosa [13] (2020). One of them is Almazy Anabara, a subsidiary of Alrosa, which produces diamond not only in the Anabar district but also in the Bulun and Olenek districts. I estimated the share of these three districts using data in USD. The data of gold are those in 2015, since data in more recent years are not available.)

## 3. Contribution of the Mining Industry to Government Budgets in Sakha

### *3.1. Contribution to the Federal and Republican Budgets*

In this section, I examine the contribution of the mining industry to government budgets in Sakha, including the federal, regional, and local budgets. In Russia, tax revenues are divided into federal and regional budget revenues, of which the regional budget includes the budget of a region and its municipalities, i.e., local budgets. Some taxes are exclusively revenues of the federal budget, including value-added tax (VAT), export and import duties, and mineral extraction tax on oil and gas, while some other taxes are those of regional budgets, including personal income tax, asset tax, and mineral extraction tax on diamond. There are some taxes in which revenues are divided between the federal and regional budgets, including corporate tax and mineral extraction tax on coal. In the case of the extraction tax on coal, 40% of its revenues are revenues of the federal budget, and 60% are those of regional budgets. Until 2006, the extraction tax on diamond was distributed the same as that on coal. By the amendment of the Budget Code (Federal Law No. 237 of 19 December 2006), 100% of these revenues became revenues of the federal budget since the beginning of 2007. According to Sakha News (21 November 2006, https://www.1sn.ru/9169.html), this amendment was made to compensate the losses sustained by the decrease in numbers of Alrosa stocks (diamond-producing monopoly) owned by the Sakha Republic in 2006. However, I was not able to confirm the change in stocks owned by the Republic in 2006 from the Annual Reports of Alrosa. The share of the Russian Federation did increase from 37% to 50.9% in 2008 [13] (various years).

In Russia, revenues from oil and gas are the largest and most important in the federal budget. In 2019, extraction tax and export duty on oil and gas accounted for 40.9% of federal budget revenues (calculated from the Federal Treasury's website). This percentage does not include VAT, corporate tax, and other taxes that oil and gas companies pay to the federal budget. Since diamond production concentrates in Sakha, the republican budget of Sakha received 82.4% of extraction tax revenues on diamonds in Russia in 2020. In other words, the existence of revenues of extraction tax on diamonds is one of the most distinct characteristics of Sakha's budget.

Table 3 demonstrates the difference in the contribution to government budgets between the oil and diamond sectors. (Note that in Table 3 and Figure 7, I regard tax revenues from other mining as those from the diamond sector since there seem to be no major minerals other than diamond in the mining industry of Sakha, excluding the fuel-energy and metal sectors.) In terms of total tax revenues, the contribution by the oil sector is 58.5%, and that of the diamond sector is 15.6% in 2019. In terms of federal budget tax revenue, the oil sector contributed almost all, owing to a large amount of extraction tax on oil. On the other hand, in terms of republican budget revenues, the contribution of the diamond sector (30.3%) is larger than that of the oil sector (18.2%). Particularly, the share of the diamond sector in corporate and asset tax revenues is significant. Since the production of diamond in Sakha is monopolized by Alrosa, the share of the diamond sector in republican budget revenues of Sakha means the contribution of Alrosa to these revenues [10]. In fact, this company almost completely monopolizes diamond production in Russia. Its share in diamond production in Russia was 90%, and its share in world diamond production was 27.5% in 2020 [13] (2021, pp. 12, 28). Alrosa also has its affiliate (diamond mining enterprise) in Arkhangelsk Oblast.

**Table 3.** Structure of tax revenues by industry in Sakha, in percentages, 2019.

| | Total Tax Revenue | Federal Budget Tax Revenue | Republican Budget Tax Revenue | Extraction Tax | Corporate Tax | Personal Income Tax | Asset Tax |
|---|---|---|---|---|---|---|---|
| All industries | 100.0 | 100.0 | 100.0 | 100.0 | 100.0 | 100.0 | 100.0 |
| Mining industry | 74.3 | 93.4 | 59.3 | 99.8 | 79.7 | 27.1 | 40.9 |
| including | | | | | | | |
| Oil and gas | 58.5 | 105.3 | 18.2 | 83.7 | 40.1 | 5.4 | 14.7 |
| Diamond | 15.6 | −0.3 | 30.3 | 12.6 | 29.5 | 12.2 | 22.7 |
| Other | 25.7 | 6.6 | 40.7 | 0.2 | 20.3 | 72.9 | 59.1 |

Note: In terms of federal budget tax revenue, the contribution of the oil and gas sector exceeds 100% because, in the other sectors of the mining industry, revenues of VAT are large negative values, which means tax refunds. Sources: Calculated by the author from the website of the Federal Tax Service.

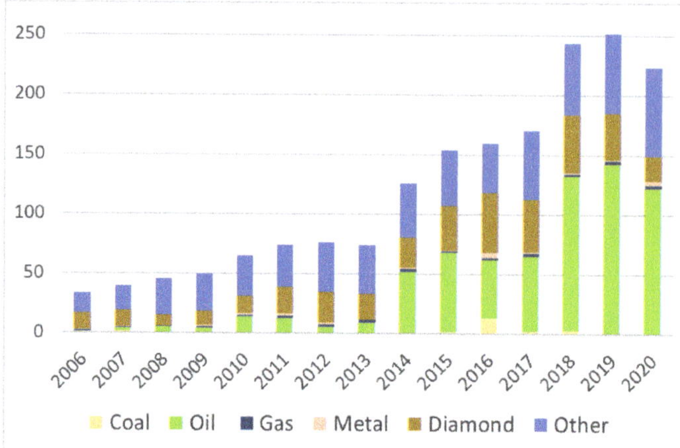


**Figure 7.** Tax revenues in Sakha by industry, in billion rubles, 2006–2020. Sources: Compiled by the author from the website of the Federal Tax Service.

Since oil production has increased rapidly in recent years in Sakha, tax revenues from the oil sector have grown as well. Figure 7 demonstrates which industry pays taxes. In 2019, for example, 74% of taxes were paid by the mining industry, including the coal, oil, gas, metal, and diamond sectors. The rapid growth of tax revenues from the oil sector is apparent from this figure.

This growth of tax revenues from the oil sector resulted in significant increases in federal tax revenues since extraction tax revenues on oil are federal budget revenues. In fact, as shown in Figure 8, federal budget tax revenues have increased tremendously in recent years. They were only 5 billion rubles in 2013 but increased to 122 billion rubles in 2019. You may ask why oil tax revenues were so small until 2013 and increased abruptly in 2014 (Figure 7), although oil production increased rather smoothly in the period 2009–2016 (Figure 4). The reason was exemptions from oil extraction taxes for new oil fields in the Sakha Republic, as well as Irkutsk Oblast and Krasnoyarsk Krai, to promote new oil field development in these regions [14] (p. 167).)

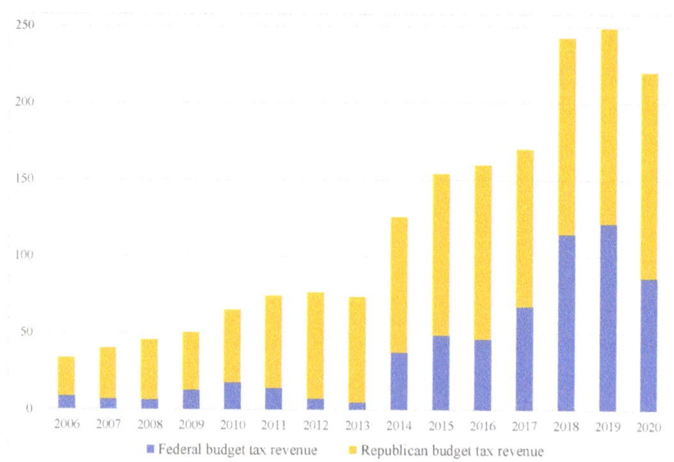


**Figure 8.** Tax revenues in Sakha, in billion rubles, 2006–2020. Sources: Compiled by the author from the website of the Federal Tax Service.

I have devised such indicators as the gross and net contribution of a region to the federal budget [15]. The gross contribution of a region is defined as federal budget tax revenues of the region, while net contribution is defined as gross contribution minus transfer. The transfer includes dotation, subsidy, subvention, and other inter-budgetary transfers that a region receives from the federal budget.

The result of the calculation is shown in Figure 9. The gross contribution of Sakha was modest until 2013, but since then, it has rapidly grown thanks to the increase in extraction taxes on oil. On the other hand, Sakha received a relatively large amount of transfer due to the high cost of public services caused by severe climate conditions. Consequently, the net contribution was negative until 2016. However, it turned positive in the following year. We confirm the great contribution of the oil sector to this change in net contribution.

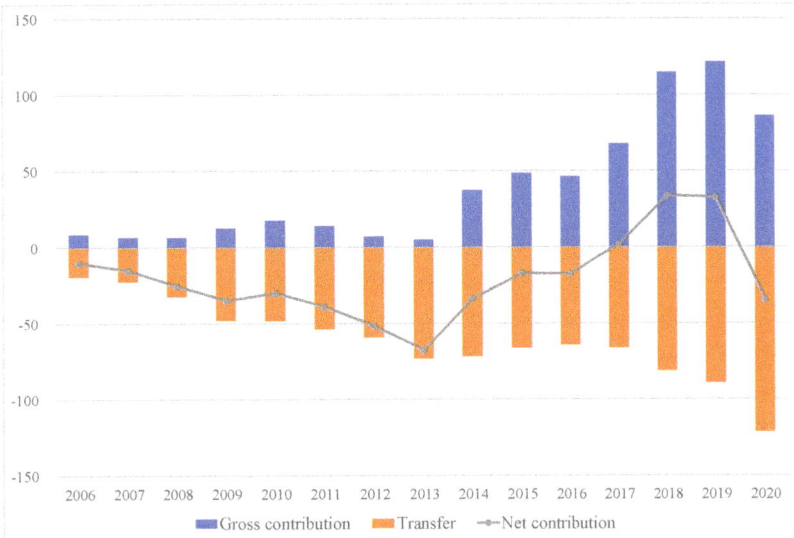


**Figure 9.** Contribution of Sakha to the federal budget, in billion rubles, 2006–2020. Sources: Compiled by the author from the websites of the Federal Tax Service and Federal Treasury.

It should be noted that this gross contribution includes only tax revenues collected by the Federal Tax Service of Russia. There are other federal budget tax revenues collected by the Federal Customs Service of Russia, including export and import duties, as well as indirect taxes (VAT and excise) on imported goods. They are exclusively revenues of the federal budget. However, the problem here is that there are no statistics that show which regions pay these duties and taxes. I attempted to make a preliminary estimate of these revenues in 2015 [15] (pp. 15–21). According to this estimate, Sakha's gross contribution increased from 49 billion rubles to 91 billion rubles (increase by 88%), and its net contribution increased from minus 17 billion rubles to 25 billion rubles. Thus, the contribution of the oil sector becomes more significant when we take into account export duties on oil. Note that there are deficiencies in this estimate. Basically, in my estimates, export duties on oil were distributed among regions in proportion to the share of each region in the production of crude oil in Russia. By adopting this method, I disregarded the special measures (exemption or reduction in export duties) in some areas taken by the Federal Government to promote the development of new oil fields [14] (pp. 166–168). Note also that export duties on diamond were abolished on September 1, 2016, since Russia promised to abolish them within four years of joining the WTO in 2012.

If we calculate gross and net contribution rates by dividing gross and net contributions by total tax revenues of a region, Sakha ranks 38th and 49th among 85 of Russia's regions in 2015 [15] (p. 17). The lower ranking of the net contribution rate of Sakha compared with the gross net contribution rate was a result of receiving a relatively large transfer from the federal budget.

### 3.2. *Contribution to Local Budgets*

The contribution of the mining industry to local budgets has two channels. One is through transfer that the republican budget provides to local budgets. The other is the dividends that Alrosa pays to eight districts.

In Sakha, as well as other regions of Russia, transfer plays an important role in the redistribution of revenues in a region [16]. As shown in Figure 10, 82.8% of local budget revenues are transfers from the republican budget in 2019. As indicated above (Table 3), 59.3% of republican budget tax revenues are paid by the mining industry, including 30.3% from the diamond sector. They are the main sources of transfer that local budgets receive from the republican budget.

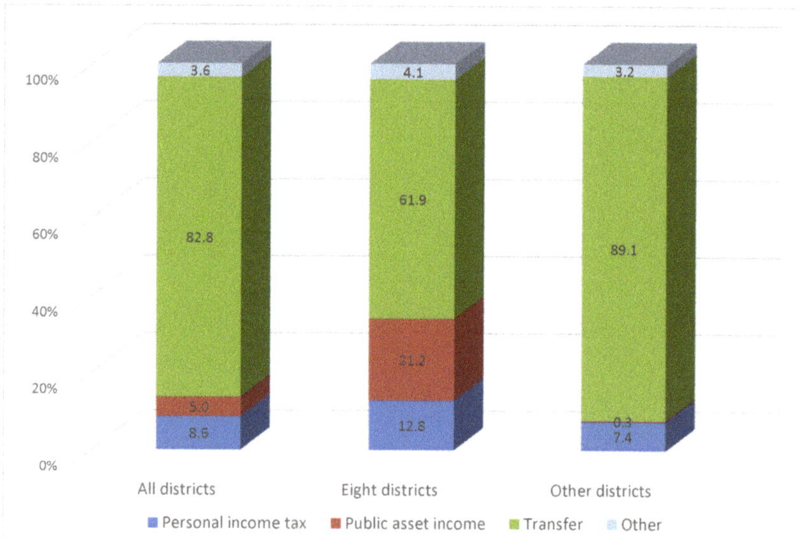


**Figure 10.** Structure of local budget revenues of Sakha, in percentages, 2019. Note: The arithmetic mean of percentages of districts is shown. Eight districts are those that receive dividends from Alrosa. Sources: Compiled by the author from Rosstat's website.

Eight districts receive dividends from Alrosa since they own stocks of this company. Alrosa is a joint-stock company; 33% of its stocks are owned by the Russian Federation, 25% by the Sakha Republic, and 8% by eight districts of Sakha, where Alrosa's production and other facilities are located. They are Anabar, Verkhnevilyuisk, Vilyuisk, Lensk, Mirny, Nyurba, Olenek, and Suntar. They have 1% of stocks each. Other stocks (34%) are owned by private entities. These shareowners received dividends from Alrosa every year. This equity information is obtained from Alrosa's website. Alrosa was transformed from a zakrytoe (closed) to an otkrytoe (public) joint-stock company in 2011. In the period from 2011 to 2016, the share of the Russian Federation and the Sakha Republic decreased from 50.9% to 33% and from 32% to 25%, respectively, and the private share increased from 9% to 34% [13] (various years).

As indicated in Figure 10, the share of public asset income in these eight districts was 21.2%, and that of transfer was 61.9% in 2019. On the other hand, in the other districts, the share of transfer was almost 90%.

These dividends caused significant differences in the revenues of local budgets. Figure 11a shows the per capita revenue of local budgets, excluding transfers in 2019. Five districts in which revenues exceed 40 thousand rubles and three of four districts in which revenues are between 30 and 40 thousand rubles receive dividends from Alrosa.

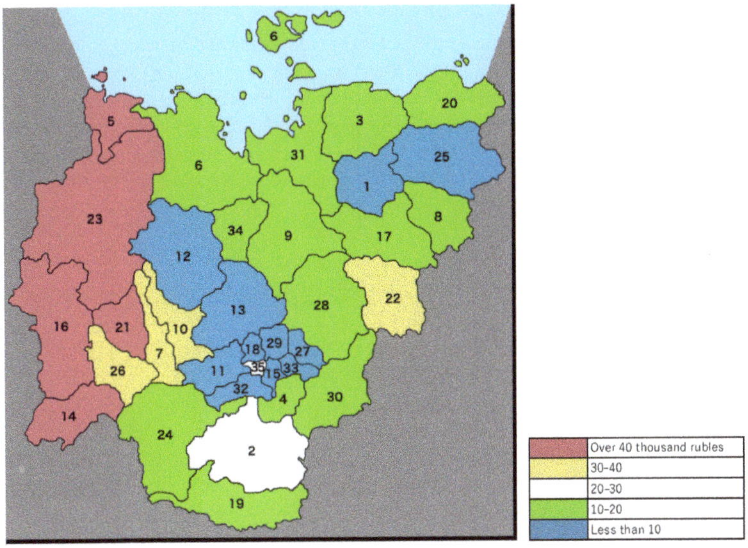


(**a**) Revenue excluding transfer.

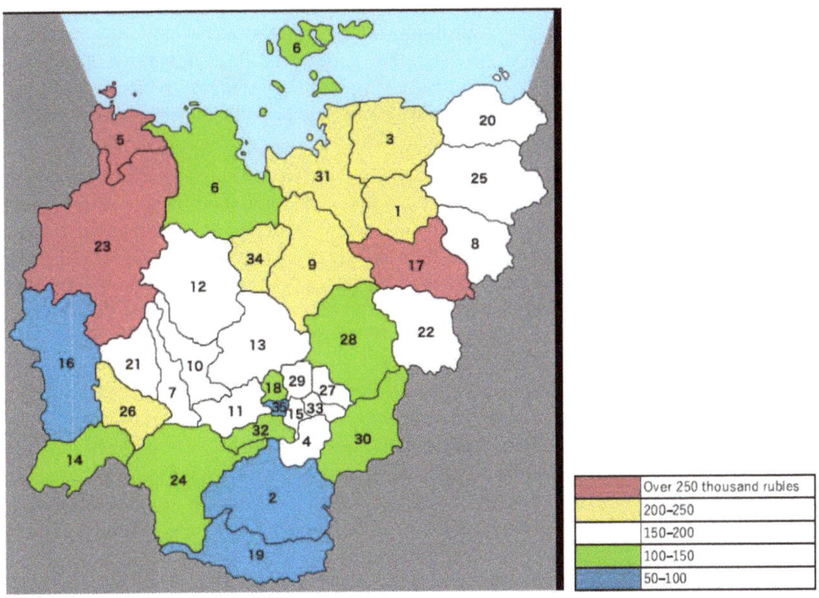


(**b**) Revenue including transfer.

**Figure 11.** Per capita revenue of local budgets of Sakha, 2019. (**a**) Revenue excluding transfer; (**b**) Revenue including transfer. Sources: Calculated by the author from websites of Rosstat and Sakha's Ministry of Finance.

The other mining productions do not seem to contribute as strongly to local budgets as the diamond sector. The correlation coefficient between the share of mining production in GMP (Figure 5) and per capita local budget revenues of a district excluding transfer (Figure 11a) is 0.65. (Note that data in Figure 5 are those from 2016. We do not have data after 2017.) It is safe to say that only diamond production has a direct and distinct influence on local budgets.

Figure 11b shows the per capita revenue after the transfer, i.e., including transfer, while Figure 11a is revenue before the transfer, i.e., excluding transfer. The correlation between these two figures is not so strong (the correlation coefficient is 0.56), which suggests that transfer plays a certain role in equalizing the revenues of local budgets. The Gini coefficient improves from 0.56 before transfer to 0.21 after transfer. Thus, the diamond sector played a prominent role both in creating differences in local budget revenues and reducing these differences through transfer.

## 4. Concluding Remarks

The findings of this article obtained by numerical and statistical analysis are summarized as follows:

1. The mining industry has been a driving force of the economic growth of Sakha in recent years. In the mining industry, the oil sector has sharply increased its presence, while the diamond sector has decreased its presence in the economic and industrial development of Sakha;
2. The mining industry is unevenly developed in Sakha, which has caused significant inequality in per capita GMP. In other words, the mining industry has considerably increased GMP in several districts;
3. Sakha's contribution to the federal budget has increased significantly in recent years due to growing oil production. The importance of the oil sector of Sakha for the federal budget revenues has been enhanced considerably;
4. Concerning the contribution to the republican and local budgets, the diamond sector is still more influential than the oil sector. While dividends of Alrosa caused considerable differences in per capita revenues of local budgets, revenues from the diamond sector account for 30% of republican budget revenues, from which transfer is provided to local budgets to equalize differences in local budget revenues.

It seems that this paper demonstrated the considerable contribution of the oil and diamond sectors to the economic development of Sakha and the appropriateness of continuing these productions in the future. This paper's main focus, however, was limited to the contribution of the mining industry to economic growth and government budget performance. There are other areas to which the mining industry contributes. For example, Alrosa's contribution to local employment is often pointed out [17] (p. 4). The influence on other welfare, including payments of salaries and other benefits, and the construction of public infrastructure and housing, requires further examination.

On the other hand, the costs of the development of the mining industry remain to be explored. They include costs to compensate for the negative influence extended by mining companies on the natural environment and other economic activities, such as livestock farming and fishery. (Ref. [18] is an excellent previous work on this topic. They also addressed pollutions by underground nuclear explosions in the Soviet era for the purpose of seismic exploration). They are also included in topics of our future joint research.

**Funding:** This work was a part of the Arctic Challenge for Sustainability II (ArCS II), Program Grant Number JPMXD1420318865.

**Institutional Review Board Statement:** Not applicable.

**Informed Consent Statement:** Not applicable.

**Acknowledgments:** My special thanks go to Tuyara Gavrilyeva for helping me gather data published by Sakhastat and relevant papers written by Russian scholars. I also wish to thank Yugo Konno, Michitaka Hattori, and Hirofumi Arai for giving me useful information.

**Conflicts of Interest:** The author declares no conflict of interest.

## Appendix A. Website Information on Data Used in This Paper

Alrosa

- Equity information (Information for shareholders): http://www.alrosa.ru/%D0%B8%D0%BD%D0%B2%D0%B5%D1%81%D1%82%D0%BE%D1%80%D0%B0%D0%BC-%D0%B8-%D0%B0%D0%BA%D1%86%D0%B8%D0%BE%D0%BD%D0%B5%D1%80%D0%B0%D0%BC/ (accessed on 9 September 2021)

Federal Customs Service of Russia

- Export data of Sakha: http://stat.customs.gov.ru/unload (accessed on 9 September 2021)

Federal Tax Service of Russia

- Tax revenues: https://www.nalog.gov.ru/rn77/related_activities/statistics_and_analytics/forms/ (accessed on 9 September 2021)

Federal Treasury of Russia

- Performance of consolidated regional budgets: https://roskazna.gov.ru/ispolnenie-byudzhetov/konsolidirovannye-byudzhety-subektov/ (accessed on 9 September 2021)
- Performance of consolidated state budgets: https://roskazna.gov.ru/ispolnenie-byudzhetov/konsolidirovannyj-byudzhet/ (accessed on 9 September 2021)

Ministry of Energy, Russian Federation

- Natural gas production in 2020: https://minenergo.gov.ru/node/1215 (accessed on 9 September 2021)

Ministry of Finance, Russian Federation

- Diamond production in Russia: https://minfin.gov.ru/ru/perfomance/jewels/KimberleyProcess/ (accessed on 9 September 2021)

Ministry of Finance, Sakha Republic

- Local budget revenues of Sakha: https://minfin.sakha.gov.ru/bjudzhet/otchetnost/godovye-otchety (accessed on 9 September 2021)

Rosstat (Federal State Statistics Service of Russia)

- GDP and GRP data: https://rosstat.gov.ru/accounts (accessed on 9 September 2021)
- Local budget revenues of Sakha: http://www.gks.ru/dbscripts/munst/munst98/DBInet.cgi (accessed on 9 September 2021)

Sakhastat (Rosstat's branch in Sakha)

- GRP: https://sakha.gks.ru/folder/32205 (accessed on 9 September 2021)
- Industrial production: https://sakha.gks.ru/folder/35778 (accessed on 9 September 2021)
- Population (year average): https://sakha.gks.ru/folder/32348 (accessed on 9 September 2021)

## Appendix B. List of Municipalities in Sakha

| | | | | | | | |
|---|---|---|---|---|---|---|---|
| 1 | Abyysky | 10 | Vilyuisk | 19 | Neryungri | 28 | Tomponsky |
| 2 | Aldan | 11 | Gorny | 20 | Nizhnekolymsk | 29 | Ust-Aldan |
| 3 | Allaikhovsky | 12 | Zhigansk | 21 | Nyurba | 30 | Ust-May |
| 4 | Amginsky | 13 | Kobyui | 22 | Oymyakon | 31 | Ust-Yansky |
| 5 | Anabar | 14 | Lensk | 23 | Olenek | 32 | Khangalas |
| 6 | Bulun | 15 | Megino-Kangalas | 24 | Olyokminsk | 33 | Churapcha |
| 7 | Verkhnevilyuisk | 16 | Mirny | 25 | Srednekolymsk | 34 | Eveno-Bytantai |
| 8 | Verkhnekolymsk | 17 | Momsky | 26 | Suntar | 35 | Yakutsk |
| 9 | Verkhoyansk | 18 | Namsky | 27 | Tattinsky | 36 | Zhatay |

Note: There are 34 districts and two cities in Sakha. The district is *raion* in Russian. It is called *ulus* in Sakha. Some of them are called national *ulus*. Two cities are Yakutsk and Zhatay. Since Zhatay is located inside Yakutsk, it is not shown in Figures 5, 6 and 11.

## References


1. Hill, F.; Gaddy, C. *The Siberian Curse: How Communist Planners Left Russia Out in the Cold*; Brookings Institution Press: Washington, DC, USA, 2003.
2. Minprirody (Ministry of Natural Resources and Environment of the Russian Federation). *Gosudarstvennyi Doklad o Sostoianii i Ispol'zovanii Mineral'no-syr"evvykh Resursov Rossiiskoi Federatsii v 2019 Gody (National Report "On the State and Use of Mineral Resources of the Russian Federation)*; 2020. Available online: https://www.mnr.gov.ru/docs/gosudarstvennye_doklady/ (accessed on 9 September 2021).
3. Kondratyeva, V.I.; Pulyaevskaya, V.L. Metodicheskii podkhod k opredeleniiu mesta munitsipal'nogo obrazovaniia v ekonomicheskom prostranstve regiona (Methodological Approach to the Determination of the Place of Municipalities in the Economic Space of the Region). *Vestn. NGUEU* **2017**, *4*, 77–85.
4. Pulyaevskaya, V.L. Valovoi munitsipal'nyi product kak pokazatel' otsenki ekonomicheskogo potentsiala raionov and gorodov (Gross Municipal Product as Measure of Estimation of Economic Potential of Districts and Cities). *Vestn. NGUEU* **2012**, *3*, 159–166.
5. Pulyaevskaya, V.L. Valovoi munitsipal'nyi product v otsenke urovnia ekonomicheskogo razvitiia Respubliki Sakha (Iakutiia) (Gross Municipal Product in Assessing the Level of Economic Development of the Republic of Sakha (Yakutia)). *Vestn. NGUEU* **2015**, *4*, 135–144.
6. Grigor'eva, E. Nalogovye i sotsial'nye obiazatel'stva predpriiatii almazno-brilliantovogo kompleksa Iakutii (Tax and Social Obligation of Diamond-Brilliant Branch of Yakutia). *Prior. Napravleniia Razvit. Nauk. i Obraz.* **2015**, *3*, 274–279.
7. Rosstat. *Regiony Rossii. Sotsial'no-Ekonomicheskie Pokazateli—2020 g.(Regions of Russia. Social-Economic Indicators—2020)*. Available online: https://rosstat.gov.ru/storage/mediabank/LkooETqG/Region_Pokaz_2020.pdf (accessed on 9 September 2021).
8. Soromotin, A. Neftegazovye resursy respubliki Sakha (Iakutiia): Sostoianie, perspektivy ispol'zovaniia (Oil and Gas Resources of the Republic of Sakha (Yakutia): State and Prospects of Use). *Vestn. SVFU* **2014**, *11*, 129–136.
9. Filimonova, I.; Moiseev, S.; Nemov, V.; Provornaya, I. Neft' i gaz Iakutii: Perspektivy i ogranicheniia (Oil and Gas of Yakutia: Perspectives and Constraint). *Heftegazovaia Vertikal'* **2019**, *20–21*, 32–42.
10. Grigor'eva, E. Predposylki k izmeneniiu struktury biudzhetnykh dokhodov regiona na osnove prognoznoi otsenki zapasov prirodnykh almazov do 2050 g. (Prerequisites for Change in the Structure of Budgetary Revenues of the Region Based on Predictive Estimates of Natural Diamond Reserves till 2050). *Seriia Vestn. SVFU* **2017**, *4*, 22–29.
11. Maksimova, K.L.; Popova, A.I. Otsenka sobremennoi podgotovki kadrov dlia granil'no-iuvelirnoi promyshlennosti v RS (Ia) (Evaluation of Modern Training for the Jewelry Industry in the Republic of Sakha (Yakutia)). *Strateg. Ustoichivogo Razvit. Reg. Ross.* **2016**, *32*, 30–36.
12. Kuboniwa, M. Estimating GDP and foreign rents of the oil and gas sector in the soviet union and present-day Russia. In *Russian Economic Development over Three Centuries: New Data and Inferences*; Kuboniwa, M., Nakamura, Y., Kumo, K., Shida, Y., Eds.; Palgrave Macmillan: Singapore, 2019; pp. 421–438.
13. Alrosa. *Godovye Otchety (Annual Report)*. Various Years. Available online: http://www.alrosa.ru/documents/%d0%b3%d0%be%d0%b4%d0%be%d0%b2%d1%8b%d0%b5-%d0%be%d1%82%d1%87%d0%b5%d1%82%d1%8b/ (accessed on 9 September 2021).
14. Tabata, S.; Liu, X. Russia's energy policy in the far east and east Siberia. In *Russia's Energy Policy: National, Interregional and Global Dimensions*; Pami, A., Ed.; Edward Elgar: Cheltenham, UK, 2012; pp. 156–181.
15. Tabata, S. Finansovye potoki mezhdu federal'nym biudzhetom i arkticheskimi regionami Rossii (Flow of Financial Resources between the Federal Budget and the Arctic Regions in Russia). *Reg. Ekon. Sotsiologiia* **2019**, *3*, 3–25. [CrossRef]

16. Alexeev, M.; Weber, S. Russian fiscal federalism: Impact of political and fiscal (de)centralization. In *The Oxford Handbook of the Russian Economy, Alexeev, M., Weber, S., Eds.*; Oxford University Press: New York, NY, USA, 2013; pp. 643–660.
17. Danilov, Y.; Leontiev, S. Dobycha almazov v Arkticheskikh raionakh Respubliki Sakha (Iakutiia) (Diamond Mining in the Arctic Regions of the Sakha Republic (Yakutia)). *Kontsept* **2016**, *11*, 1–6.
18. Yakovleva, N.P.; Alabaster, T.; Petrova, P.G. Natural Resource Use in the Russian North: A Case Study of Diamond Mining in the Republic of Sakha. *Environ. Manag. Health* **2000**, *11*, 318–336. [CrossRef]

*sustainability*

MDPI

*Article*

# Climate Change and Unalakleet: A Deep Analysis

**Tero Mustonen [1,*] and Brie Van Dam [2]**

[1] Snowchange Cooperative, 81235 Lehtoi, Finland
[2] Snowchange Cooperative, Fairbanks, AK 99703, USA; brievd@gmail.com
* Correspondence: tero@lumi.fi

**Abstract:** This multi-disciplinary science and Indigenous knowledge assessment paper reviews over 20 years of research materials, oral histories and Indigenous views on climate change affecting Unalakleet, Alaska, USA and Norton Sound. It brings a historical review, statistical analysis, community-based observations and wisdom from Unalakleet Iñupiaq knowledge holders into a critical reading of the current state of climate change impacts in the region. Through this process, two keystone species, Pacific salmon and caribou, are explored as indicators of change to convey the significance of climate impacts. We rely on this historical context to analyse the root causes of the climate crisis as experienced in Alaska, and as a result we position Indigenous resurgence, restoration and wisdom as answers.

**Keywords:** Norton Sound; Alaska; ecosystem change; Iñupiaq knowledge; Bering Sea

**Citation:** Mustonen, T.; Van Dam, B. Climate Change and Unalakleet: A Deep Analysis. *Sustainability* **2021**, *13*, 9971. https://doi.org/10.3390/su13179971

Academic Editors: Stephanie Pfirman and Gail Fondahl

Received: 6 August 2021
Accepted: 31 August 2021
Published: 6 September 2021



## 1. Introduction

*"I was born free. I won't die free. What happened to us?"*—An Elder from Unalakleet, 20th Century.

Alaskan climate change in all of its context has not been thoroughly discussed as a socio-ecological whole. Studies of northern climate change and, consequently, the impacts on Indigenous people in Alaska and beyond have emerged in the past 20 years as a major research topic [1–10]. Studies of Indigenous knowledge and concerns have also included sectoral approaches in the Bering Sea, such as around transport [11].

Our focus area is Norton Sound, within the Bering Sea, and more specifically the village of Unalakleet, Alaska. We, a Finnish human geographer and a non-Indigenous Alaska-based scientist, present a multi-disciplinary assessment of climate change from the focus of this location building on two decades of engagement. We explore questions that go beyond monitoring and observation to include an assessment of the meaning of climate change from various viewpoints and what are argued to be the root causes and implications of the present day and future changes. The article, therefore, includes a strong equity—and Indigenous rights—focus. We recognize all Indigenous communities do not agree on the degree of climate change or the causes and solutions.

The majority of present-day climate change and Indigenous knowledge studies operate from a uniform, geographically-fixed base of contemporary social and cultural locations. A challenging view is to take the Indigenous historical experience into account; this has also been called an "endemic approach" [12,13]. Iñupiaq knowledge holder Herbert O. Anungazuk [14] (p. 189) has called this the "unwritten law of the sea": *"The lifeways of the Iñupiaq people cover an entire spectrum, a spectrum so wide and profound that it continues to astound the Western mind."*

In this article we try to position climate change as observed, experienced and interpreted in the community into an Iñupiaq-tradition informed matrix that included Indigenous Nations self-governing over their lands and seas [12]. These Nations also decided on issues following their specific socio-cultural, political, cosmological and spiritual processes [14,15]. The traditional land uses and social institutions were built on an intimate understanding of nature and its cycles. Burch [16] (p. 40) points to the historical fact that "each of the Iñupiaq nations discussed both claimed and asserted dominion over a distinct territory having clearly defined border ... When people crossed the border into another nation's territory, they were either trespassers or guests, depending on the particular circumstances attending their passage."

According to Ray [17,18] each of the distinct units had a "chief/'omelik'" who served as the leader of the *kazgi,* the community house where decisions and actions were taken. These chiefs were mostly male. The village with a *kazgi* was the central political location for a nation. Other villages would belong to the nearest autonomous *kazgi* or Nation with distinct borders. A central political-social method of maintaining power, social relations, trade and governance was the messenger feast in the villages [18] (p. 224). It also assisted in adaptations to change and disruptions.

Northwest Alaska was a homeland of several Iñupiaq Nations self-governing their assets and natural resources according to customary governance until the time the process of colonization began, first by Russia (beginning in the 1700s) and then subsequently the United States of America (from 1867 onwards). The loss of self-governance was sped up by the introduction of dependencies on firearms [19], cash economies and large epidemics [15,16,20–23] and political takeover of lands, resources, language and social spaces.

Pratt [24] (p. 105) has described that the Unalakleet region with the Iñupiaq, Yupiaq and the Athabascans provided an important exception to Burch's understanding of Indigenous Nations. According to him, this was due to trading and that the groups "had a stable, friendly relationship and had friendly borders" as opposed to conflicts elsewhere in Alaska. Ray [17,20] also identified specific historical reasons for the Norton Sound region as an exception to the rule of the Nations and their territories.

Our work has been informed by and builds on Napoleon's [15,21] Indigenous evaluation of the events that have transformed the Iñupiaq and Yupiaq self-governing nations into present day modern communities. According to him the colonial process, especially the internal loss of culture and society as a result of the "Great Death", i.e., the epidemics, has not been understood to this day and manifests in the present day as a transference of post-traumatic stress of social ills and collapse of specific nation-based governance, culture, languages and social realities across Alaska.

We have chosen this approach (founded on [21]) to look at current climate change in greater depth and to view it in context as a historical and as an equity process in Alaska. We do not claim that we know a comprehensive view nor that the space allows for an exhaustive review of the situation in Unalakleet. Instead, our attempt is to position the events underway into the losses and partial resurgence experienced in Unalakleet. We conduct this in order to explore the root causes of current change by combining oral histories, written Indigenous knowledge statements, science and governmental reports.

## 2. Materials and Methods

We combine Iñupiaq knowledge, in oral histories and written statements, with the latest natural sciences view of climate change in Unalakleet and, subsequently, the Norton Sound and connected Bering Sea. In the studies of Alaska, there is a trend to include "observations" [5] of change from Indigenous co-researchers, but the deeper contexts and frames of knowledge, histories and cultures are often left out. Instead, Iñupiaq knowledge has been seen at best as Indigenous literature [22,25] or cultural production, but the narratives and statements by Indigenous Alaskans linking a historical and endemic approach have not often been included as a study of climate or ecological change. On the other hand, oral histories and communal lore have been of great importance at the community level [12].

Hykes-Steere [26] has called attention to the fact that the internal dimensions of the Iñupiaq are only slowly being discovered: *"Our world is so completely different than most of you can imagine. A world without time because our world is timeless .... We are taught that you can live in a moment a whole lifetime and words are sacred."* With this realization, the understanding of Unalakleet, Norton Sound and the whole Bering Sea should be re-assessed from a new viewpoint and methodology.

More specifically, the Indigenous history and peoples of Unalakleet have been described, using an "endemic view" by Ticasuk in detail [27] (Ticasuk's views have also been under critical review locally). Historic governmental reports on the early modernisation of Unalakleet [28] provide a demographic and analytical view of the post-World War II situation. Community-based oral history work between 2002 and 2019 ([19] and Snowchange Unalakleet Oral History Archives) contains the words and knowledge of our co-researchers, the people of Unalakleet and their voices from the first decades of the 21st century. This is also the period of intensifying climate change impacts in the region resulting in the present-day crisis on land and at sea.

Victoria Hykes-Steere (an Iñupiaq woman originally from Unalakleet, presently in Anchorage) shares her written statements from 2002 to 2020 [29,30] on climate change issues and their links to the history of injustice in Alaska. We also include written statements of the Kawerak Inc. regional tribal consortium [31] as well as Iñupiaq individuals from Alaska who have shared their views on climate change in public [6,32].

We use two species (Pacific salmon (*Oncorhynchus* sp.) and caribou (*Rangifer tarandus*)) as key socio-ecological species for the community. One of these indicators, Pacific salmon, transects between the ocean and the river ecosystem, i.e., they are anadromous. This is, in a way, symbolic for the shape and role of Indigenous knowledge as a fluid and connected method of knowing. We also include open-ended sections on weather change and Indigenous wisdom. We treat Indigenous knowledge (sometimes known as traditional knowledge, Iñupiaq knowledge, local-traditional knowledge, Indigenous wisdom and also 'endemic knowledge' [13]) and associated streams of information (oral histories, paintings and written statements) being of equal and independent value as a method of knowing.

The methods used include Indigenous evaluation [29], endemic ways of knowledge [13], oral history [33] and narrative analysis [34] to position the observations, views and Indigenous wisdom of climate change into a frame relative to Unalakleet. Secondly, we use community-based monitoring [35] as a vehicle to position and understand the issues from the materials.

These methods have been complemented with field visits during 2002–2019, community based workshops to document and understand the Unalakleet climate change situation, literature reviews, place name analysis and youth engagement (2002–2019). All oral histories, interviews and workshops have been collected and conducted using the principles of free, prior and informed consent.

For the western scientific understanding of climate and ecological change in Unalakleet, we used a literature review and publicly available data sets from the region to position the speed and scope of change into a useful framework. Cartographic and satellite data interpretations, field visits during 2002–2019 and regional ecological monitoring of key indicators and species are also used. In the conclusion, we position the Indigenous knowledge and science trends on the two keystone species into a dialogue to point to potential divergences, disparities and commonalities followed by a final assessment using the long-term Indigenous knowledge of the situation.

## 3. Results

### *3.1. Historical Context*

It is impossible to investigate the present-day experience of climate change without first reviewing the historical context of Indigenous peoples of Unalakleet as a tapestry of Alaska. Unalakleet is a village located on the Norton Sound at the mouth of the Unalakleet river in Alaska (Figure 1 provides a map of the Norton Sound region, with place names as described below). It is located approximately 630 km from Anchorage and is a fly-in community (https://kawerak.org/our-region/unalakleet/) (accessed on 1 December 2019). The earliest archaeological evidence from the present-day village site dates back to 200 BC (also see [32]). Emery, Redlin and Young [36] (p. 482) say that the site would have been used for "15,000" years but do not provide a more exact source for their data.

**Figure 1.** Map of Norton Sound region.

According to Ray [20], place-name analysis is a method of understanding occupation and presence over long periods of time in Indigenous Alaska. On the Norton Sound, according to her, the Iñupiaq and Yupiaq systematized their travel and camp sites as well as dwelling sites on place names. Ray [20] (p. 256) and [16,24] point out that, today, most of the Indigenous place names have been lost as they ceased to be used for the most part in the end of the 1800s: "(place names) once held a tribal territory together, provided mnemonic guides for travel and utilisation of resources and forged a permanent and identifiable bond with the land."

Pratt [24] identifies the community to have been a major historical node for pan-Indigenous trading and migrations. According to Pratt [24] (p. 94), the Indigenous peoples in the region at the time of the European contact were the Koyukon Athabascan and Unalit Yupiaq peoples. Some sources recall earlier Iñupiaq presence already in 1830s (Snowchange Unalakleet Oral History Archive 2019). Ray [20] also reports occasional Iñupiaq "travellers" down to the Kuskokwim river in the early 1800s. Until around 1800, Unalakleet was a major border of Iñupiaq and Yupiaq languages. After that, Ray [17,20] says that it became a trilingual community of Malemiut Iñupiaq, Kauwerak Iñupiaq and Yupiaq (see critical view of the Malemiut concept in [16]). This change was triggered by a loss of caribou in the Iñupiaq home areas further north and increasing trade [18,37]. Burch [23] places the overall crash of this Nulato Hills caribou herd (see below) between 1870 and 1900 (see critical assessments of his analysis in [38,39]). Ray [18,37] refers to this as a major economic revolution in the region. The destination of the increased trade was the Russian-established trading market of Anyui on the Kolyma river delta in Siberia in the Chukchi homeland [18].

Burch [16] (p. 319) offers a very important analysis of the arrival of what he calls a "new social system". He writes that the combination of loss of caribou, small-pox, trade and other drivers all contributed to the establishment of the Iñupiaq presence on Norton Sound. According to him, the social structure of the Iñupiaq allowed strangers to join with a specific Nation. He says Unalakleet received the survivors of famine even from the Kivallinigmiut people [16] (p. 321).

The documented place names of Iñupiaq and Yupiaq origin on the sea coast around the community reflects the seasonal rounds of hunting, fishing and gathering (see [20] for a full list), including those listed in Table 1.

**Table 1.** Selected older Iñupiaq, Yupiaq, Unalit and Athabascan place names of the Unalakleet catchment area as in Pratt [24] (p. 112).

| Iñupiaq and Yupiaq Place Names | Translation |
|---|---|
| *Kungikuchuk* | Norton Bay |
| *Ingektuk* | A (good) mountain where it was good to pick blueberries |
| *Choatulik* | Blueberry place |
| *Igikpait* | Big mountain where caribou hunting took place, see more on this in Burch 2012: 75, placing this place name in the 1840s |
| *Pitikshuit* | Place of shooting caribou, see Burch 2012: 75 |
| *Putulgit* | Rocks with a hole, a way of measuring number of people passing by adding a rock |
| *Kikiktuk* | Island (whale island) for whaling between Unalakleet and Shaktoolik, today Besboro |
| *Paimiut* | Unalit name for the river mouth and associated fish camp area |
| *Nigukmuthluk* | Cache area for meat |
| *Sikseriak* | Where to go for hunting squirrels |
| *Angakuksharak* | Old lady's camp |
| *Kaglik* | Place of seining on the Unalakleet river |
| *Mekliktlik* | Place or a stream of good water close to the ocean |
| *Nagoyumkuti* | Sandbar where the sea gulls congregate on the Unalakleet river |
| *Naplathlasit* | Location of the Sámi / Lapp reindeer herders, also an old fishing camp |
| *Nunamitkoa* | End of the world, a river camp site where the Iñupiaq fish camps ended in 1800s (today, the river is used more extensively) |
| **Yup'ik Place Names** | **Translation** |
| *Ungalaqliq* | "south/south wind [village]" or "one river to the south" or "from where the south wind blows" or "the way the [Unalakleet] river flows south to the ocean" or "where the Unalit live". Ray (1971: 253) quotes Rasmussen as to "farthest south". |
| *Kuiggavluaq* | "Swift river" or "little river" |
| *Iktigalik* | "Possesses Indians" |
| *Ulukaq* | "Woman's stone knife"; stone formerly used to make women's knives |
| **Lower Koyukon Place Names** | **Translation** |
| *Yoonłe* | "The distant perimeter" |
| *Ses Tseegé* | "Ochre-colored bear" |
| *Kk'aadoleekkaakk'et* | "Mouth (of Kk'aadoleet Nó)" |
| *LeggUyh No'* | "White river" |
| *Edemełek Denh* | uncertain/unknown |
| *Tiyh T'oh* | "Where the trail lies at the foot of a hill" |
| *Haatoghee'o Denh* | "Place where/to which the water reaches" |
| *Ses Tseegékkotno'* | "Little [Ses Tseegé (ochre- colored bear)]" |
| *Too Kk'utl* | "spring water" |
| *Kk'aadoleet Nó* | "Water flowing on (a downward slope to the sea)"; "river flowing on top" ["i.e., on the portage"] |
| *Yoonłe Tene* | perimeter trail |

Challenging, in part, Ray [17,20] and other scholarship, Pratt [24] argues that the Unalakleet river catchment area and the upstream river have historically been Koyukon Athabascan-occupied, as reflected in the place name analysis. Burch [16] (p. 12) also adds to Ray's analysis that there are names of individuals reflected in the Norton Sound place names. Toponyms such as *Haatoghee'a Denh* are evidence of the large use and occupancy by the Athabascans [24] (p. 100).

The village has been a significant pre-historical trade area between the Athabascan, Yupiaq, known as Unaligmiut (considered the "original inhabitants" of the community in pre-history), and "northern" Iñupiaq (also speaking Malemiut and other Iñupiaq dialects, such as Kauwerak) peoples in Norton Sound connected by overland route to the Yukon River in the East. This route is known as the Kaltag portage (see map in Figure 2) and positioned local nations into a geopolitically central role as mediators of trade up and down the Yukon River.

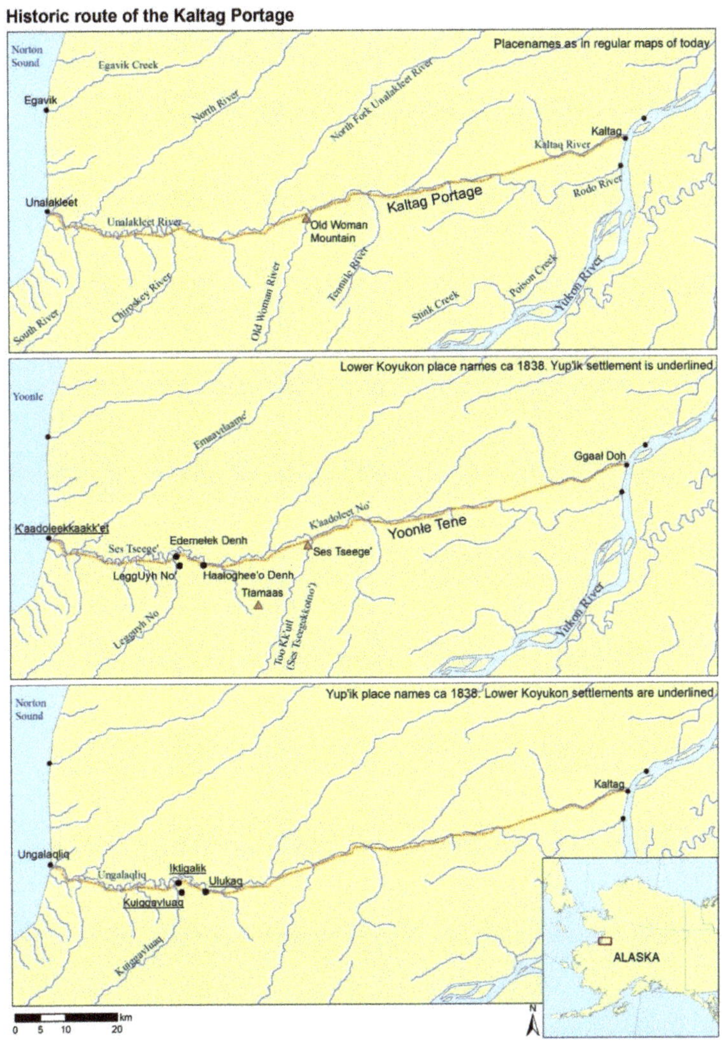


**Figure 2.** Historic route of the Kaltag portage.

The place name Unalakleet has been defined as "From the South Side" (some sources earlier translated it as "Where the East Wind Blows", but this is today considered inaccurate ([19,24]). Ray [20] also says "One river to the south, or where the south wind blows"). Ray [20] and Pratt [24] offer the historical evolution of the place name: From "Uliakagmiut/Ungalaklik/Unalaklik" (this referred specifically to the Yupiaq community south of the river mouth. The Iñupiaq had their settlement north of the river) into "Ulukagmiut" (in 1800s) into the present-day Unalakleet (these toponyms have also referred to the ethnic communities in place across time). Pratt [24] (p. 112) says that Ungalaqliq referred to "from where the south wind blows" amongst other meanings.

Ticasuk [27] (p. 29) says that the Kuvunmute Iñupiaq people arrived at Unalakleet from the north from the Kobuk River area. She [27] (p. 49) points also to the Yupiaq influence in the community associated with the proliferation of trade following the contact with the settlers (Russians) in the 1700s. Pratt [24] (p. 95) provides that the first contact happened in 1778 (while Ray [20] (p. 252) points out that King Island, Ukivuk, was reported by Daurkin in 1765). He says [24] (p. 95) that the first European to visit Unalakleet was Andrei Glazunov in 1833. Ray [37] and [16,23] report that the first firearms arrived in the region in 1819. Pratt [38] offers criticism of this theory and links these herds to a larger Western Arctic caribou herd (WAH), pointing that Burch's assessment of the firearms was vague. According to them, the introduction of firearms may not be a valid argument in losses of caribou populations in the region. Instead, other reasons such as cyclical population dynamics may be at play. Mager [39] has also continued the assessment but comes close to [23] genetic evidence.

The earliest recorded notes from Unalakleet are from the Russian-American company who constructed a post there in 1830s. Ray [18] points to the number of ships associated with the search of the Franklin expedition that caused major social impacts in 1849–1853. In 1892–1898, reindeer herders were brought into the region to introduce this northern trade to Alaska [23] (p. 17). These herds dwindled after 1936 for the most part, with some exceptions such as in Stebbins [40].

Ticasuk [27] offers a view on the pre-contact and early-contact Unalakleet society as observed by the peoples themselves and "handed down" from the past using Iñupiaq knowledge transfer. She provides a view of the royalty, kinship and a description of the endemic [13] decision-making processes in the community. Her account describes the Iñupiaq traditional royalty from the region, including Alluyagnak I and II ([18] she spells them as Alluiyanuk) and Queen Masu. Ray [18] says that majority of the Iñupiaq population in Norton Sound is related to Alluiyanuk. Hykes-Steere [41] describes these royalties in her recent reflection: "Dorothy Jean Ray did field work in Unalakleet in the 1960s and stayed with my great-grandmother, my grandmother's mom and our last traditional Queen. Her father was Chief Nashalook (see also [27])—the last one Unalakleet had. His two brothers, (the oldest) and Paniptchuk were co-Chiefs, but Nashalook, the youngest brother, was made Chief when he was 20 or 21. Their dad was Chief. There are five Iñupiaq languages in Unalakleet, so unlike most villages we were not all related."

According to Ticasuk [27] (p. 27), the Norton Sound Iñupiaq practiced their seasonal rounds according to ecology and the species and food available. Freeze-up and ice-melt defined the uses of the marine and terrestrial areas for hunting, fishing, gathering, trading and other activities [27]. The authors of [20,28] point to the fact that, historically, the Unaligmiut had been living as far west as in Golovin Bay, influencing the linguistic situation there.

A major social change event was the arrival of the Reverend Axel E. Karlson (originally from Sweden) who became a significant advocate of Christianity in the village and in the region. He arrived into the Norton Sound Region in 1887 and from there on started to expand his missionary work [27] (p. 95). This has been summarized as follows: "On June 25, 1887, Evangelical Covenant missionary A.E. Karlson arrived in St. Michael. In St. Michael, he happened to meet Nashoalook of Unalakleet, who spoke some English and Russian. Nashoalook, a medicine man and the last traditional chief of Unalakleet, was one of five tall, dark brothers who had moved to Unalakleet from Malemiut country, which is east of Kotzebue. Nashoalook invited Karlson to Unalakleet, and travelled there on July 12, 1887. In Unalakleet, not all of the locals were as friendly to Karlson as Nashoalook was, and a plan was hatched by three men to kill Karlson. Upon learning of this murderous plan, Nashoalook hid Karlson in his home for three months until the angry men could be persuaded not to kill Karlson" [42].

Ticasuk [27] (p. 97) describes that there used to be a "kargii" (kazgi is the spelling cited in [17]) surrounded by sodhouses in the area, reflective of the Iñupiaq customary rule and governance. She points to a fact that many earlier residents of Unalakleet had "died" before the re-settlement was underway in the 1880s. Ray [20,28] confirms this and points to a small-pox epidemic of 1838–1839, which left the community with only "13 survivors".

This indicates the massive role of the epidemics of the region [15,16]. According to Ray [18], subsequent epidemics included measles and pneumonia in 1900 and the 1918 influenza epidemic. She also refers to the 1830 tsunami that affected coastal villages. A major outside driver in the region was also the number of gold rushes between 1898 and 1900.

Ulukagmiut also entered into regional conflicts with the Athabascans, including the Nulato Massacre in 1846–1851 ([24] p. 97). However, at the time of the 1838 epidemic, there was an Iñupiaq settlement on the north side of the river (the one Karlson saw), which amalgamated the 13 survivors from the Unalit (Snowchange Unalakleet Oral History Archive 2019).

Ray [28] (pp. 150–151) says that the Iñupiaq arrival to the Unalakleet region was also driven by the accelerating status of the Russian fur trade, which was prominent between 1836 and 1868 ([24] p. 97). By establishing personal kinships in the region with the Unaligmiut, the Iñupiaq were able to establish their presence in the area (see also in [27] on the personal kinship relations of Norton Sound). Ray [28] (p. 151) says that the key period for co-existence of the Iñupiaq and the Unaligmiut happened around 1865–1867. The amalgamation process sped up with the missionary work by Karlson.

Karlson was involved also in the construction of the "new" village of Unalakleet [27]. The area had been used for a long time as an Indigenous trading and occupancy site, yet the 1887 events saw the fixing of the village site to its present location. At this time Russian was a common trading language in the area due to the history of the Russian-American company trading and settling since 1700s. In later years, English became dominant.

The question of land rights has persisted in Alaska beginning with the 1867 transfer officially to the US from Russia [29]. A range of legislation, including the Dawes Act and the Wheeler–Howard Act, advocated either for recognition of "Native lands" or extinguishment. A central driver of these political-legal actions was the question of how the lands of Alaska could be utilized. The people themselves were equally active early on. Already in 1912, the Alaska Native Brotherhood struggled for recognition of Indigenous rights. Between 1942 and 1946, Unalakleet was included in a group of seven reservations, similar to the mainland United States. Unalakleet was the smallest with a land base of only 870 acres [43] (p. 87).

In modern history, Alaska was established as a state in 1959. The twin drivers of unsettled Indigenous rights and the discovery of oil and gas as well as other resources led the U.S. Congress to pass the Alaska Native Claims Settlement Act (ANCSA) of 1971 (see critical views in [29,44] and supporting views in [43]). This Act transferred Indigenous lands and governance areas into ownership by regional and village for-profit corporations—a solution that has not been used elsewhere. The city of Unalakleet was incorporated in 1974 (https://kawerak.org/our-region/unalakleet/) (accessed on 15 January 2020).

According to Hykes-Steere [29] the ANSCA transformed the land governance away from the traditional governments into the hands of "western"-defined corporations. Hykes-Steere [29] (p. 384) says that "congress ... forced (Indigenous hunters, fishers and gatherers) into the market economy and the 20th century, the largest land grab in U.S. history and a marvel of social engineering designed to destroy the social fabric of our communally-based societies."

### 3.2. Climate Change

According to Indigenous observations from Unalakleet, climate change impacts began to emerge primarily during the post-ANSCA era. We call this period "modern Unalakleet". Climate change impacts, as perceived by the people in the community, started in the early 1990s. Major visible climate change impacts as perceived locally include coastal erosion, sea level rise and storm surges. The community has been chosen as one of the most vulnerable in Alaska because of this by NOAA [45]. Emery, Redlin and Young [36] (p. 481) estimate that 86% of the Alaska Native villages in Alaska will face "destruction" because of these drivers.

Aronson [45] (p. 7) stresses the spirit of survival in the community. This has led the community to take a range of actions overall, including erosion monitoring with the U.S. Army Corps of Engineers, sales of uphill land lots to residents to have new housing areas, construction of a sea wall and wind power installations to diversify community energy sources.

We have chosen one marine and one terrestrial species of central importance to the Unalakeleet residents as keystone species (Pacific salmon and caribou). This allows for an ecosystem-crossing view and to position and discuss the community-based Indigenous observations from several co-researchers between 2002 and 2019, several of whom have passed on. We wished to allow a rigorous dialogue with scientific indicators of the same species to take place and, hence, chose only two species, a terrestrial and an anadramous example, even though, as is often said, Indigenous knowledge does not separate or uplift certain species over others. Two final categories refer to weather change and an open-ended space for "Indigenous wisdom" where the undefined free expression of Iñupiaq knowledge can be outlined.

Even this method of expressing observations remains limited. The space here does not allow for linguistic scaling, Indigenous evaluation [29] or a full oral history disclosure of the nuances and details of change. Our purpose is, however, to position observations into a twenty-year frame and meaning in order to offer a coherent view of change.

Table 2 lists key observations, an interpretation of meaning in the given context and the source material where the observation can be referenced for each indicator species listed. Brief summaries from the observations listed in the table are in the sections below.

**Table 2.** Key observations, meaning and source material references for Pacific salmon and caribou.

| **Pacific Salmon (*Oncorhynchus* sp.)**<br>**Key Observation** | **Meaning** | **Reference** |
|---|---|---|
| Salmon harvested for dry fish; important seasonal food | Major dependency on Pacific Salmon | [19] |
| Salmon populations started to dip in 1992; warmer ocean | Moreover, birds and salmon dip in numbers. The ocean is warmer | [19] |
| Ocean bottom vegetation is increasing and affects salmon | Warmer trend noticed in 2002 (nets) | [19] |
| Overharvesting of salmon in the Bering Sea (international) | International boats take too much in 2002 | [19] |
| Salmon have more lesions, 2002 | More disease in salmon | [19] |
| Everything depends on the nutrients of the salmon on river | Salmon is the key to river health | [19] |
| King Chinook fishery was very big in 1980s; now it has dipped | Chinook numbers down in 2002 | [19] |
| King salmon drops "first noticed in 1980s" | Chinook decline first observed in 1980s | [6] |
| Silver (Coho) salmon plentiful in 1980s; now it has dipped 2002 | Coho numbers down in 2002 | [19] |
| 1997–2002 "considerable drop in salmon" | Numbers dwindling | [19] |
| In 1983, daily catch was 20–45 tonnes commercially; now it is low | Major drop compared to 1983 | [19] |
| "Ten times more king, chum and coho" in 1983 than in 2002 | Collapse of salmon | [19] |
| King Chinook arrives later than usual 2002–2003 | Warmer ocean? | [19] |
| Silver (Coho) salmon arrives late 2002–2008 | Changes in water temperature | [19] |
| Silver Coho salmon made a comeback in 2008 | Increased Coho numbers | [19] |
| Many dead fish at the ocean bottom 2002 | Hook fishery by-catch or catch and release? | [19] |

**Table 2.** *Cont.*

| | | |
|---|---|---|
| Abundance of Pink Salmon in 1940s | Loss of plentiful stocks | [19] |
| Pink salmon years are "even" | Pink salmon plentiful in even years; cyclic nature of pink salmon stocks | [19] |
| Average catch 5–6 King Salmon in early 2000s | Major collapse compared to 1980s | [19] |
| "Half of the salmon" lost | Major salmon collapse | [19] |
| River very low; warmer temperatures 2002–2008 | Prevents salmon spawning in time | [19] |
| Fish deformations more common after 2000 | Sick fish increasing | [19] |
| Pacific Salmon also occasionally returns back to ocean | Individual behaviour reported | [19] |
| Pinks, Chums, Silver and Kings stay in the area all year | Resident salmon stays | [6] |
| Chum (dog) salmon used for dog food in the past | Change in modes of transport | [19] |
| International fleet harvesting millions of tons | Lack of Indigenous rights | [5,19] |
| Pink salmon in dire straits since 1992 | Collapse of Pink Salmon | [19] |
| Biggest threat is the loss of subsistence salmon | Cultural and food security threat | [19] |
| King (Chinook) numbers on the decline 2008 | King Salmon loss | [19] |
| King salmon has a 24-year salmon run | Cyclic runs are long | [19] |
| Large numbers of jellyfish | New event; kept the silver salmon away | [6] |
| Salmon spawning areas disturbed by people 2002–2009 | Human interference on the spawning | [19] |
| In 2007, the industrial trawlers took 120 tonnes | 1982–2019 severe impact on chinook from trawling | [46] |
| Salmon and tomcod health linked | Tomcod harvest decreasing, loss of ice | [47] |
| Salmon and herring health linked | Link between herring decrease | [47] |
| Silver main commercial ocean fish in Unalakleet 2009 | Central to the community | [6] |
| Seining, rod and reeling for silver in the river | Subsistence use high | [6] |
| Subsistence closures due to overfishing in Unalakleet | Major commercial harvests | [6] |
| Trout harvest Pink salmon fry and smolts for food | Trout affecting Pink | [6] |
| King salmon is smaller than in the past 2015 | King salmon size loss | [6] |
| Pollock fishery affects King salmon as a side catch | Pollock harvest affects king numbers | [6] |
| King salmon eats more herring or does not eat | Changes in King diets | [6] |
| Kings first to arrive 1 June; arrival dependent on ice cover | May be delayed if a lot of ice | [6] |
| King salmon do not have a whitish "tip" anymore in 2014 | Changes in King salmon nose | [6] |
| King salmon decline due from predation of trouts | Predation | [6] |
| Habitat changes cause less King salmon | Habitat degradation potential | [6] |
| Commercial King catch ended | Voluntary moratoriums in place | [6] |
| Silver salmon do not have "noses" in 2014 | Changes to bodies of silvers | [6] |
| Silver salmon has skin diseases and problems | Quality of the silver down | [6] |
| Drop in Pink salmon numbers in 2014 | Pinks affected, numbers down | [6] |
| Red (sockeye) salmon rare in Unalakleet, but now increase | Not often observed in the past | [6] |
| Chum salmon comeback in 2014 | Chum doing better in 2014 | [6] |
| Increase in beaver dams may affect salmon spawning | Dams of beavers affecting salmon | [6] |
| River temperatures too warm causing salmon death | Fish death events in Summer 2019 | [31] |
| Salmon arrives earlier, fry leaves later | Changes to salmon cycles | [31] |
| Pink Salmon pre-spawning death events | Water temperatures high | [48] |
| Thousands of Pink Salmon dead in Unalakleet in 2019 | Water temperatures high; will impact mammals and birds | [40] |
| King salmon returned plenty in summer 2019 | Have been mostly gone for 15 years; closures have worked | [40] |
| So many dead fish that youth could not jump from the bridge to swim | Salmon deaths noticed by all age groups | [49] |
| Best King harvest since 2014, but mostly small fish | Improved king situation, but size is small | [49] |
| **Caribou (*Rangifer tarandus*)** | | |
| **Key Observation** | **Meaning** | **Reference** |
| "Immense herds in Unalakleet in May 1867" | Caribou (possibly NHCH) plentiful | [23] |
| Caribou stocks collapsed in Northwest Alaska in mid-1800s | Redistribution of Iñupiaq peoples | [24,28] |
| Caribou had further decrease in 1870s | Redistribution of Iñupiaq peoples | [18] |
| Communal drives for the caribou have ended in 1900s | End of a communal harvest | [50] |
| Caribou hunt remains central to food security 2002 | Caribou still available in 2002 | [19] |
| Caribou re-appeared close to village in 1980s | In 1940–1960, few caribou | [19] |
| Quota of 5 caribou/day in place in 1980–1990 | Food security guaranteed | [19] |
| Between 1997 and 2002, Caribou have not come to town | Animals stay further out | [19] |
| Caribou hunting is performed usually in winter, January | Seasonal round observed | [19] |

**Table 2.** *Cont.*

| | | |
|---|---|---|
| Sámi reindeer mixed with the caribou since 1891 | Reindeer and caribou mix | [19] |
| 17000 caribou mixing with reindeer 1989–2005 | Reindeer and caribou mix/Seward | [51] |
| Caribou have not really been here since 2001 (2008) | Disappearance of caribou | [19] |
| Caribou migration close to town ended in 2004 | Disappearance of caribou | [36] |
| Caribou are "128 miles" away | Disappearance of caribou | [36] |
| Caribou are far away from the village | Continued absence of caribou | [52] |
| One family had to travel over 300 miles to caribou (2018) | Caribou far away from the village | [52] |
| Hunting trips extended to Buckland and SE Buckland 220 miles; 450 miles total | Caribou far away from the village | [46] |
| Caribou used to be harvested in January, now in March | Travel on the land becomes hard due to conditions | [46] |
| Hunters have to travel to Koyuk for caribou | Caribou far away from the village | [40] |

#### 3.2.1. Pacific Salmon (Various *Oncorhynchus* spp.)

Key summary: The authors of [6] (p. 127) describe the fish species important in the community (including Pacific herring, King salmon, Chum salmon, Pink salmon, Silver salmon and trout species), and how they are fished and used. Pacific salmon species are central for food security and culture. Major declines in numbers occurred between 1983 and the 2002–2007 period. After this, Silvers rebounded for a while, whilst King salmon numbers were very low until 2019 when they made a return. Pink salmon mass death events triggered widespread concern in 2019, alongside changing ocean and river conditions. Industrial harvests (by "foreign pirates", pollock by-catch, major fisheries in the Alaska Peninsula and Aleutian Islands Management area and. to a much smaller extent, the local commercial fishery) affected the subsistence catch. Trout are increasingly affecting salmon smolts.

#### 3.2.2. Caribou (*Rangifer tarandus*)

Key summary: Early crashes of caribou stocks in Northwest Alaska triggered a migration of some of the Iñupiaq further south [23]. In 1867 there were large numbers of caribou in Unalakleet. During the early 1900s, the caribou were not as plentiful, and reindeer introduced by Sámi herders in 1891 [51] eventually joined the caribou herds. Caribou were abundant and provided for the village during 1980–2000, but since then numbers have dwindled, and animals have moved further away from the village ([28] gives a base number of 200,000 caribou in the Bering Strait region, with 15,000 annually on average).

#### 3.2.3. Weather

Table 3 shares key observations related to weather observations and changes, including references for the source material. The information includes how Elders and earlier generations could predict weather accurately, and some people could even influence the wind and weather, but those skills are being lost as a community. Snow amounts and storms have been observed to change significantly. The timing and process of break up and freeze up have also been affected in recent years. Continuing coastal erosion is a concern, as are algal blooms in the ocean. There is less sea ice, and it is thinner. Major storms and lack of sea ice have occurred more recently in 2010s.

**Table 3.** Key observations related to weather and source material references (including summarized observations and direct quotes; direct quotes are indicated with quotation marks).

| Weather<br>Key Observation | Reference |
|---|---|
| 1830: A major tsunami event in Bering part of Alaska. | [18] |
| Only in 1892, Bering Strait could be crossed on ice. | [18] |
| 1892: Winter was unusually cold, fall was late and was spring early. | [18] |
| Shaktoolik suffering from erosion (caused by river) already in 1931. | [18] |
| In 1950s, weather was "normal". | [6] |
| In 1960s, ocean conditions started to be "rougher". | [6] |
| Freeze-up happened from September to October in 1960s. | [6] |
| Weather was more consistent. | [6] |
| Change started in 1970s. | [6] |
| Ocean frozen solid last time in 1970s; polar bears close by. | [49] |
| [When asked about weather prediction] "I... you know we grew up without grandparents, the ones that have grand- parents are the ones that... they sure know how to predict. There's one lady we used to go camping with a boat down the shore for salmon berries; she'll predict the weather in the evenings. She'll go out and put her hands inside her parka and stand around and look at the clouds, and look at the hills, and some evening she'd come in and say: 'tomorrow will be nice,' other days then, 'tomorrow won't be too good, either rains, or be real windy.' If it's windy they enjoy it, because of the mosquitoes." | [19] |
| "I would say that there's gonna be years with really heavy snow, and that gives us a reminder of when I grew up. But it seems like it was like that every year. And now, maybe once in a decade we'll have a really good heavy snow year. In fact, several of the last ones have been pretty dry, not as much snow as I could remember. But it seems like when I grew up we had lots of snow every year. But it does not seem to be the case now." | [19] |
| "I moved in to my grandfather's home back in 1939, and there's a lotta change since. It's noticeable. We used to have, you know, the break-up of the Unalakleet River used to occur late in May and in June. And now it's late April or first week of May It's much earlier and much warmer, than it used to be. And when I was a kid, during the middle forties, it was a fun game for us to jump from ice cake to the water, and it'd be middle of June!" | [19] |
| "The first time I really can remember having rain in winter time was when I was in grade school, and then it froze the next day, and everybody went skating all over the place. It's a long time ago. Around 1951, or somewhere around there." | [19] |
| "It just doesn't seem like we've had too much of the blizzards, like we used to have when I was younger. And long period of bell blizzards in, you know, not the one-day, two-day kind. I mean it used to blow seven to ten days in a row, you can't see anything, that's why we had the big snow banks. But it seems a little bit warmer." | [19] |
| "There has been changes since I've lived here. I came here in 1978, and I've noticed that one of the big changes in the winter is that there is no more big huge snow banks in town. They were very, very high above buildings each winter. Now there is practically none. And when I sit around with people, and visit maybe the older ladies, they talk about how the summers are hotter than when they were young. Or ever since I've been here maybe it seems like the summers are hotter." | [19] |

**Table 3.** *Cont.*

| Weather<br>Key Observation | Reference |
|---|---|
| "We notice these things, so maybe the global warming may have affected them. It wasn't that noticeable. One other thing that I noticed is that even the snow... that may not be natural, I'm almost certain it's because of global warming. Because in the fall, we used to get snow all the way from October and December-January, and the snow used to be all over, you know, and the snow banks. But during the last few years the snow doesn't seem to come until January, or so late. And last year, we got quite a bit of snow, and when the warm weather come, and the snow just dissolved more or less." | [19] |
| "My step grandfather said years ago the winter was severe. They used to go dog teaming to go to St. Michael's, and they used to make straight cut. But now they say the weather is changing on account of the wind, and this keeps it from forming solid [ice where] the wind would blow. We do have a lot of east wind during winter months. And it is not as solid as it used to be, to go right straight to St. Michael's." | [19] |
| "I've hunted seals ever since I was a little boy. My dad took me up when I was so small, and I could not see the land in any direction, and I was worried. I was hoping that he'd know the way back, because I didn't know where I was. He taught me how, and he taught me wind direction and how to catch the seals. You know, when I am looking for seals, cause minor problem, but if I'm not looking for the oogruk [bearded seal], I need the leave the shore-fast ice and get out there, you know, twenty-three miles out on the ocean. And if I am looking for walrus I gotta go even further.<br>But he taught me how and basically where to find them, and how to catch them and how to take them a part. And what to bring home and what not to bring home; there was not too much we left behind." | [19] |
| "It's warmer! Yeah it's warmer, and I'm told that the scientists are seeing a lot of different types of algae growing out there. And blue whale... blue whale is right in it. They're not supposed to be up here! But they're out here on the Bering Sea, in that green stuff. So, I don't know maybe they're just using it for shade or something, but... there's a lot of change in the ocean! When they first started reporting [changes in birds]... that's when our salmon started taking a dip in population. It was about ten years ago; they said that birds are dying out there. And it showed here, in the fish; we get less fish." | [19] |
| "The sea ice is thinner in the winter- time. Here it blows out frequently, cause we have east winds, and after freeze up when the wind blows the ice off, and that might be a contributing factor too. I've noticed that the ocean ice is considerably thinner. Because of it's thinner it breaks up sooner. " | [19] |
| "You can also use the wind and the tide, somehow I can't think of how you can do it right now. But you can tell which way the wind is going to blow by just looking at the tide. Also, when we are out berry picking and it's very calm and all the mosquitoes are there, we usually whistle. You know, and it'll bring the breeze. So we still do that." | [19] |
| 2012: Extratropical cyclone produced storm surged up to 3–4 m. | [45] |
| 1994: A really high flood on the river. | [45] |
| Really early winter, 1992. | [45] |
| There is more rain now affecting the drying of fish in 2014. | [6] |
| Floods affect the salmon spawning in 2014. | [6] |
| Decreased snowfall makes the river freeze to the bottom affecting fish and spawning survival. | [6] |
| New storms cause substantial erosion. | [6] |
| Breakup in 2014 was noted to be different than in the past, ice melts on the place, no bangs. | [6] |

**Table 3.** *Cont.*

| Weather<br>Key Observation | Reference |
|---|---|
| Freeze-up has been delayed in the autumn 2014, now it happens October–November. | [6] |
| Sea ice is less, with more open water areas. | [6] |
| Loss of dog teams means that the fish are not used as much. | [6] |
| Sea ice formation usually in October; in 2018, open water well into Autumn. | [32] |
| Disappearance of the Bering Sea "cold pool". | [31] |
| Projected disappearance of sea ice by 2037. | [31] |
| Hurricane in Savoonga 2017 damaged over 60 homes. | [31] |
| Energy infrastructure affected. | [31] |
| Four feet of ice in June during seal hunt in 1970s. | [46] |
| No sea ice and hard sealing in 2018 on March 14th, 60 miles one way for hunt trip. | [46] |
| 3rd August 2019: massive storm with flood events. | [46] |
| Lack of sea ice and 35 mph winds cause significant erosion. | [46] |
| Permafrost thaw events on the coast reported and documented; sinkholes. | [46] |
| New storms are so massive that people need to evacuate. | [40] |
| Plants bigger because of warmer weather. | [40] |
| Winter 2018–2019 no sea ice at all. Seagulls stay longer. | [40] |
| Algae forming in bay and caught in nets. | [49] |
| Elders saw the changes in visions beforehand. | [49] |

### 3.2.4. Indigenous Wisdom

Table 4 provides space for the open-ended sharing of Indigenous wisdom that occurred in interviews and materials. By highlighting these specific statements, we are pinpointing what constitutes as meaningful as perceived by Indigenous evaluation methods. In summary, climate change is observed to be caused by greed and loss of traditional Indigenous values. Survival and well-being in Unalakleet is understood to be deeply connected to the health of animals and fish and the environment. The "ownership" of Alaska was described to have been wrongly transferred to parties that are not managing the environment well. The Alaska Native Claims Settlement Act has not delivered the promised rights or solved the crisis in the villages.

**Table 4.** Open-ended Indigenous wisdom, including source material references.

| Indigenous Wisdom<br>Key Observation | Reference |
|---|---|
| "If the global warming really affects the salmon and I think it does, salmon and all these other resources that natives subsist and rely on, when those are gone, it will be hard on a lot of people here in our area". | [19] |
| "Our ability to survive as native people been depends a lot on that fish and it's staple in our diet." | [19] |
| "We grew in a times, where we were the only ones here and we owned the whole state. You know, the native people did. That forty million acres, sound [like] an awful lot of land that they say they give us. But there's 364 million acres in the State of Alaska. What happened to the other 328 million, that was divided up between the federal government and state government? Our land, and they have taken the land. They've taken the money derived from that land. They've taken the money derived from the oil, and they've spent it in urban centers. While our communities go without water and sewer yet." | [19] |

**Table 4.** *Cont.*

| Indigenous Wisdom<br>Key Observation | Reference |
| --- | --- |
| "Native people have to be a part of the society. They have the right to be here, and we have a right to be here. And I'd like to keep fishing and hunting, and pass on what my dad taught me, plentiful, bounty of the ocean, and the bounty of our land that we live in." | [19] |
| "I worry if they kill the resources, that not only we depend, but the marine mammals depend on. If those fish species are gone, what are they gonna eat? If they change the cycle of life, that we depend on, there is a missing part of that cycle that [reverberates] throughout the whole cycle. You end up with nothing. It's kind of scary." | [19] |
| "I have really deep concerns about oil and gas development. I think it benefits primarily the huge companies and the nation as whole. And they don't take care of the local people, I mean, they say they do. And they do indirectly through taxes and the employment. But based on their experience with the North Slope development, there were very few native people that got hired up there permanently. There's some. You know, there's no question that there's benefit there. But the majority of people didn't really have an opportunity. Most of the people employed were outside people, imported to do the work. And that's been common in Alaska since the beginning of world trade." | [19] |
| "And then how long does oil pollution last? It's really... it's really hard. You know, there's always the argument of scientists. Western scientists versus aboriginal people, and how they know their area. And generally the scientists are written-language based, and document -based. And they don't believe anybody unless several people saw it. And when they're dealing with fish and game, and wildlife, there's no way, you know, they're always behind schedule by fifty years. I summarized Western science being 'oops' science." | [19] |
| "If you judge by the fish, no the sea is not healthy, but there's something going wrong, there's something going on." | [19] |
| "I think it [knowledge of weather] changes gradually. Right now you got satellites and television, and they... they keep communications going.<br>But, I think some people can read weather very well, and for what they do. I mean, they take care of themselves, like the fishermen and hunters. And they know what's going on, and they can tell what is gonna happen. So, they're prepared for what's coming up. I know, one of my friends told me that he was suspicious about his weather forecasting ability, because of the change. He said, 'The ice isn't as thick as it used to be, the currents are different, and the weather patterns changed a bit.' So, when you see the different cloud formations in relation to the hills, you know there is a change. It changed a bit. It's harder to tell, how the weather's going to act with his knowledge. So, we, the younger ones we are a bit more dependent on... they're using the Internet now. And there's more observation in points that's formalized. A long time ago you just depended on yourself. You know, where you are at and you had to read the weather for your own benefit, and your memory of what happened before. You don't have to think about it, you just have to be ready to go and do the stuff. And now that weather forecasting has become more formal and more [dependent on] modern technology for... I don't know... It's just different! It's not mind based it's... technology based with mind interpreted." | [19] |
| "I remember Elders like saying, like, I don't remember exactly, but you can see if it's gonna rain on you in twenty minutes or not. You can see [it from the ways the] clouds are going and see what kind of clouds are coming. Sun mostly comes with calm weather, or you can somehow see if it is going to be there within the next few hours. But I don't know how to predict the weather." | [19] |
| "Etok (Charles Edwardsen from Barrow) and the others who began fighting for land rights in Alaska where when the oil companies joined them in their fight to clear title for the right-of-way to build the Trans Alaska pipeline to access the oil in Prudhoe Bay. Most of the leaders, including Etok, accepted credit cards to pay for their travel to DC. Their travel had been funded by bake sales, raffles, bingo and often they ended up sleeping at the airport, which at the time had showers in DC. This lead them down a very precarious road and in the end we lost our hunting and fishing rights, the land went to corporations and the villages lost control of the land." | [53] |
| Given the speed of climate change, tribes cannot rely on state or federal action for decisions. | [54] |
| "Though the Earth changes, it is still giving. Providing. Nurturing. Inuqtaq (a small boy in 2018) will still learn respect for what gives life. I hope the rest of the world quickly adapts and also respects the Earth—as we have for milleniums and will continue to do so." | [32] |

**Table 4.** *Cont.*

| Indigenous Wisdom Key Observation | Reference |
|---|---|
| "What scares me more than that, and the observation, is the effects it's having on our resources. And this summer is the scariest I've ever seen. I'm really scared of what's happening in our ocean. The Bering Sea is the richest sea in the entire world, or was. We are now seeing die-offs from the killer whale to the krill. Zooplankton, you know, like the bowheads eat. Without the ice, you know, I don't know what's happening. When you see even killer whale die from something out there, all the way, we've got Killer whale, belugas, bearded seals or oogruks, ring seals, spotted seals, harbor seals, you know, all dying. Birds, auklets, murres, you know, puffins, a lot of them just belly up dying. This summer as a commercial fisherman I've seen more fish come belly up, floating out of our river. Normally they spawn, and then they rot trying to go upriver, and they feed the little fish, as they, you know, when they decay and fall or sink to the bottom. The bears eat them, everything becomes part of the food chain. But if they die and they float out into the river, then they're not part of the food chain. And I don't know if that, um, is a way that God kinda controls things, you know, if we got too much in our river, and there's only so much oxygen, and if you got 7 million humpies out there then naturally there's gonna be some natural death. But this looked unnatural." | [46] |
| "For the first time in all my years of fishing, at the end of the season my nets were so stinky I had to strip them, and they're no more good. And it scares me, with the ocean warming 10–15 degrees in Norton Sound this summer they didn't catch their crab. The crabbers couldn't find them. I worry that the food chain is tremendously impacted by climate warming. And I worry about the climate warming, but even more worrisome is the effect of Fukushima. I studied nuclear waste. Nuclear waste lasts for thousands of years. And, so it's been two years maybe, so it's got another 998 years to kill, you know, long afterward. And then even before Fukushima up north was Chernobyl. And they just dump their nuclear waste into the ocean, as I understand it. And that was, what? Thirty years ago. So it's still got 970 years to kill. And it's killing today, yet." | [46] |
| "Basically if we kill what we depend on to sustain life, then we're basically cutting our own throat. And the powers that be need to realize that this is not a red or yellow, black or white thing. It's an everybody thing. And money is not gonna be able to pay your way back to a healthy ecosystem, which we need. But, those are my observations. No snow, no ice, the platform for our marine mammals, where they feed and raise their young, is not there anymore, so they're hauling out on land. Our actions are having effects on the walrus that have to haul out on land, which makes it more dangerous." | [46] |
| "For me as an Alaskan Eskimo, born and raised in Unalakleet, Alaska, I have a tremendous respect for my land. I don't pollute it. Same way, tremendous respect for my ocean I don't pollute it. I clean up, and teach my children to clean up. If you go camping, make it cleaner than when you get there." | [46] |
| "The changes are happening. We see it coming, with social media and everything. In terms of climate change, it's happening and it's happening really fast. And it's accelerated. It's accelerated to the point where we had no longer have the ice to keep the people out of our Bering Sea. The super tanker ships that are going through our northwest passage. Before it was quiet, there was no noise pollution. There was no wastewater, gray water pollution, there was no oil being dumped. But with this global warming and the climate change, our northwest passage is open. Which opens up another avenue of pollution, another avenue of disruption to the habitat that we, our resources depend on, and we depend on these resources." | [46] |
| "Things have changed. You have to go far. We don't get walrus anymore, at all. Zero. That's been gone for many years. When I first came here in the late 70s people used to hunt walrus and eat walrus, but it's no longer available cause there's no more ice." | [40] |
| "It's crazy how we're losing so much. Even the river's changing. The river is changing, eroding, so much that it's changed, it's changing its path it's like it's alive, you know. And then the coast, the erosion is so huge. And then we have, if you fly over and you land towards the village you can see these huge craters, and they look very bizarre. All along north and south, just melting of permafrost and just, it's very ugly and it's kinda dangerous for, like, berry pickers. Yeah. A lot of mammoth tusk being exposed. (laughs) Very weird. It's very bizarre." | [40] |

### 3.3. *"Children's Voices"*

Listening to the voices of the youth of a community provides an important perspective on the present and future and what it means for them to experience and adapt to the present changes. Mustonen and Mustonen [19] collected a range of oral histories from the youth of the early 2000s. In these materials, the presence of communal distribution of catches and the presence of a number of cultural practices, such as feeding the seal salt water after a hunt, are still prominent. Young boys recounted harvesting beluga whale with nets and the knowledge to cut the whale using Indigenous patterns.

Between 2008 and 2018, the school of Unalakleet and two schools in the Eurasian North, Selkie in Finland and Norya in Udmurtia, Russia, initiated a traditional knowledge education programme and exchanges. This revitalisation of educational and cultural aspects has included activities on the land, weather observation and prediction, painting and dancing, to name some examples.

In 2015, the children in Unalakleet school created animated stories about a "School of Dreams" (Available online: http://www.snowchange.org/2015/02/school-of-dreams-produced-animations-and-a-look-to-the-future-in-urals-bering-straight-and-boreal-finland/) (accessed on 14 February 2015). The topics of the animations were focused around a faith in a technological future, including Antigravity, Flying Books, Jet Pack, Robot Girl, Time Traveller and Village in a School [55].

In 2019, the third grade class in the Unalakleet school shared their visions of the future of Unalakleet through a painting exercise (Available online: http://www.snowchange.org/2020/02/children-of-unalakleet-alaska-paint-their-future-2020/) (accessed on 14 February 2020). Themes in the paintings incorporated aspects of observed climate change and adaptations, as well as village changes and desires. For example, students painted other families moving to the hills above town, a rising ocean, snow, warming temperatures, more trees in town, a water park and airplanes.

### 3.4. Science Results

Information from western scientific methods indicate that the Arctic is undergoing a period of unprecedented change [56]. It is well established in the scientific literature that greenhouse gas emissions from the industrialized world and land use changes are causing far-reaching and accelerating change to the climate and ecosystems of the Circumpolar north. Temperature increases across Alaska vary greatly by season and location, with the fall and winter periods in the western and northern regions of the state experiencing the greatest temperature increase [57–59]. Figure 3 shows the trend in annual average surface air temperature at Unalakleet for the available record, from 1950 to 2019 ([60]; NOAA NCEI ISD dataset as described in [61]). There has been an approximately 2 °C change in the annual average surface air temperature in Unalakleet since 1950, based on a least-squares linear fit.

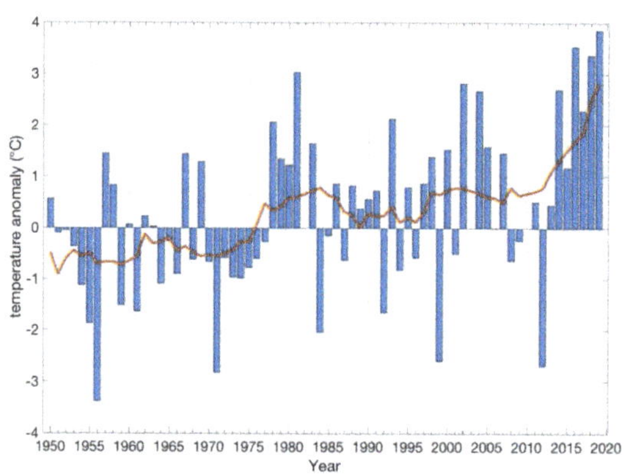


**Figure 3.** Annual surface temperature trends at Unalakleet, AK, 1950–2019. Grey bars are annual mean atmospheric temperature anomalies at Unalakleet, AK, calculated relative to a base period 1961–1990. Air temperature data are from a meteorological station at Unalakleet, from the NOAA National Center for Environmental Information (NCEI) Integrated Surface Database (ISD), a quality-controlled global repository of surface meteorological observations from governmental and meteorological organizations worldwide [61]. The data were accessed through the IMIQ Data Portal [60]. The red line is a 10 year moving average.

Seasonal sea ice in the Bering Sea is crucial for coastal communities, marine and terrestrial ecosystems and regional climate. The annual variation in sea ice extent between 1850 and 2020 in the Bering Sea is shown in Figure 4. The winters of 2017–2018 and 2018–2019 represent an unprecedented deviation from the over 160 year record of sea ice extent in the Bering Sea. The record low sea ice extent observed in the Bering Sea during these recent winters is an integral part of the massive ecosystem shifts underway in this region due to the importance of sea ice as a control on ocean temperature and by initiating primary production in spring through the provision of algae that forms under the ice in winter [62–64]. The conditions that contributed to ice loss during these years include unusually warm southerly winds during the winters that caused substantial ice retreat, delayed ice arrival in the Northern Bering Sea due to late freeze-up of the southern Chukchi Sea in 2017 and 2018 and warm ocean temperatures [65]. Work by Thoman et al. [66] indicates that the 2018 extreme low sea ice extent in the Bering Sea may become the mean level by the 2040s.

Typically, shore-fast sea ice in Norton Sound dampens the impact of late fall and winter storms. The lack of sea ice combined with winter storms during the 2017–2018 and 2018–2019 winters resulted in coastline erosion, infrastructure damages and unprecedented winter flooding events in villages along the northern and western coasts of Alaska, including Unalakleet [59,67,68]. In other regions of coastal Northern Alaska, the lengthening sea ice free season is resulting in an extended fall storm season with more destructive waves and damage later in the year, including increased flooding and erosion [69].

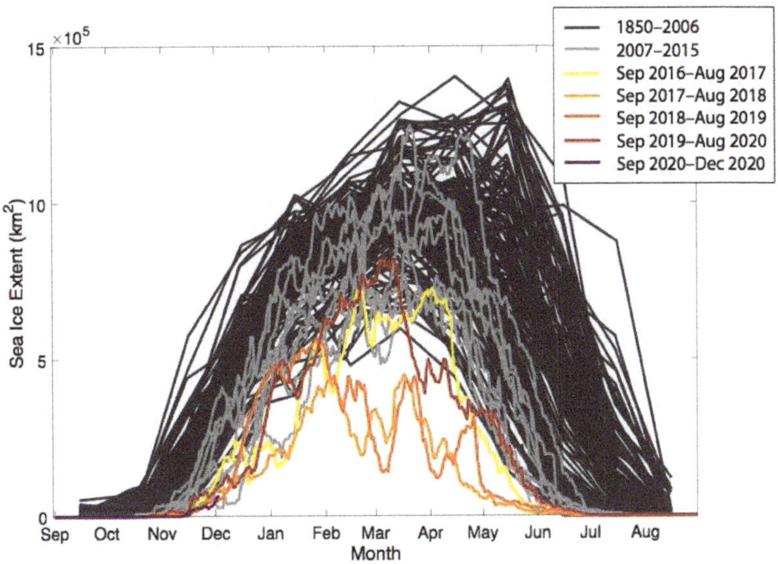


**Figure 4.** Annual cycle of Bering Sea Ice Extent 1850–2020. Dark grey lines indicate data from the Historical Sea Ice Atlas, 1850–2007 [70]. Light grey (2007–2015), yellow (winter 2016–2017), light orange (winter 2017–2018), dark orange (winter 2018–2019), red (winter 2019–2020) and purple (through Dec 2020) lines indicate NSIDC MASIE-NH sea ice extent data (https://doi.org/10.7265/N5GT5K3K) (accessed on 14 February 2015). The [70] data provide the Bering Sea ice extent value on the 15th of each month, whereas the NSIDC MASIE-NH data provide a daily sea ice extent value.

Summer sea surface temperatures along Alaska's west coast were 2.5 to 6 °C warmer than average during the summer of 2019 [58]. Historically, the thermal barrier that existed between the southeastern region of the Bering Sea and the Northern Bering Sea was an important division between the linked but separate ecosystems in these regions [64,71].

The location and extent of the "cold pool" (water temperature less than 2 °C) at the ocean bottom is primarily determined by the southern extent of sea ice during the preceding year and its melt-out date. This cold pool is a critical aspect of the ecosystem of the Northern Bering Sea, as it provides habitat and refuge for cold-tolerant species such as Arctic cod (*Boreogadus saida*) [72,73] and acts as a barrier for adult subarctic fishes such as walleye pollock (*Gadus chalcogrammus*) and Pacific cod (*Gadus macrocephalus*) [74,75].

The extent of the cold pool was substantially reduced in the summers of 2017–2019, and temperatures in the Northern Bering Sea near the seafloor have been above 0 °C in general for an increasing amount of time [64]. The loss of sea ice, lack of thermal barrier and warm ocean and atmospheric temperatures have triggered massive ecological shifts 2017–2019. Increases in Pacific cod, walleye pollock, and several flatfish species were observed in the Northern Bering Sea [76]. Work by Thorson et al. [72] demonstrates a northern shift and reduced area of the Arctic community assemblage (which includes Arctic cod and other species) between 2010 and 2018. Genetic analyses indicate that the increase observed in Pacific cod biomass in the Northern Bering Sea in 2017 was a result of northward migration from their historical range in the Southeastern Bering Sea during the anomalously warm summer conditions [77].

In addition to fish community assemblage and species range shifts as described above, ecosystem changes have reverberated elsewhere throughout land, ocean and air. Harmful algal species that are responsible for toxic algal blooms are expanding in the Arctic alongside warming ocean conditions [78,79]. Seabird surveys and monitoring demonstrated low seabird abundances at sea and low reproductive success, partially due to low forage fish abundance [63]. An increase in seabird die-off events in 2018 in the Northern Bering and Chukchi Seas continued through summer of 2019, with preliminary studies indicating starvation as the likely cause of death [80,81]. In 2019, NOAA declared an UME for three of the four Alaskan ice seal species in the Bering and Chukchi Seas (bearded (*Erignathus barbatus*), ringed (*Pusa hispida*) and spotted (*Phoca largha*) seals were affected; ribbon seals (*Histriophoca fasciata*) were not) [82].

### 3.4.1. Pacific Salmon

Pacific salmon species are experiencing various changes across Alaska. To the south, climate change and habitat loss threaten their distribution. While in the north, a warming Arctic has provided opportunity for range expansion of some species [83]. In the Norton Sound region, the size of chum salmon has been declining steadily since 2000, potentially impacted by factors such as a warming climate, changing ocean ecosystem and fisheries-induced evolution [84]. Juvenile pink salmon, measured as catch per unit effort during trawl surveys in the Northern Bering Sea, increased markedly in 2017 [64].

A major event was the high atmospheric and water temperatures in July 2019, which is the likely cause of a large death event in pre-spawned pink salmon in multiple river systems in the Norton Sound Region [31,85]. This event is of great concern for salmon species in the region, as the observed high temperatures are likely to become more frequent under future climate change.

### 3.4.2. Caribou

Specific to Northwest Alaska, Burch [16,23] outlines the existence of the now-disappeared Nulato Hills caribou herd (NHCH) that partly overlaps with today's Western Arctic Herd (WAH; November to March overlap); there might have been seven specific herds before the 1800s in Northwest Alaska. Burch [23] says that after the population crash of the 1800s, only the WAH and Porcupine herds remained, with the NHCH disappearing by 1900. A herd can be identified based on their calving area. Burch [23], by documenting the changes on caribou in 1850–2000, identifies the dynamics of fluctuation and says that the key driver of the territorial dispersal is the size of the herd, with the calving area as the center of dispersal (see the seasonal herd dynamics in [23] and Figure 5).

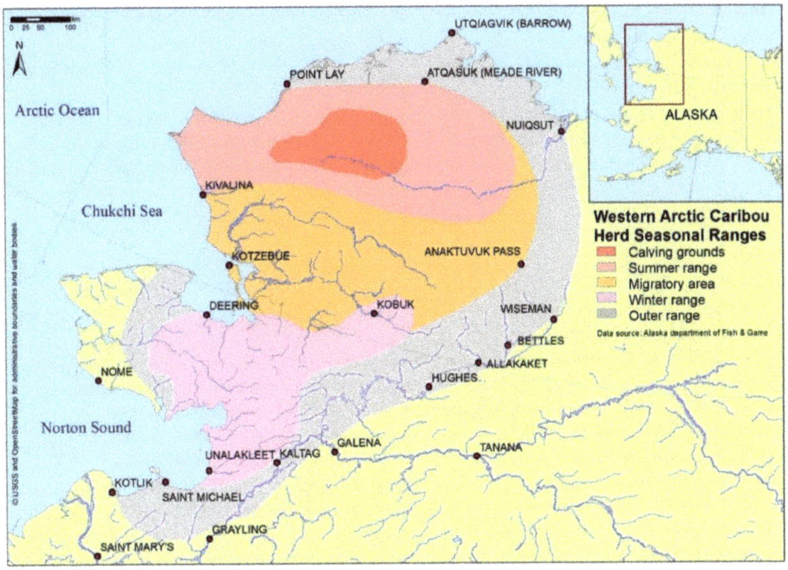


**Figure 5.** Seasonal ranges of the Western Arctic caribou herd (WAH).

The WAH calves in the Utuktok highlands [50,51]. According to scientists, the herd fluctuates on a decadal scale. Burch [23] (p. 67) says that the herd was at around 400,000–500,000 in 1840–1860; in his words, "immense herds" occupied and passed Unalakleet in 1867. There was a recent low of 75,000 animals in 1976 which then rebounded to the present 259,000 (peaking at 463,000 in 1996 and 493,000 in 2003, [50,51]). This also indicates that Unalakleet is at the southern edge of the herd, meaning a loss of caribou near the community if numbers go down ([23,50,51] see also continued and critical assessment of this model in [38,39]). Across their circumpolar range, caribou and reindeer herds have declined in response to factors related to climate warming and anthropogenic landscape changes [86,87].

The mixing of reindeer and caribou in the earlier parts of the 1900s was also a cause of changes [23]. Caribou behaviour has been puzzling both to local Iñupiaq reindeer herders and scientists in trying to determine why the WAH caribou use the Seward peninsula irregularly [51]. The author of [54] points to the knowledge that the loss of caribou cascades in the community, including lack of intergenerational knowledge sharing.

## 4. Discussion: Positioning Indigenous Knowledge and Wisdom with Science from Unalakleet

Western science demonstrates the main drivers of climate change to be the burning of fossil fuels by industrialized societies and land use changes [88]. Indigenous knowledge from community voices takes this understanding a step further: the messages of the changing ocean, lands and waters of Unalakleet are understood to be caused by greed and the industrial (mis) use of resources, especially oil and gas [19,49]. This link has been established by individuals [29,32,46] as well as numerous community participants [19,49].

The impacts of climate change have accelerated, especially the Bering Sea warming that is reflected in the lack of sea ice, health and numbers of animals and other key indicators of a sub-Arctic ecosystem that is now suffering from the lack of its key components. These processes manifest both in western science and Indigenous knowledge materials (Table 5). We recognize that the understanding of the "why" and "how" of climate change is not the same across all coastal communities in Alaska.

**Table 5.** Summarizing Indigenous knowledge and western science related to changes in the two keystone species and observed weather.

| Keystone Species | Indigenous Knowledge Summary | Western Science Summary |
|---|---|---|
| Pacific Salmon | Major losses in salmon from 1983 to 2002–2007; high summer temperatures in July 2019 led to mass death event in pre-spawn pink salmon. Impacts from industrial harvests. | High summer temperatures in July 2019 led to mass death event in pre-spawn pink salmon. |
| Caribou | Caribou have declined and moved further from the village since early 2000s. | WAH has declined; at least in part due to impacts of climate change and anthropogenic landscape changes. |
| **Weather Patterns** | **Indigenous Knowledge Summary** | **Western Science Summary** |
| | Snow amounts have changed a lot and increased storms. Break up and freeze up of ocean and river are very different. Coastal erosion is intensified. Algal blooms out in the ocean. Less sea ice, major storms in recent years. Death events in marine mammals and seabirds and salmon. | Increased temperatures. Increased storms, flooding events and coastal erosion. Sea ice drastically reduced in recent years, with cascading effects throughout marine ecosystem, including harmful algal blooms and seabird and marine mammal deaths. |

Hykes-Steere [29] (on her criticism of the corporation models and see also [44]) says that a central value for the Iñupiaq traditional society was to shy away from greed. One of the remedies for the misuse and overuse of natural resources, therefore, might be to re-establish an Iñupiaq-based value and management system of the traditional kind that would mind the limits and value of these resources (see also [16]). Weaver Ivanoff already identified in the 1980s how the new layers of government were complex in terms of the relationship to the traditional governance of Unalakleet (in [44] p. 149).

Pointing towards this direction, Hykes-Steere [29] (p. 384) recounts her grandmother's words stressing that each person is born with "gifts". This points to the presence of an Indigenous sense of the cosmos and the world still relatively in place in the early 1900s. Later the grandmother, reflecting on her life experiences of the tumultuous 20th century, had said: *"I was born free. I won't die free. What happened to us?"* ([29] p. 384).

## 5. Conclusions

By tracing the histories of Unalakleet, we have observed a place that has been continuously inhabited at least for thousands of years. There have been waves of abandonment such as the epidemics of the 1800s [24,28]. The area used for trading and settlement has welcomed many Indigenous Nations in the past, including the "original" Yupiaq Unaligmiut and "northern" Iñupiaq as well as other Yupiaq and Athabascans before the European settlement. The people had Indigenous self-governance and customary justice system of their lands, including royalty [27].

A tumultuous era starting in the 1830s introduced small-pox and other epidemics to the Unaligmiut and other peoples of the region, resulting in what Napoleon [15] has called the "Great Death". The impact over generations of this massive tragedy has only in recent decades been discovered, as observed from within the cultures [15]. The disruption of endemic, Indigenous governance and sheer loss of populations allowed the United States to settle Alaska in a speedy manner, following the early Russian influence.

A central driver for the settlement of land rights was the desire at the US Federal level to secure Alaska oil and gas for industrial uses. Western science and Indigenous knowledge shared from the Norton Sound region agree that the burning of such fossil fuels is the root cause of climate change and, thus, the present-day collapse of animal, fish and bird populations as well as the major changes on land and out at sea.

This has led many of the Unalakleet knowledge holders to question whether survival is possible [19] at all in the worsening conditions and the present-day economic model, which has brought short term gains to some in the village but has fundamentally altered what has been known as the "community-based culture of Unalakleet" [29].

Martha Itta from Nuiqsut, further north in Alaska, expresses a similar view and is currently involved in a divestment process for her community in order to get away from the oil production and dependency [89]. Jerry Ivanoff [46] produced the following overall assessment: *"When I was a little boy, we were the power that was. My community, as a tribe, we controlled what we did in our village, in our tribal area .... What happened to us, you know? We owned everything from zero to 200 miles and all the land that we can run around in our tribal area, our tribal jurisdiction. And if you cross my land, and you're a [ ... ] Indian then we killed them. Or if we crossed their land and we were Eskimo on their land they killed us, you know. It was just territorial, tribal and Nation, not for money. But for resources, hunting on my land ... .They didn't even know what we were doing you know, in terms of land ownership. We owned it all. In terms of the river and the rights to all the resources around it. If we took care of it, it was all ours, you know. Everything in our river valley was ours. If you go into our river system today, it's all clean, clear and cold. It's pristine."*

At the same time, the community members are determined to thrive and adapt, as Donna Erickson [40] summarized: *"Despite all the climate changes, we move and change with the climate changes, try to adapt."* At the community level, a number of actions have been taken in Unalakleet to address climate change impacts, including those described in this article and in Aronson [45].

Hykes-Steere [29] provides a number of drastic measures as answers. She outlines a pathway to Indigenous re-territorialisation and a vehicle for survival under climate change [15]. According to her [29] (p. 392), Alaska should be formally listed as a territory under the United Nations. This would open the door to a future decolonisation process. According to her analysis, the Alaska Native Claims Settlement Act of 1971 is a unilateral Act and does not extinguish the prior existing claims of the Indigenous Alaskans.

Berger [44] is in alignment and calls for self-determination and the advancement of the Tribal governments of Alaska as a vehicle to correct the perceived mistakes of ANCSA. Napoleon [15] outlines the pathway for the social transformation of the Alaska Indigenous societies. According to him, a positive path for the future can be secured if the past "Great Death" is finally acknowledged. Then, the pathologies and the enduring cross-generational impacts and social ills should be reflected on as the reason for them is finally clear. Then, Napoleon [15] concludes that the Indigenous Alaskans can go on having undergone a decolonial self-reflection and acknowledgement for the past to build a new future; this time on their own terms.

Combined with a decolonisation process at the legal-political level, Hykes-Steere [29] points to allowing the climate change drivers to be addressed according to Iñupiaq values and knowledge, resulting in a new pathway of survival. This view is also reflected in a statement by Ivanoff [32]: "Though the Earth changes, it is still giving. Providing. Nurturing. Inuqtaq [a small boy in 2018] will still learn respect for what gives life. I hope the rest of the world quickly adapts and also respects the Earth–as we have for millenniums and will continue to do so" [32].

**Author Contributions:** Conceptualization, T.M.; methodology, T.M. and B.V.D.; validation, T.M. and B.V.D.; formal analysis, T.M. and B.V.D.; investigation, T.M. and B.V.D.; resources, T.M. and B.V.D.; data curation, T.M. and B.V.D.; writing—original draft preparation, T.M. and B.V.D.; writing—review and editing, T.M. and B.V.D.; visualization, T.M. and B.V.D.; supervision, T.M.; project administration, T.M. Both authors have read and agreed to the published version of the manuscript.

**Funding:** This research was internally funded by the Snowchange Cooperative.

**Informed Consent Statement:** This project adheres to the principles of Free, Prior and Informed Consent of Indigenous peoples.

**Data Availability Statement:** Publicly available climate data sets were used in this study. These sources include the IMIQ hydroclimate Database and Portal at http://imiq-map.gina.alaska.edu/ (accessed on 1 December 2020) (Unalakleet surface temperature data); the Historical Sea Ice Atlas at https://nsidc.org/data/g10010/ (accessed on 1 December 2020) (1850–2007 Bering Sea ice extent; doi:10.7265/jj4s-tq79); and the National Snow and Ice Data Center MASIE-NH sea ice extent data at https://doi.org/10.7265/N5GT5K3K/ (accessed on 1 December 2020) (2007–2020 Bering Sea ice extent). Oral histories are available through sources referenced in the test, and through the Snowchange Oral History Archives by request only. Snowchange Cooperative follows a number of guidelines regarding Oral Histories and Indigenous Knowledge materials, including: Free, Prior, Informed Consent; GIDA CARE principles (https://www.gida-global.org/care/ (accessed on 1 December 2020)); and the Ottawa Indigenous Knowledge Principles (https://www.arcticpeoples.com/knowledge#indigenous-knowledge/ (accessed on 1 December 2020)). Not all of the materials in the Snowchange Oral History Archives are available publicly, though they can be made available upon request in most cases.

**Acknowledgments:** The authors would like to thank Kaisu Mustonen, Victoria Hykes-Steere, Nancy Persons, Bethany Fernstrom, teachers at the Unalakleet schools (Bering Strait School District), and the residents of Unalakleet for their contributions and help over the two decades of this work.

**Conflicts of Interest:** The authors declare no conflict of interest.

## References

1. CAFF. *Arctic Biodiversity Assessment*; CAFF: Akureyri, Islanda, 2013; Available online: https://www.caff.is/assessment-series/233-arctic-biodiversity-assessment-2013 (accessed on 1 August 2021).
2. ACIA. *Arctic Climate Impact Assessment Overview Report*; Arctic Council. ACIA Overview Report; Cambridge University Press: Cambridge, UK, 2020; p. 2005.
3. Hunt, G.L.; Coyle, K.O.; Eisner, L.B.; Farley, E.V.; Heintz, R.A.; Mueter, F.; Napp, J.M.; Overland, J.E.; Ressler, P.H.; Salo, S.; et al. Climate impacts on eastern Bering Sea foodwebs: A synthesis of new data and an assessment of the Oscillating Control Hypothesis. *ICES J. Mar. Sci.* **2011**, *68*, 1230–1243. [CrossRef]
4. Whiting, A.; Griffith, D.; Jewett, S.; Clough, L.; Ambrose, W.; Johnson, J. *Combining Iñupiaq and Scientific Knowledge: Ecology in Northern Kotzebue Sound, Alaska*; Alaska Sea Grant College Program, University of Alaska Fairbanks: Fairbanks, AK, USA, 2011.
5. Huntington, H.P.; Braem, N.M.; Brown, C.L.; Hunn, E.; Krieg, T.M.; Lestenkof, P.; Noongwook, G.; Sepez, J.; Sigler, M.F.; Wiese, F.K.; et al. Local and traditional knowledge regarding the Bering Sea ecosystem: Selected results from five indigenous communities. *Deep Res. Part II Top. Stud. Oceanogr.* **2013**, *94*. [CrossRef]
6. Raymond-Yakoubian, B.; Raymond-Yakoubian, J. *"Always Taught Not to Waste": Traditional Knowledge and Norton Sound/Bering Strait Salmon Populations*; Kawerak, Inc.: Nome, AK, USA, 2015; Available online: https://kawerak.org/wp-content/uploads/2018/04/TK-of-Salmon-Final-Report.pdf (accessed on 1 August 2021).
7. Raymond-Yakoubian, J.; Raymond-Yakoubian, B.; Moncrieff, C. The incorporation of traditional knowledge into Alaska federal fisheries management. *Mar. Policy* **2017**, *78*, 132–142. [CrossRef]
8. Meredith, M.; Sommerkorn, M.; Cassotta, S.; Derksen, C.; Ekaykin, A.; Hollowed, A.; Kovinas, G.; Mackintosh, A.; Melbourne-Thomas, J.; Muelbert, M.M.C.; et al. Polar Regions. In *IPCC Special Report on the Ocean and Cryosphere in a Changing Climate*; Portner, H., Roberts, D.C., Masson-Delmotte, V., Zhai, P., Tignor, M., Poloczanska, E., Mintenbeck, K., Nicolai, M., Eds.; University of Alaska Fairbanks: Fairbanks, AK, USA, 2019.
9. Crate, S.; Cheung, W.; Glavovic, B.; Harper, S.; Jacot Des Combes, H.; Ell Kanayuk, M.; Orlove, B.; Petrasek MacDonald, J.; Prakash, A.; Rice, J.; et al. Cross-Chapter Box 4: Indigenous Knowledge and Local Knowledge in Ocean and Cryosphere Change. In *IPCC Special Report on the Ocean and Cryosphere in a Changing Climate*; Portner, H.O., Roberts, D.C., Masson-Delmotte, V., Zhai, P., Tignor, M., Poloczanska, E., Mintenbeck, K., Eds.; IPCC: Geneva, Switzerland, 2019.
10. Jeff Birchall, S.; Bonnett, N. Thinning sea ice and thawing permafrost: Climate change adaptation planning in Nome, Alaska. *Environ. Hazards* **2020**, *19*, 152–170. [CrossRef]
11. Huntington, H.P.; Daniel, R.; Hartsig, A.; Harun, K.; Heiman, M.; Meehan, R.; Noongwook, G.; Pearson, L.; Prior-Parks, M.; Robards, M.; et al. Vessels, risks, and rules: Planning for safe shipping in Bering Strait. *Mar. Policy* **2015**, *51*, 119–127. [CrossRef]
12. Apassingok, A. *Lore of St. Lawrence Island: Echoes of Our Eskimo Elders*; Bering Strait School District: Unalakleet, AK, USA, 1987.
13. Mustonen, T. Endemic time-spaces of Finland: Aquatic regimes. *Fennia* **2014**, *192*, 120–139. [CrossRef]
14. Anungazuk, H. An Unwritten Law of the Sea. In *Words of the Real People: Alaska Native Literature in Transition*; Fienup-Riordan, A., Kaplan, L., Eds.; UAF Press: Fairbanks, AK, USA, 2007; ISBN 978-1-60223-004-0.
15. Napoleon, H. *Yuuyaraq: The Way of the Human Being*; Madsen, E., Ed.; Alaska Native Knowledge Network: Fairbanks, AK, USA, 1996; ISBN 978-1877962219.
16. Burch, E. *The Iñupiaq Eskimo Nations of Northwest*; University of Alaska Press: Fairbanks, AK, USA, 1998; ISBN 0-912006-95-1.
17. Ray, D.J. Land Tenure and Polity of the Bering Strait Eskimos. *J. West.* **1967**, *3*, 371–395.
18. Ray, D.J. *Ethnohistory in the Arctic: The Bering Strait Eskimo*; Limestone Press: Kingston, ON, Canada, 1983; ISBN 9780919642980.

19. Mustonen, T.; Mustonen, K. *It Has Been in Our Blood for Years and Years That We Are Salmon Fishermen: A Book of Oral History from Unalakleet, AK, USA*; Snowchange Cooperative: Kontiolahti, Finland, 2010.
20. Ray, D.J. Eskimo place-names in bering strait and vicinity. *Names* **1971**, *19*, 1–33. [CrossRef]
21. Napoleon, H. Yuuraraaq: The Way of the Human Being. In *The Alaska Native Reader: History, Culture, Politics*; Williams, M., Ed.; Duke University Press: Durham, NC, USA, 2009.
22. Williams, M. *The Alaska Native Reader: History, Culture, Politics*; Duke University Press: Durham, NC, USA, 2009; ISBN 978-0822344803.
23. Burch, E. *Caribou Herds of Northwest. Alaska, 1850–2000*; University of Alaska Press: Fairbanks, AK, USA, 2012.
24. Pratt, K.L. Reconstructing 19th-century Eskimo-Athabascan boundaries in the Unalakleet River drainage. *Arctic Anthropol.* **2012**, *49*, 94–112. [CrossRef]
25. Fienup-Riordan, A.; Kaplan, L. *Words of the Real People: Alaska Native Literature in Transition*; University of Alaska Press: Fairbanks, AK, USA, 2007; ISBN 978-1602230057.
26. Hykes-Steere, V. *We Used to Sing: An Arctic Elegy*; Animation film, 2011. Available online: https://vimeo.com/19500715 (accessed on 10 January 2020).
27. Ticasuk (Emily Ivanoff Brown). *The Roots of Ticasuk: An Eskimo Woman's Family Story*; Alaska Northwest Books: Portland, OR, USA, 1981; ISBN 978-0882401171.
28. *Alaska Natives and the Land.*; Federal Field Committee for Development Planning in Alaska; Federal Field Committee: Washington, DC, USA, 1968.
29. Hykes-Steere, V. Interpreting the Journey: Where Words, Stories Formed. In *Indigenous Pathways into Social Research: Voices of a New Generation*; Mertens, D., Cram, F., Bagele, C., Eds.; Left Coast Press: Walnut Creek, CA, USA, 2013; ISBN 978-1598746969.
30. Hykes-Steere, V. Narratives on Climate Change 2002–2019. In *Snowchange Unalakleet Archives*; Snowchange Cooperative: Kontiolahti, Finland, 2019.
31. Kawerak Inc. Climate Change Letter to Senate Indian Affairs Committee. Available online: https://kawerak.org/climate-change-letter-to-senate-indian-affairs-committee/ (accessed on 2 October 2019).
32. Ivanoff, L. *The Bearded Seal My Son May Never Hunt*; New York Times: New York, NY, USA, 2018.
33. Macdonald, J. *The Arctic Sky: Inuit Astronomy, Star Lore, and Legend*; Nunavut Research Institute: Iqaluit, NU, Canada, 2000; ISBN 978-0888544278.
34. Cortazzi, M. Narrative Analysis in Ethnography. In *Handbook of Ethnography*; Atkinson, P., Coffey, A., Delamont, S., Lofland, L., Lofland, J., Eds.; Sage Publications: London, UK, 2001; pp. 384–394. ISBN 9780761958246.
35. Johnson, N.; Alessa, L.; Behe, C.; Danielsen, F.; Gearheard, S.; Gofman-Wallingford, V.; Kliskey, A.; Krümmel, E.M.; Lynch, A.; Mustonen, T.; et al. The contributions of Community-Based monitoring and traditional knowledge to Arctic observing networks: Reflections on the state of the field. *Arctic* **2015**, *68*, 28–40. [CrossRef]
36. Emery, M.; Redlin, M.; Young, W. Native Leadership and Adaptation to Climate Change. In *Environmental Leadership: A Reference Handbook*; Gallagher, D., Ed.; Sage Publications: Thousand Oaks, CA, USA, 2012.
37. Ray, D.J. Early Maritime Trade with the Eskimo of Bering Strait and the Introduction of Firearms. *Arctic Anthropol.* **1975**, *12*, 1–9.
38. Pratt, K.L.; Ganley, M.; Slaughter, D.C. New Perspectives on the Late Nineteenth-Century Caribou Crash in Western Alaska. In *Arctic Crashes: People and Animals in the Changing North*; Krupnik, I., Crowell, A., Eds.; Smithsonian Scholarly Press: Washington, DC, USA, 2020.
39. Mager, K.H. Ghosts of the Caribou Herds Past: Evaluating Historical Caribou Crashes in Alaska Using Genetics. In *Arctic Crashes: People and Animals in the Changing North*; Krupnik, I., Crowell, A., Eds.; Smithsonian Scholarly Press: Washington, DC, USA, 2020.
40. Erickson, D. Oral History Interview, 17th October 2019. In *Snowchange Unalakleet Archives*; Snowchange Cooperative: Kontiolahti, Finland, 2019.
41. Hykes-Steere, V. *Snowchange Unalakleet Archives*; Unpublished manuscript; Snowchange Cooperative: Kontiolahti, Finland, 15 September 2019.
42. Bering Straits Native Corporation. Historical Spotlight: Nashoalook and Karlson. *The Agluktuk: Winter/Spring 2016*. 2016, p. 2. Available online: https://beringstraits.com/historical-spotlight-covenant-high-school/ (accessed on 20 June 2021).
43. Arnold, R. *Alaska Native Land Claims*; Alaska Native Foundation: Anchorage, AK, USA, 1976; ISBN B0006CVEIE.
44. Berger, T. *Village Journey: The Report of the Alaska Native Review Commission*; Inuit Circumpolar Council: Utqiaġvik, AK, USA, 1985.
45. Aronson, R. Adapting to Climate Change in Unalakleet, Alaska. Ph.D. Thesis, University of Washington, Washington, DC, USA, February 2014.
46. Ivanoff, J. Oral History Interview, 18th October 2019. In *Snowchange Unalakleet Archives*; Snowchange Cooperative: Kontiolahti, Finland, 2019.
47. ICC Alaska. *Bering Strait Regional Food Security Workshop: How to Assess Food Security from an Inuuit Perspective: Building a Conceptual Framework on How to Assess Food Security in the Alaskan Arctic*; Inuit Circumpolar Council: Anchorage, AK, USA, 2014.
48. Hovey, D. Warmer waters investigated as cause of pink salmon die-off in Norton Sound Region. Available online: https://www.adn.com/alaska-news/rural-alaska/2019/07/12/warmer-waters-investigated-as-cause-of-pink-salmon-die-off-in-norton-sound-region/ (accessed on 13 July 2019).
49. Van Dam, B. Oral History Notes from Community Documentation, October 2019. In *Snowchange Unalakleet Archives*; Snowchange Cooperative: Kontiolahti, Finland, 2019.

50. Joly, K. History, Purpose and Status of Caribou Movements in Northwest Alaska. Available online: https://www.researchgate.net/profile/Christopher-Sergeant/publication/325768480_Future_Challenges_for_Salmon_and_the_Freshwater_Ecosystems_of_Southeast_Alaska/links/5b22e5de0f7e9b0e374800fc/Future-Challenges-for-Salmon-and-the-Freshwater-Ecosystems-of-Southeast-Alaska.pdf#page=51 (accessed on 13 July 2019).
51. Rattenbury, K. Reindeer herding, Weather and Environmental Change on the Seward Peninsula, Alaska. Ph.D. Thesis, University of Alaska, Fairbanks, AK, USA, 2006.
52. Wilson, J. Jewell's Hunt; Provider Films. 2019. Available online: https://www.youtube.com/watch?v=_pz88CumPp4 (accessed on 1 October 2019).
53. Hykes-Steere, V. *Snowchange Unalakleet Archives*; Unpublished Manuscript; Snowchange Cooperative: Kontiolahti, Finland, 20 May 2019.
54. *Alaska Inuit Food Security Conceptual Framework: How to Assess the Arctic from an Inuit Perspective Report*; Inuit Circumpolar Council—Alaska, 2015. Available online: http://iccalaska.org (accessed on 10 December 2019).
55. Mustonen, T. *Selkien Perinteestä 7*; Snowchange Cooperative: Kontiolahti, Finland, 2015; Available online: http://www.snowchange.org (accessed on 1 December 2019).
56. Box, J.E.; Colgan, W.T.; Christensen, T.R.; Schmidt, N.M.; Lund, M.; Parmentier, F.J.W.; Brown, R.; Bhatt, U.S.; Euskirchen, E.S.; Romanovsky, V.E.; et al. Key indicators of Arctic climate change: 1971–2017. *Environ. Res. Lett.* **2019**, *14*, 045010. [CrossRef]
57. Markon, C.; Gray, S.; Berman, M.; Eerkes-Medrano, L.; Hennessy, T.; Huntington, H.; Littell, J.; McCammon, M.; Thoman, R.; Trainor, S. Alaska. In *Impacts, Risks, and Adaptation in the United States: Fourth National Climate Assessment, Volume II*; Reidmiller, D.R., Avery, C.W., Easterling, D.R., Kunkel, K.E., Lewis, K.L., Maycock, T.K., Stewart, B.C., Eds.; U.S. Global Change Research Program: Washington, DC, USA, 2018; pp. 1185–1241.
58. Thoman, R.; Walsh, J.E. *Alaska's Changing Environment: Documenting Alaska's Physical and Biological Changes Through Observations*; International Arctic Research Center: Fairbanks, AK, USA; University of Alaska: Fairbanks, AK, USA, 2019.
59. Walsh, J.E.; Brettschneider, B. Attribution of recent warming in Alaska. *Polar Sci.* **2019**, *21*, 101–109. [CrossRef]
60. Cherry, J.; Jacobs, A.; Heinrichs, T.; Fisher, W.; Delamere, J.; Haase, C.; Martin, P.; Bradley, J.; Jenkins, J.; Grunblatt, J.; et al. Imiq Hydroclimate Database at the University of Alaska Fairbanks. Available online: http://arcticlcc.org/projects/imiq/ (accessed on 21 January 2021).
61. Lott, J.N.; Vose, R.S.; Del Greco, S.A.; Ross, T.R.; Worley, S.; Comeaux, J.L. The integrated surface database: Partnerships and progress. In Proceedings of the 24th Conference on Interactive Information Processing Systems for Meteorology, Oceanography, and Hydrology, New Orleans, LA, USA, 20–24 January 2008.
62. Stabeno, P.J.; Kachel, N.B.; Moore, S.E.; Napp, J.M.; Sigler, M.; Yamaguchi, A.; Zerbini, A.N. Comparison of warm and cold years on the southeastern Bering Sea shelf and some implications for the ecosystem. *Deep Res. Part II Top. Stud. Oceanogr.* **2012**, *65*, 31–45. [CrossRef]
63. Duffy-Anderson, J.T.; Stabeno, P.; Andrews, A.G.; Cieciel, K.; Deary, A.; Farley, E.; Fugate, C.; Harpold, C.; Heintz, R.; Kimmel, D.; et al. Responses of the Northern Bering Sea and Southeastern Bering Sea Pelagic Ecosystems Following Record-Breaking Low Winter Sea Ice. *Geophys. Res. Lett.* **2019**, *46*, 9833–9842. [CrossRef]
64. Huntington, H.P.; Danielson, S.L.; Wiese, F.K.; Baker, M.; Boveng, P.; Citta, J.J.; De Robertis, A.; Dickson, D.M.S.; Farley, E.; George, J.C.; et al. Evidence suggests potential transformation of the Pacific Arctic ecosystem is underway. *Nat. Clim. Chang* **2020**, *10*, 342–348. [CrossRef]
65. Stabeno, P.J.; Thoman, R.L.; Wood, K. Recent warming in the Bering Sea and its impact on the ecosystem. In *Arctic Report Card 2019*; Richter-Menge, J., Druckenmiller, M.L., Jeffries, M., Eds.; NOAA: Washington, DC, USA, 2019.
66. Thoman, R.L.; Bhatt, U.S.; Bieniek, P.A.; Brettschneider, B.R.; Brubaker, M.; Danielson, S.L.; Labe, Z.; Lader, R.; Meier, W.N.; Sheffield, G.; et al. The record low Bering sea ice extent in 2018: Context, impacts, and an assessment of the role of anthropogenic climate change. *Bull. Am. Meteorol. Soc.* **2020**, *101*. [CrossRef]
67. Hovey, D. Sea ice almost gone in Norton Sound; Conditions "uncannily similar" to last March. Available online: https://www.knom.org/wp/blog/2019/03/01/sea-ice-almost-gone-in-norton-sound-conditions-similar-to-last-march/ (accessed on 2 March 2019).
68. Shallenberger, K. *Winter Storms Flood Houses in Nunapitchuk and Kotlik*; Alaska Public Media: Anchorage, AK, USA, 2019.
69. Fang, Z.; Freeman, P.T.; Field, C.B.; Mach, K.J. Reduced sea ice protection period increases storm exposure in Kivalina, Alaska. *Arct. Sci.* **2018**, *4*, 525–537. [CrossRef]
70. Walsh, J.E.; Chapman, W.L.; Fetterer, F.; Stewart, S. Gridded Monthly Sea Ice Extent and Concentration, 1850 Onward, Version 2. Available online: https://nsidc.org/data/g10010 (accessed on 3 December 2019).
71. Sheffield, G.; Hauser, D.; Thoman, R. Rapid Change: 2019 in Northwest Alaska's Oceans and Impacts to Ecosystems and People. Available online: https://uaf-accap.org/event/rapid-change-2019-in-northwest-alaskas-oceans-and-impacts-to-ecosystems-and-people/ (accessed on 3 December 2019).
72. Thorson, J.T.; Fossheim, M.; Mueter, F.J.; Olson, E.; Lauth, R.R.; Primicerio, R.; Husson, B.; Marsh, J.; Dolgov, A.; Zador, S.G. Comparison of near-bottom fish densities show rapid community and population shifts in Bering and Barents Seas. In *Arctic Report Card 2019*; Richter-Menge, J., Druckenmiller, M.L., Jeffries, M., Eds.; NOAA: Silver Spring, MD, USA, 2019.
73. Marsh, J.M.; Mueter, F.J. Influences of temperature, predators, and competitors on polar cod (*Boreogadus saida*) at the southern margin of their distribution. *Polar Biol.* **2020**, *43*, 995–1014. [CrossRef]

74. Boldt, J.L.; Buckley, T.W.; Rooper, C.N.; Aydin, K. Factors influencing cannibalism and abundance of walleye pollock (*Theragra chalcogramma*) on the eastern Bering Sea shelf, 1982–2006. *Fish. Bull.* **2012**, *110*, 293–306.
75. Ciannelli, L.; Bailey, K.M. Landscape dynamics and resulting species interactions: The cod-capelin system in the southeastern Bering Sea. *Mar. Ecol. Prog. Ser.* **2005**, *291*. [CrossRef]
76. Stevenson, D.E.; Lauth, R.R. Bottom trawl surveys in the northern Bering Sea indicate recent shifts in the distribution of marine species. *Polar Biol.* **2019**, *42*, 407–421. [CrossRef]
77. Spies, I.; Gruenthal, K.M.; Drinan, D.P.; Hollowed, A.B.; Stevenson, D.E.; Tarpey, C.M.; Hauser, L. Genetic evidence of a northward range expansion in the eastern Bering Sea stock of Pacific cod. *Evol. Appl.* **2020**, *13*, 362–375. [CrossRef]
78. Alaska Sea Grant. 2018/2019 Bering Strait/Chukchi Sea: Alexandrium Algae, Saxitoxin, and Clams. 2019. Available online: https://seagrant.uaf.edu/bookstore/pubs/MAB-75.html (accessed on 15 January 2020).
79. Anderson, D.M.; Richlen, M.L.; Lefebvre, K.A. Harmful Algal Blooms in the Arctic. In *Arctic Report Card*; NOAA: Silver Spring, MD, USA, 2018.
80. Siddon, E.; Zador, S. *Ecosystem Status Report 2018: Eastern Bering Sea*; NOAA Fisheries, 2018. Available online: https://www.fisheries.noaa.gov/resource/data/2018-status-eastern-bering-sea-ecosystem (accessed on 9 May 2019).
81. US Fish and Wildlife Service 2019 Alaska Seabird Die-off Fact Sheet 508C. Available online: https://www.nps.gov/subjects/aknatureandscience/upload/9Sep2019-Die-Off-USFWS-Factsheet-508C-revised-29Aug.pdf (accessed on 10 January 2020).
82. NOAA. Fisheries Diseased Ice Seals: Unusual Mortality Events for Ice Seals in the Bering and Chukchi Seas. Available online: https://www.fisheries.noaa.gov/alaska/marine-life-distress/2018-2021-ice-seal-unusual-mortality-event-alaska (accessed on 10 January 2020).
83. Nielsen, J.L.; Ruggerone, G.T.; Zimmerman, C.E. Adaptive strategies and life history characteristics in a warming climate: Salmon in the Arctic? *Environ. Biol. Fishes* **2013**, *96*, 1187–1226. [CrossRef]
84. State of Alaska Salmon and People (SASAP). The Declining Size and Age of Salmon, Norton Sound Region. Available online: https://alaskasalmonandpeople.org/topics/the-declining-size-and-age-of-salmon/ (accessed on 10 December 2020).
85. Norton Sound Economic Development Corporation (NSEDC). 2019 Norton Sound Pink Salmon Die Off Event Likely Linked to Temperature. Available online: https://www.nsedc.com/norton-sound-pink-salmon-die-off-event-likely-linked-to-temperature/ (accessed on 10 December 2020).
86. Vors, L.S.; Boyce, M.S. Global declines of caribou and reindeer. *Glob. Change Biol.* **2009**, *15*, 2626–2633. [CrossRef]
87. CARMA Western Arctic Caribou Herd. Available online: https://carma.caff.is/herds/538-carma/herds/626-western-arctic. (accessed on 3 February 2020).
88. NASA. Global Climate Change: Vital Signsof the Planet. In *Facts: The Causes of Climate Change..* Available online: https://climate.nasa.gov/causes/ (accessed on 5 January 2020).
89. Eilperin, J. *Alaska Climate Change: Facing Catastrophe, This Town Can't Quit Big Oil*; Washington Post: Washington, DC, USA, 2019.

sustainability

MDPI

Article

# Return of Nimat?—Wild Reindeer as an Indicator of Evenki Biocultural Systems

**Tero Mustonen [1,*], Tamara Andreeva [2], Vyacheslav Shadrin [2] and Kaisu Mustonen [1]**

1 Snowchange Cooperative, 81235 Lehtoi, Finland; kaisu.mustonen@snowchange.org
2 Institute for Humanities Research and Indigenous Studies of the North, Siberian Branch of the Russian Academy of Sciences (IHRISN SB RAS), 677000 Yakutsk, Russia; taan2001@mail.ru (T.A.); odul_shadrin@mail.ru (V.S.)
* Correspondence: tero@lumi.fi

**Abstract:** This paper reviews oral histories and established scientific materials regarding wild reindeer (Rangifer tarandus spp.) in the Southern Sakha-Yakutia, in the Neriungri district and surrounding highlands, river valleys and taiga forest ecosystems. Wild reindeer is seen as an ecological and cultural keystone species through which environmental and social changes can be understood and interpreted. Oral histories of Evenki regarding wild reindeer have been documented in the community of Iyengra between 2005 and 2020. During this 15-year-co-researchership the Southern Sakha-Yakutian area has undergone rapid industrial development affecting the forest and aquatic ecosystems. The wild reindeer lost habitats and dwindles in numbers. We demonstrate that the loss of the wild reindeer is not only a loss of biodiversity, but also of cultural and linguistic diversity as well as food security. Our interpretative and analytical frame is that of emplacement. Socio-ecological systems have the potential and capacity to reconnect and re-establish themselves in post-extractive landscapes, if three main conditions are met. These conditions for successful emplacement include (1) surviving natural core areas, (2) links to cultural landscape knowledge and (3) an agency to renew endemic links.

**Keywords:** Siberia; rewilding; Evenki; Sakha-Yakutia; Rangifer

**Citation:** Mustonen, T.; Andreeva, T.; Shadrin, V.; Mustonen, K. Return of Nimat?—Wild Reindeer as an Indicator of Evenki Biocultural Systems. *Sustainability* **2021**, *13*, 12107. https://doi.org/10.3390/su132112107

Academic Editors: Stephanie Pfirman, Gail Fondahl and Grete K. Hovelsrud

Received: 1 September 2021
Accepted: 26 October 2021
Published: 2 November 2021



## 1. Introduction

Evenki live over a vast geographical range in East Siberia, Far East of Russia, Northern China and Mongolia [1–3]. Wure'ertu [4] provides us with an Evenki-authored view of the distant past and the deep memories the Evenki have over large spatial scales, essentially spanning the continent. Yet, each region and community has a unique heritage of the Evenki culture [1,2,5].

It is estimated that the entire Evenki community has approximately 36,000 members today, and ca. 7000 speak the language. Traditional livelihoods have revolved around nomadic reindeer herding, hunting and fishing [6]. There is also an Evenki community in China. This research focuses on the Evenki community of Iyengra and the surrounding taiga camps, in the Southern Sakha-Yakutia (spelling also for example Iakutia).

Collaborations and co-researcher -ship have emerged in the past 10 years in different parts of the Evenkia (here referring to the larger Evenki home realm, see more of this concept for example in [7] and other parts in [8]), where reindeer herders and women from the taiga camps have been positioned as equal partners in research. This structure has been in place in our long-term research from the very beginning. The emergence of the Evenki landscape studies [9–11] have offered similar approaches in research. More precisely, the people in the community shared their oral histories over the past years to form an oral history corpus. Individual contributors own their knowledge and are acknowledged by name unless they have requested anonymity. Two of the authors of this article are

Indigenous (Evenki and Yukaghir) and have led the research effort in the community and in the republic for 15 years.

Our case community, the Evenki of Iyengra, has faced a continuous, though varying over time, pressure of displacement since the days of the first encounters with Russians in the early 17th century, including forced settlements in the 1920s. We summarize these community changes briefly below and throughout the article but our primary focus is on wild reindeer as a keystone indicator of change.

One of the first encounters with Russians took place in 1619 when Petr Albychev and Cherkas Rukin enslaved Evenki nobility Iltik. Since this event 400 years ago, the Evenki have interacted with the Indo-European peoples. Prior to contact with Russians the Evenki had met with the Sakha (Yakut) people, with whom the relationship had ranged from trade to war. During those days, the Evenki were also aware of and had connections to a range of Siberian nations, such as the Chukchi, Even and Yukaghirs [12]. For example the place names surrounding Iyengra reflect in part the meshed landscape and influence of the Sakha on the Evenki homeland [6].

Neriungri, the regional district where Iyengra is located, is a southern region of the Republic Sakha-Yakutia in the Russian Federation, located close to the Chinese border. The population of the district is about 75,000, most of it being urban. Neriungri, the capital city, is also the centre for coal-mining operations. The district produces up to 75% of the 10 million tons of coal that is produced in Sakha annually [12–14].

Southern Sakha-Yakutia is part of the continental climatic zone. Siberian larch, other coniferous trees and birch cover much of the taiga. The village of Iyengra is located in the southern part of the Sakha. Mountains and large hills dominate the landscape, combined with shallow rivers flowing in the valleys. For example, the Aldan catchment area, a subcatchment of Lena River, flows through the district. Winters are usually very cold with small amounts of snowfall. Temperatures can plummet down to −50 °C and below.

Springtime is often short with snowmelt already well under way in April. Summers are continental and hot. Autumn brings the first frosts, often in September or October. In addition to terrestrial hunting and herding economies, fishing for salmonid fish such as trout, grayling and whitefish is important. This is reflected in many Evenki place names around Iyengra [6,15]. They play an important role in the traditional food security of the community. Evenki have used these salmonid fish as cultural bioindicators to assess the degradation of river health and change over time [6,16]. They also place significant cultural value especially to local trout as a culturally relevant species.

Under the Soviet regime and during the post-Soviet years, the Evenki lands and waters became a target of expansive industrial operations [14,17,18]. Mining for coal and gold, hydropower construction, energy pipelines [19] and many other infrastructure projects mark the landscape [14]. This expansion widely and thoroughly altered the post-Ice Age landscapes of the southern Sakha-Yakutia. During Soviet times, most of the Evenki in the area were officially relocated into the village of Iyengra, which was founded on the catchment area of the river of the same name in 1926, along the traditional nomadic routes of the Evenki [6].

According to the 2018 census (which is the most recent in public records), the population of Iyengra is 918, of which the majority (over 800) is Evenki. Some of the Iyengra Evenki spend most of the year on the taiga but are registered in the village (Other hunter-herders, like Kolesov, shared their oral histories with willingness and consent to be named. If this has been the case, we have quoted a person direct with names.). Other nationalities in the village include for example Even, Karelian, Russian and Sakha-Yakut peoples. Evenki children go to school in Iyengra, which functions as a residential school where the children of the reindeer herders spend the winter season while their parents are working in the taiga.

We, the authors of this paper, have been working in the community to document and co-research Evenki knowledge between 2005–2020. The original invitation to work with the Evenki of Iyengra originated amongst Evenki scholars and leaders in Yakutsk. The original group was led by Evenki Galina Varlamova, Evenki Tamara Andreeva, Evenki Anna Myreeva and Yukaghir Vyacheslav Shadrin. The aim was to seek new collaborations to strengthen Indigenous knowledge preservation, document climate and ecological observations in the village and surrounding taiga forest ecosystems and to position the rapid development plans for the region into a broader context of the Evenki culture, way of life, food security and traditional professions.

In the present paper, we focus on available science and Indigenous Evenki knowledge of the wild forest reindeer (*Rangifer tarandus* Linneaus, 1758, Krivoshapkin and Mordosov [20] name the southern reindeer to be *Rangifer tarandus valentinae* Flerov, 1933 and speculate further that the southern herds would be actually close to the Ohota subspecies *R. tarandus phylarchus* Hollister, 1912 [21] in Southern Sakha-Yakutia).

Reindeers are overall central to the Evenki in terms of food security, cultural and linguistic knowledge, beliefs and to the maintaining of social order, justice and knowledge of nature across the Evenki living space [5,22–24]. Vorob'ev [24] (p. 34) quotes G.M. Vasilevich, a specialist in Evenki language and culture, sharing "that the Evenki considered that moose and (wild rein)deer were sent by a spirit—the master of the taiga: The ancient belief in their resurrection survived for a long time. After processing the meat and hides, hunters were supposed to collect all the bones of the skeleton and to rest them on a platform [gulik], so that dogs could not drag them away".

According to Vorob'ev [24] in Krasnoyarskii krai the northern wild reindeer is known as as severnyi dikii olen. The word *dikovka* is derived from *dikii*, Russian for "wild." Vorob'ev [24] (p. 36) continues that the "seasonal hunt for migrating northern wild deer called *dikovka*, as in most regions of Siberia, underscoring the distinction between the northern wild deer and its domesticated cousin. The wild deer hunters are called *dikovshchiki*." Shubin [24] (p. 34) reports that in Buryatia the meat and appreciation of wild reindeer meat as food comes behind for example moose, even though overall wild reindeer are important to the culture.

Overall there are at least 71 distinct endemic concepts for the reindeer, domestic and wild (same biological animal, different stocks and uses for the Evenki) [25]. When accounting for dialects and synonyms, this equates to hundreds of specific reindeer-related concepts. The language differentiates the animals according to age characteristics, fur color, as well as their character and behavior. To offer some examples:

- "sonnga" = newborn calf
- "ukoto" = nursing calf
- "epchakan" = female reindeer, one to two years old
- "ektana" = bull reindeer, two to three years old
- "semeki" = female reindeer that does not let people approach it during calving
- "arkichan" = old riding (on saddle) reindeer
- "kongnomo", "kongnorin" = black color and fur color of reindeer
- "igdiama", "igdyama" = ginger fur color
- "kurbuki" = reindeer that has become wild
- "sungnaki" = restless reindeer (summarized from [26])

Specifically for wild reindeer, people in the oral history materials pointed for example to *kuraika*, which can be translated as 'autumn male wild reindeer'. Accordingly *neŋ cheng* is the 'spring wild reindeer' (see discussion in [10] on tensions between 'official transliterations' and vernacular spelling of Evenki).

Given the vast industrial developments, infrastructure process and mining in the Southern Sakha-Yakutia space, the wild reindeer are disappearing and losing their ancestral ranges [20]. We explore the key implications, observations and ultimately, solutions, to these negative drivers.

Several earlier summary reports and publications have been released from this on-going knowledge documentation and Indigenous knowledge work (see for example [6,27–29]).

The Evenki have been studied at length over the past years, with a primary focus on their belonging to the taiga forest ecosystem [7,9,10,15,22,24,30–32]. Already during the historic, and later Soviet times, the Evenki were a target of anthropological and geographical interest, with 'classical' desciptions of their lives, human societies and ways of life documented. During the Russian and Soviet era, the Evenki were targets of scrutiny and ethnographic investigations that have provided an outsider view of the transitions and changes of the Evenki societies across their vast home area.

Authors such as Arkadii Anisimov, Sergei Shirokoroff, Anna Sirina and perhaps most importantly, Glafira Vasilevich [e.g., see English summaries in [9,24], researched the Evenki customs and lifeways from the late 1800s to the mid- and late-1900s. Evenki Alexander Sergeevich Shubin (born in 1929, see collections of works in [23]) from the Khabarzhan area of the Barguzinsky region produced a large collection of Buryat Evenki histories and change during the Soviet era. He has written (during Soviet times and being affected by them, see [5,22,23] from an '(Soviet) internal viewpoint on the Evenki historic development and cultural issues, in some ways, much like Varlamova [1].

Emerging from the 1990s, the concept of Indigenous communities of Northern Russia being in a position of co-interpreting and researching with outside scholars started to gain a foothold (for notions of inherent, endemic knowledge of a situated heritage, see [27], and for the persistence of the 'Evenki as other' see [33]). For example Safonova and Sántha [34] have reviewed the 'classical' intepretation of 'hunter gatherer Evenki' and advanced a more broad interpretation. We follow such overall positioning in our present work and have also included the voices of the Evenki academic scholars, such as the late Galina Varlamova [1], Wure'ertu [4] and Anna Myreeva [25], as well as our Evenki co-researcher Tamara Andreeva in Yakutsk. Overall, we use this co-interpretation (see below) and 'novel research space' in our approach of emplacement of the Evenki, described in more detail below.

The empirical section of the article begins with a description of the traditional concepts, cosmological narratives and aspects of the Evenki and thereafter focuses on the wild reindeer and change in the community over our review period.

## 2. Materials and Methods

We frame our article on two theoretical understandings. Following Montonen [35] and in part from Vorob'ev [24], we build our case around the understanding that the wild forest reindeer (Rangifer) for the Evenki is a cultural keystone species [26]. Montonen [35] demonstrated in the case of the Sámi and Finnish wilderness communities that the wild forest reindeer maintained the land uses, oral histories, traditional songs, hunting patterns and food security of several forest communities until it was over harvested and the stocks collapsed in NE and Eastern Finland, early 1900s and in 1928, respectively.

This sped up, according to Montonen [35], the erosion of traditional knowledge. The practise and the time-consuming actions of hunting wild reindeer in the forest fell out of use, either then replaced by other activities or species, such as increased hunting pressure on moose, or a switch to larger scale reindeer herding. More recent scholarship, such as Frainier et al. [26] agrees, and emphasizes the triangulation between Indigenous knowledge, linguistic and biological diversity on which the people depend.

Second, as we explore the potential for maintaining and fostering the return of wild forest reindeer into the Southern Sakha-Yakutia we draw on Mustonen and Lehtinen [36]. They argue that whilst modernity and associated natural resources extraction has altered, often permanently, natural systems, they contain elements of a renewed emplacement despite the damages.

They [36] investigated how community emplacement functions. Here we can summarize that it is an emerging spatial understanding of severed, preserved and reconnected belongings to a place. We wish to offer a renewed emplacement approach as a mechanism to understand the complex reality (and potential future pathways) of the present-day Evenki life in Iyengra. More precisely, investigations of emplacement require the following components:

- some remaining natural ecosystems having enough carrying capacity and production that nature-based traditional livelihoods have maintained "core" areas, despite extractive / altered impacts from human uses
- a traditional knowledge corpus of a cultural engagement with a landscape of former but recent (often discontinued or abandoned) practices that is still within community's reach
- a concentrated willingness and actionable process to maintain, and where emplacement allows, revitalize nature-based livelihoods, cultural practices and specific engagements with the surround ecosystems (summarized from [36], with elements of scaling the loss from [24]).

We need to be clear also that if industrial and extractive land uses have created conditions where emplacement does not meet these three criteria, evident and permanent loss may happen. Emplacement is not an automatic solution space or to be taken for granted. It is a concentrated, combined effort of emerging and surviving potential of socio-ecological systems combined with a determined agency [37] that is willing to implement and position emplacement as a realistic potential future pathway.

We recognize the important descriptive and interpretative early works building on Russian and Soviet ethnography (Arkadii Anisimov, Sergei Shirokoroff, Anna Sirina, Glafira Vasilevich and others). We also acknowledge the present ethnographical works that have partnered with the Evenki to understand change, continuity and loss [7,9,10,15,24,30,31] and identities [32] and provide valuable descriptions and precise documentation of the Post-Soviet years.

Our emplacement conversation builds to conceptualize a 'third way' in the science-Indigenous nexus, where the Indigenous peoples self-articulate and co-produce the situational analysis and potential solution spaces. In order to realize Indigenous rights and authorship, we also have to critically self-reflect on the established academic practices and the othering that has been one of the underlying drivers of Russian and Siberian ethnography [38].

In our understanding emplacement offers a situated, post-colonial and localized assessment of a post-industrial situation [36] where the empirical situation 'as-it-unveils' provides the best window to the issues. This has also implications to the future of the Arctic and Indigenous studies and how we understand change, belonging, memory and landscape studies.

In the context of Iyengra, we frame this work to build on the fact that the Indigenous forest lifestyles and presence have continued across and in the middle of the industrial developments [14]. For example, continued fish harvests and reindeer herding demonstrate a range of qualities and adaptations to the ecological alterations that do produce cultural continuums even if in the middle of industrial upheaval [14].

We can see these examples constitute endemic (self-defined, self-articulated) acts of emplacement [27,36]. This rests on tensions between preserved traditional lifeways of the taiga and the impacts of industrialization. They constitute also acts of co-intepretation. This emerges in oral histories and choices of narration Evenki themselves priorize—overall in our study period with the community members the fate of wild reindeer as a overall theme and the specific drivers was raised and resulted in the analytical frame of this article.

Co-interpretation comes in many forms, one of the most central being agency [37]. It allows outside researchers to refine their view and on the other hand addresses the power relations in scholarship where the local/community/knowledge holders have an even space for their priorities as opposed to outside, prior and set expectations in research. Huntington et al. [37] value the inherent value of community choice, knowledge and interpretation as unique—providing a tension line between general and specific. In this research we address this tension by allowing the co-intepretation to flow from the oral history corpus which is then further reflected on by T.A., Evenki co-author to maximize the voices and knowledge of the Evenki as they see it. All interpretation is always an interpretation. However, as Huntington et al. [37] show, co-intepretation advances the scholarly process in marked ways and it is applied here with rigor.

Traditional skills do not necessarily disappear during industrial modernization. Instead, in certain encouraging conditions, they can re-emerge and pave routes to endemic futures. In the conclusions we discuss these aspects of potentials of preservation of wild reindeer stocks (and connection to Evenki) in Southern Sakha-Yakutia.

As a culturally-positioned solution we refer to such an (future) act of emplacement as *nimat*. Nimat refers to a process, ceremony and distribution system by which successful hunters share their wild reindeer catches amongst the Evenki community [1]. The purpose of this distribution of assets is linked with community cohesion and food security—those who cannot hunt, i.e., the women, Elderly, children, are also given food by the providers. In *nimat*, wild reindeer provides for the Evenki.

We can see the forest ecosystem services culminating in *nimat* for the Evenki—what is taken, is shared, and overharvesting is strictly observed and forbidden. McNeil [2] highlights the role of the revival-focused concepts amongst the Evenki. We therefore use the concept of *nimat* as a concept in this vein. We following co-interpretation priorities from the community envision rewilded Iyengra territories to be a *nimat* of the future—an act of emplacement, a return and a comeback to the restored taiga and restored beings in the forest. Whilst this may seem unrealistic at present, the situation has changed in the recent past in a speedy manner [38].

For capturing the Indigenous knowledge of the Evenki regarding the southern stocks of the wild reindeer, we have reviewed over 80 primary oral history files that have been translated from Evenki or Russian (depending on the case) into English, as well as diary entries, written statements communicated by email (especially for 2019–2020 season) from Iyengra on the situation of the wild reindeer over the decades. Our primary focus is on the 15 years between 2005 and 2020, but we add Indigenous knowledge of past changes on wild reindeer where applicable, also reviewing the literature (for example on losses of knowledge at [24]).

In 2007, a large international conference "Traditions of the North" was organised in Iyengra to discuss the midpoint results, natural resources extraction of the region and further cooperation points. This allowed the community members participating to hear of the results and discoveries until 2007. Between 2012 and 2019, the materials of the oral histories were summarised into an online Atlas (see [6]) that has been made available, based on the wishes of our co-researchers, in Russian in January 2020 (an Evenki version may be a future option).

The primary oral history tapes were recorded by the authors (Tero Mustonen, Kaisu Mustonen, Tamara Andreeva), mostly in the reindeer camps around Iyengra, or in the community itself. Consent forms were used. Anonymity was offered and respected when requested. Some summaries and examples of the oral histories have been shared in Mustonen [27], Lehtinen and Mustonen [28], Mustonen and Lehtinen [29] and in Evenki Atlas [6].

The primary tapes are located in oral history archives in Yakutsk and in Finland. Vyacheslav Shadrin led this process. Each tape was transcribed and analyzed for contents and thematized. Evenki translations of interviews were carried out by Indigenous co-authors Yukaghir Vyacheslav Shadrin and Tamara Andreeva, an Evenki herself. Interview tapes exist in WMA and MP3 format in the archives.

Given the thematic focus on wild reindeer we concentrate on offering the trends and key messages from the oral history documentation and other sources, with carefully chosen oral history quotes where applicable for the results. These quotes were a question of co-intepretation of needs—the losses of wild reindeer, observations and ultimately impacts of the losses emerge strongly from the oral history corpus.

For a broader temporal-analytical frame, building on Montonen [35] and Mustonen and Lehtinen [36] we use three conceptual-temporal themes

- baseline science on wild reindeer in general in Sakha-Yakutia
- the situation from pre-industrial era into first industrial changes until 1990 (displacement)

- shifts from 1990s to the second wave of industrial land use 2020s (emplacement potential emerges)

These will service as a mechanism to discover the status, trends and intertwined impacts of the wild reindeer numbers, populations and their observed or assumed disappearance. Then the fourth and the largest analytical section "Evenki knowledge of the reindeer" highlights these implications for the Evenki (see the loss of knowledge in another Siberian region in [24]).

## 3. Results

### *3.1. Understanding the Baseline of Wild Reindeer in Southern Sakha-Yakutia*

An independent survey conducted by Argunov [39] explains that the wild reindeer stocks in Sakha-Yakutia can be divided roughly into tundra wild reindeer (populations of Leno-Olenok, Indigirka and Sundrun populations) and then taiga, or forest, reindeer stocks that are often understood to consist of Western taiga, Southern Sakha and mountainous taiga populations. An earlier, separate study was conducted by Krivoshapkin and Mordosov [20]. We use these as a baseline and indicators of change in summary form from scientific sources.

In this paper we focus, according to the categorisation of Argunov [39], on the Southern Sakha wild reindeers and more specifically on those herds which have occupied and continue in part to occupy the Neriungri region. We should note that despite some of the monitoring efforts presented in this paper, the exact populations on 20 km × 20 km scales remain still a gap, and there is a need for further research and documentation.

### *3.2. From Pre-Industrial Populations into the First Wave of Industrial Land Uses 1975–1990*

Argunov [39] presents a fairly recent historical assessment of the herds of the wild reindeer in the region. His results build on aerial and on-the-ground -surveys of stocks as well as hunter inventory data from the ministerial sources in Yakutsk. The large stocks of 1960s referred to by Argunov in Krivoshapkin and Mordosov's [20] opinion are based on the re-organisation and liquidation of many wilderness and rural communities in the 1960s (so-called Liquidation of Villages with no Prospects). This provided more spatial extent and habitat to the wild reindeer.

Aerial surveys of wild reindeer in the taiga parts of Sakha-Yakutia can be summarized in the following:

- 1960: 100,000 animals
- 1975: 57,000 animals
- 2001: 24,500 animals
- 2010: 24,500 animals (based on [39], partly complemented by [20])

Rather positively Argunov [39] presents also data from the ground monitoring which has been re-started in 2010s and summarized below:

- 2013: 70,000 animals
- 2014: 70,000 animals
- 2015: 85,000 animals
- 2016: 63,500 animals
- Late 2010s: 85,000 animals (based on [39])

Argunov [39] says that the southern taiga herds of the wild reindeer would number around 7900 animals. He points out the rather massive (75%) collapse in the aerial survey numbers of the southern wild reindeer over the span of 50 years, despite the aerial and spatial extent being rather secure.

He also points out that the southern herds have moved in the more recent 2000s into the middle parts of the Republic and have merged with herds there—a potential consequence of the industrial development in the south. Krivoshapkin and Mordosov [20] mark the southern herds to number around 8000 in 2001 and they list the overall forest reindeer numbers to be around 12,000 animals in 2001 (significantly lower than presented by Argunov for the same period).

This period in the lives of the Evenki was a massive transformation, and an often forced process. The self-governance of the life in the taiga changed into a mixed, and more controlled Soviet era, when the settlement of Iyengra was founded in 1926. Natural ecosystems remain relatively intact until 1970s—we can also see the preservation of cultural knowledge of wild reindeer to have survived better, given that we can assume the herds were bigger, closer and healthier.

From 1970s onwards the advancement of several industrial actions, including the Baikal-Amur railroad and proliferation of coal mining started to affect the Evenki nature-based life. Many Evenki had been working with mineral exploration crews as guides. The building of the city of Neriungri in 1975 sped up the infrastructure and road construction in the region.

In 1991 the Evenki faced the dissolution of the state farm -based reindeer herding. Lack of financial assets triggered what Pika [40] has called "neotraditionalism", or the reawakening of modes of life building on traditional taiga life. We can see this as a first era of a renewed emplacement [36] —wild reindeer stocks also benefitted from reduced industrial activities. Some Evenki established obshchinas, Indigenous kin-based community-cooperatives as a basis of private reindeer herding. For example Gonam is one of these communities [6]. The conceptualisation of nomadic schooling also emerged from within the freedoms of the 1990s [30].

### *3.3. 1990s into the Second Wave of Industrial Land Uses 2005–2020*

As a whole Argunov [39] determines that the populations of the taiga wild reindeer are rather stable on a republic-wide level. Krivoshapkin and Mordosov [20] disagree to certain extent in those data sets where their observations overlap (until 2007). They are concerned about the loss of numbers of animals evident in the aerial survey data and have an estimation of 24,500 wild reindeer in all of Sakha-Yakutia. They name the main drivers of losses as:

- wildfires and the long recovery time of lichen pastures after fires
- industrial land uses, more specifically pipelines, diamond and coal mining, oil field developments
- forest logging
- predation by wolves

Krivoshapkin and Mordosov [20] also add to the drivers of change the expansion of domestic reindeer herds that have overtaken pastures from the wild reindeer. Montonen [35] provided similar observations from northern Finland, where he says the domestic reindeer and wild reindeer herds tend to avoid each other.

Specifically for Neriungri region, Krivoshapkin and Mordosov [20] say that the decline of wild reindeer has been one of the sharpest in the Republic in the 2000s (density from 0.31 individuals per 10 square kilometers in 2001 to 0.18 individuals per 10 square kilometers in 2007) due to the construction of the East Siberian Oil Pipeline [14]. Argunov [39] does not go this this level of detail. According to Krivoshapkin and Mordosov [20] this pipeline construction has destroyed pasture lands, changed the seasonal migrations of the reindeer and instigated direct loss of wild reindeer life (due to poaching, road kills and better access to the forest areas).

For the Evenki, this reawakened extraction era ended many of the promises of 1990s on Indigenous rights and freedoms. Industrial activities expanded significantly but also divided families, clans (see more on [32] on the complex question of clans and the Evenki) and obshchinas with some receiving financial compensations [19] and others advancing rights-based discourses.

Others in Iyengra observed the impacts but did not either care or participate in these public narratives. The window for enabling a renewed emplacement in various spatial (territorial) or social levels (obshchinas, families, and whole communities) diminished due to the proliferation of pipelines and other extractive developments [14].

All this implied that those who had carried the practice and wisdom traditions associated with the wild reindeer and practices like *nimat* faced difficulties passing on the knowledge. As Frainier et al. [26] describes, the link between biodiversity and cultural practices is sensitive, contextualized and fragile. Mustonen [27] describes how some of the herders tried to negotiate on the uses of the lands with gold mining crews in the taiga, but on a landscape level, there was little the Evenki could do to even maintain the knowledge nexus they had before. Next we explore and investigate the elements of this nexus, especially centering on wild reindeer.

### *3.4. Centering the Wild Reindeer in the Knowledge Systems of the Evenki of Iyengra—Cultural, Linguistic and Ecological Knowledge*

According to Sirina [41] and for Iyengra especially, Mustonen [27] and Frainier et al. [26] say the Evenki use their reindeer for transport, handicrafts, and food security. Wild reindeer populations are prized as food, a source of handicrafts and game. We need to be mindful, however, that this is not the case for all Evenki communities (see [23,41,42]), nor has knowledge survived in all locations [24] and therefore this applies here mostly for the context of our study region in Southern Yakutia. Lavrillier and Gabyshev [10] explore the Evenki landscape and terrain concepts in detail, demonstrating the close proximity of the language and landscape forms and details both for the Khabarovsk and Sakha Evenki.

Unlike many other reindeer herding peoples of Eurasia, the Evenki do not (in most cases) eat their domestic herd animals, rather, wild reindeer is hunted and considered to be of significance in terms of cultural and food security purposes [1,20]. Shubin [23] demonstrated that in some regions, in this case in Buryatia, the Evenki mixed wild reindeer stocks with domestic herds for improvements in strength and blood lines. Amongst the oral histories presented here no direct evidence of that practice has been found in Iyengra.

Vorob'ev [24] in his study of Chirinda Evenki relations with the wild reindeer observed also significant "erosion" and losses of cultural practices, noting the vulnerability of customs if the hunting is no longer practiced. In Iyengra this cultural role of the wild reindeer is still prominent especially amongst older residents, as is evident in some hunters observing that they dream about the wild reindeer and the hunt (OR 2005).

Evenki have developed strict guidelines and Indigenous customary law systems for their relationship with the taiga and cosmos. These include rules on how to live with nature, how to travel on Evenki land [1,10], and how hunters should behave when they are given certain animals [7,9,31]. (Navigation in the forest depends on the rivers as main pathways of transport, and we can summarize the Evenki to be living with a cultural landscape of the forest, one that they know intimately [6,10].

In 2005 Vladimir Kolesov, a reindeer herder and a hunter, explained some of the Evenki principles on these issues:

> *We say: Earth mother. If we go past large rivers, we hang a piece of cloth there. Close to the mountains we do that too. We hang a piece of cloth there. You are not allowed to leave pieces of firewood lying around. It is not allowed to cut more wood than what is needed. When you are someplace, for example hunting, don't leave pieces of wood crosswise. Everything needs to be in order. Don't throw bones around. I make a shelter, and all bones are put there. So that nothing is out of order. It is also because the reindeer come and bite the bones and suffocate.*
>
> *Clean and safe. To keep the reindeer from harm.*
>
> *You fish only as much as you need. If the next day you need more, you go fishing again then.*
>
> *If no one would buy the sable skins, it would not be hunted as much. If you need a hat, it is only then you're allowed to hunt. If there was no need, it would not be killed. This goes for all of the animals I think.*

*And trees too. If you need wood for sleds, then you take but otherwise no. If there is no need, nothing will be cut.*

(as quoted in [27])

Galina I. Varlamova, or "Keptuke" as she is known to the Evenki, was a daughter of a spiritual leader (in the cultural context, a shaman). Some consider her to have been a shaman herself. Unfortunately, she passed from this world in 2019. All of her life she wrote and researched as well as practised the Evenki traditions related to nature. According to her [1], the functioning principles of Evenki civilization are based on enforcing the moral system of what the Evenki call *Ity* through *odjo*: a set of principles of taboos and rules of human behaviour:

In every part of life, be it material or cultural, there are reflections of relationship between the Evenki and nature. This relationship that was formed and reformed across centuries was a basis for general understanding for justice, traditions and moral guidelines. These are reflected in the system for ecological law, Ity. They are also reflected in the prohibition-taboos, named Odjo. Evenki oral tradition is not just folklore and traditional poetry but includes many other cultural texts that offer teachings for life in nature and in social family and tribal system. Traditions, fixed rituals and ancient rites that have survived to this time have all been subordinate to experience of living in nature, which is Evenki homeland and Buga—Mother god. Evenki place nature at the highest level [43].

These rules and human lives operate in the universe of *Buga*, or God, Galina Varlamova explained further (see a larger treatment of her synthesis of the Evenki beliefs in English in [1]). See Sirina [24], [41] and [44] for similar cultural and ethical codes in different parts of the Evenki home area. According to [1] the most important role and significance in the Evenki universe is placed in nature. Evenki see nature as the highest god, Buga. Everything is created by Buga/nature.

Already in the 1990s, Anatoli Ivanovits Lasarev shared the teachings of his father with Galina Varlamova in Iyengra:

*My father spoke like this, and I think like this and I tell you this now.*

*Buga gives life to all kinds of scraps on earth, including humans. Buga sees everything, warms everything with its inner warmth, makes us human.*

*Buga does not like badness. You should not be selfish and greedy, but share.*

*Buga gives it for everyone according to the nimat custom.*

*Buga has prescribed this law for every living thing.*

(in [1])

Buga is (was) celebrated, according to Varlamova [1], in the seasonal rituals and festivities. In spring, the Ikenipke Festival, which is a celebration of renewing of life [2], Buga has to be remembered and respected through singing and dancing. While the Ikenipke Festival has been transformed into a summer festival and lost most of its original procedure [2,10] it is still an important marker event of the year for some Evenki.

These cultural practices and concepts define what we refer to as a renewed emplacement potential for the Evenki of Iyengra [2]. *Nimat* then becomes the cultural expression of this emplacement process. The forests and taiga lands and waters around Iyengra are not "wilderness" for the Evenki [15]. Rather they are a cultural landscape filled with history and presence [6,28]. They are defined as the Evenki homeland. The forest contains "close proximity use areas"; for example, past and present camp sites and nomadic routes that are a transitional space between human and natural realms. This is clarified by Lavrillier (e.g., in [7,9]) and Mustonen [27] in their studies of the spatial organisation of the taiga among the Evenki.

In contrast, the deep forest and remote hunting areas are "for the nature" [10], only to be visited by Elders who are aware of a proper behaviour and/or for occasional hunting trips. Mustonen [27] explains how the time and dwelling in the village of Iyengra differs from the taiga memories and presence—many people in the settlement long for the freedom, self-autonomous organisational and spatial order of the deep forest and nomadic camps whereas the town is seen to be ordered and governed by Russian state norms.

Evenki navigation skills are rather well known. Lavrillier [15] stresses the role of local streams and rivers for the Evenki navigation, especially during winter (see [45] for discussion on Nenets and rivers for comparison). Rivers emerge as "highways" on which connections can be made easily using reindeer. Wure'ertu [4] points that clan names of the Evenki are based on the rivers they occupy.

One of the most central of animals in this cosmological order of the Evenki is the wild reindeer [24]. Nature and its phenomena reflect this relationship based on oneness, unity. There are other important animals too, but wild reindeer occupies a strong position in the cultural whole. For example hunters may also associate sacred places in the taiga, such as ancient burial sites with hunting luck of wild reindeer, pointing to the spatial, endemic concepts of the forest the Evenki maintain [44].

Some of the Indigenous knowledge materials from 2005 to 2020 emerge to highlight the role of wild reindeer in the Iyengra Evenki culture. We present them here in summary and anonymous form respecting these herders' wishes (Other hunter-herders, like Kolesov, shared their oral histories with willingness and consent to be named. If this has been the case, we have quoted a person direct with names.). For example, one oral historian (anonymous OR) from the village reflected in 2020 that a meeting of white wild reindeer brings good fortune. If it is hunted the skin is preserved and considered to bring luck to the owner.

Summary from the OR corpus indicates that the overall, appreciated role of the wild reindeer is shared amongst most of the people hunting and living in taiga. Little or no differences can be found on the main role. Divergence of experiences emerges from the spatial-territorial obshchina areas—each of these kin-based communities focus their hunt on their own territories.

The question of rut season harvest remains a complex and partially unanswered whole. Several knowledge holders indicated that the hunt stops for the rutting season in September. Yet, individual statements refer to some harvests especially in the past (when the stocks were plentiful) even during rut (see below on 2005 harvest). As no systematic, reliable reporting in the community exists, exceptions may happen (for example when the opportunity emerges).

Rutting is also the period when the wild and domestic reindeer populations may encounter each other due to biological reasons. Montonen [35] reflecting on Finnish wild forest reindeer and domestic population relations disagreed. He pointed to major avoidance patterns between the wild and domestic herds. Rut events, access and avoidance and associated hunt windows will constitute one of our future focus points should the wild reindeer survive in the taiga forests surrounding Iyengra.

Variations of engagement and maintainence of traditions emerged from individual oral histories. We offer carefully selected samples below, from the consented OR materials, to illustrate some of the gendered, and nuanced practices documented.

An oral history by the late knowledge holder Oktyabrina Naumova linked the wild reindeer liver with good eye sight. She conveyed in 2005 that the 107-year-old spiritual person (shaman), Matryona Kulbertinova, had used the liver for her health and maintaining of eyesight in the forest. Oral histories also refer to the teeth of the wild reindeer to be used as talismans for babies during the nomadic travels, to protect them from evil spirits.

Naumova also conveyd that then the hunting was initiated, people would leave offerings to trees and river crossings to seek good fortunes. Stories of wild reindeer luring domestic reindeer have been plentiful in the documented materials. In summary the wild reindeer is seen to have an agency, a determination and awareness of human behaviour and practices.

Customary cultural practices, besides nimat, are still known in the camp. For example, when the wild reindeer has been harvested, its bone marrow should be sacrificed on the campfire before eating the animal. Also a part of the tongue and kidneys are fed to the fire before humans can start to eat.

In a reindeer camp in 2005 one of the more active hunters of the reindeer brigade 4 mentioned that he averages 10 hunted wild reindeer in a winter season. According to him *"wild reindeer are completely different from reindeer"*. He differentiates for example by comparing the ears, tail and nose to be very distinct.

The scientific materials pointed out that the years 2000–2007 were a source of a large decline of stocks in the area. Hunters agreed in 2005: *"In the past, during the rut, even women could go out and kill a wild reindeer, they were plentiful. Now there are too few of them."* Many other oral histories confirmed this observation in town and in the camps in Winter and Spring 2005. In an oral history from 2020 an anonymous male hunter (quoted above) stressed that in 2020s the rutting season is now a time when the hunt is suspended—a potential self-imposed measure to allow the stocks of reindeer to have a chance to replenish.

Some positioned the decline to have emerged already in 1970s: *"When the Baikal-Amur railway was built (in the 1970s), the number of animals decreased."* Vorob'ev [24] highlighted the nexus of increased wild reindeer stocks and diminishing of reindeer herding in another region as one of the drivers of change. We mark this important regional discovery but the evidence from Southern Sakha points more towards a combination of human-induced and potentially climatic changes, given the speed of losses of stocks reported both by scientists and the Evenki themselves.

In the oral history corpus several hunters also referred to climate change impacts as a driver of loss of wild reindeer. According to them the autumnal freezing rains had caused the wild reindeer to move to Amur region, away from the southern Yakutian pastures. This would be in some contrast to the science deduction by Argunov [39] that the movement of wild reindeer is more towards north, central Yakutian wild herds.

To position this in a dialogue with science, we can determine that climate change has already started to manifest both in the oral histories and the science data from the regions (see Figure 1). Mean annual temperature in the region, according to Russian Academy of Sciences data, has increased up to 0.5 °C per decade between 1930 and 2015 using a linear least squares fit to the data. Evenki views on climate change and oral histories have been explored elsewhere at length [27]. For the wild reindeer displacement analysis it is sufficient to position the weather and climate change to be major drivers this century that are already affecting the nature and livelihoods (see documentation of snow quality changes in [10]).

In Spring 2020 in the oral histories the multiple issue, present crisis of Iyengra manifested clearly in the words of one of the female reindeer herders, who prefers anonymity [46]:

> *"Spring is the season of a new life. New calves come into this world. For us, for Evenki, this is the happiest and the most exciting season. Reindeer calving is very important for it increases the number of reindeer. There are domestic and wild reindeer. Today we will talk about hunting of wild reindeer. In past, long time ago when I was a child, out reindeer herd was located about 110 kilometres away from the settlement. An all-terrain-vehicle delivered food two or three times a year. That is why, our main food was wild reindeer meat. There was enough wild reindeer in the area both for us and for other predators (wolves, bears, etc). There was enough wild reindeer. We harvested as much as Seveki (C—one of the names of the Evenki god-creator) would give us—Evenki do not take too much, they have to share with others—'Nimat' (ā—process of sharing harvested animal with others (a ritual)). Sometimes in the morning you would come out of your tent and reindeer were going in circles and were jumping away from something within kure (ē—coral for reindeer)—turned out that a wild reindeer came with domestic reindeer. Nowadays, it does not happen often. In the course of time, the climate has changed—fires, droughts. Because of the fires the number of wolves and bears has grown; they have chased away wild reindeer and got at our domestic reindeer. We had to move closer to the settlement. At present there are horrible times for reindeer herders. Gold-diggers are*

*everywhere, destroying the land. But reindeer have to eat lichen and drink clean water. Everywhere is the noise of machines, there are no wild reindeer left, they migrated to pristine areas. Because there is no wild reindeer, Evenki food supply is disrupted. In summer, if you are lucky, you can harvest one wild reindeer and the probability of it is 1 to 100. And in winter, reindeer herders go away for a month in order to harvest some meat for their families. When gold-diggers have appeared, nothing alive was left in the forest."*

Summarizing, much in line with Krivoshapkin and Mordosov [20], the Evenki of Iyengra, based on the observational visits to their camps and home area 2005–2020 as well as documenting their oral histories, agree about the losses of the wild reindeer populations. In the materials we can see a speeding up of the situation especially between 2001 and 2007 due to the pipeline construction, which corresponds to the Krivoshapkin and Mordosov [20] data of major losses of pastures, ecosystems and reindeer themselves during this period.

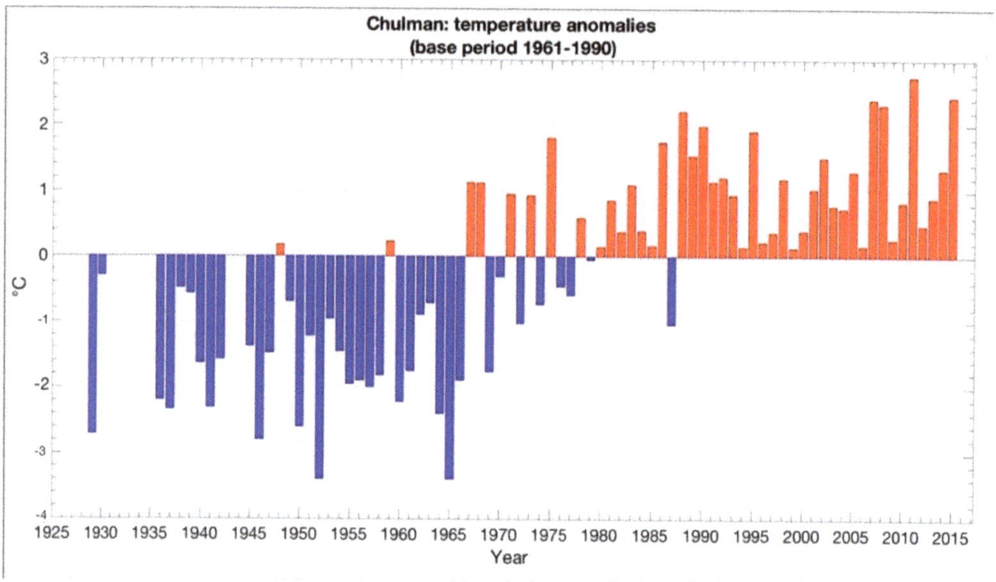


**Figure 1.** Climate change summarised on the basis of regional weather station data from Chulman (close to Iyengra). Bars indicate the temperature anomaly for each year 1929 to 2015, compared with a base period of 1961–1990. The anomaly is calculated as the annual average temperature minus the 1961–1990 mean temperature. Red bars indicate positive temperature anomalies (the annual average temperature is higher than the baseline mean) and blue bars indicate a negative anomaly (annual average temperature less than the baseline mean). These data were derived from Russian Academy of Sciences ground temperature stations with subsequent statistical analysis by Brie Van Dam, PhD, Snowchange.

Complementing the potential human-induced drivers, the Evenki point to impacts of climate change that manifest for example in the freezing of reindeer pastures in the Autumn. Diverging from the scientific understanding some oral histories point to Amur region as the destination of the wild reindeer when they move away from Iyengra and Neriungri area. Argunov [39] pointed that the taiga wild reindeer would amalgamate and populate central Sakha-Yakutia territories. Further research is needed to determine these directions better.

By late 2000s Evenki hunters reported in oral histories that the overall hunted amount of wild reindeer over the whole winter had fallen to 10 animals, in a good year, around Iyengra. "In the past", especially during the Autumn rut, wild reindeer had been perceived to be so plentiful that everyone received enough.

## 4. Discussion

Krivoshapkin and Mordosov [20] state clearly that the wild reindeer populations of Sakha-Yakutia, especially the southern herds, are in severe decline, mainly from three main reasons: industrial development, forestry and the loss of land access.

Wild reindeer according to Frainier et al. [26] are a cultural keystone species for the Evenki of Iyengra. Relations with, and the hunt and uses of wild reindeer maintain a large, even ancient cultural complex that is the source of Evenki food, linguistic diversity, customary law and relationality with the taiga forest landscapes [24]. We may determine this to be endemic to the specific dialect of Evenki spoken in the area. From the viewpoint of potentials of a cultural and ecological emplacement, we can see an even stronger role of the wild reindeer as a keystone species.

More precisely the industrial land use in the surroundings of Iyengra have had a deep impact on the traditional land use, culture and economies of the Evenki overall. Railroads, hydroelectric stations, and coal and gold mines have gradually dominated the landscape. In addition, the construction of the East Siberia-Pacific Ocean oil pipeline and Power of Siberia gas pipeline during the 2000s and 2010s [14,19,27] have largely modified the lands and lives of the people in the region.

The first railroad, the Baikal-Amur railway (The so-called BAM track) was already constructed across the taiga in the 1970s. Megachanges in nature and society have affected all of the Evenkia—in 1927 they possessed approximately 49,000 domestic reindeer, in 1968 up to 63,800, and in the 2010s, the numbers have fallen to perhaps a low of 3000 animals [47].

The Iyengra co-researchers in their priorized co-interpretations linked with the author deductions repeatedly returned to the abrupt link between railroad construction, loss of wild reindeer and reindeer herding. When the first tracks were built, accounts of several young reindeer herders committing suicide were reported [48]. It is believed that they could not come to terms with the imposed dramatic changes in their lands and lives. Such upheavals have produced contexts where traditional skills and values have lost meaning and their whole world was turned upside down.

With the new-found freedoms of the post-Soviet Siberia, Evenki were able to establish kin-base communities. Across the region a cultural resurgence brought pride and value to traditional practices and cultures. Pika [40] called this the 'neotraditional' era. Out of food security necessity and due to lack of resources and funds, many families and herders in Iyengra maintained and even renewed modes of life in the taiga that can be seen as acts of emplacement, even though we can question of course in hindsight, how much of an agency [37] was realised and how much happened because of daily needs.

The cumulative impacts of past land-use changes, and especially those associated with the industrial megaprojects of the 2000s and 2010s are as immense as they are difficult to assess. In summary, mining, energy and infrastructure projects have thoroughly altered the Evenki land and lifescapes in the following ways [13,14,19,27]:

- Major hydrological regimes and aquatic ecosystems have been efficiently transformed by mining extensions and the construction of hydropower.
- Smaller streams and old-growth forests have been contaminated by oil pipelines, as well as mercury release and land churning by artisanal gold mining.
- Changes in forest cover, fish stocks, water colour and quality have in turn exhaustively affected fishing livelihoods and reindeer herding.
- Mammals, birds and other fauna that are dependent on post-Ice Age pristine old growth taiga forests have suffered and retreated elsewhere, making hunting and subsistence economies harder to maintain.
- Soviet introductions of species such as sable, to name one example, are cases of biomanipulation that affected the region early on.
- Climate change-induced droughts and unsafe fire management have affected Evenki capacity to maintain seasonal rounds. Forest fires have also turned more frequent due to increase of tourist hunters.
- Major transport corridors have sliced the taiga around Iyengra.

- The amount of waste water released from the city of Neriungri has become alarming. (summarized also from [6,29]).

Such losses suffered by the wild reindeer populations are therefore affecting the Evenki too [24]. They make the potential for a sustained or a realistic emplacement process extremely hard. Unique, priceless cultural practices, songs, customs and ways of life become endangered as the wild reindeer disappears. As Frainier et al. [26] demonstrates, the loss of biodiversity cascades simultaneously into a loss of cultural and linguistic diversity. Vorob'ev [24] researched in Krasnoyarskii krai where Evenki are also living and discovered that the renewed herds of wild reindeer instigated a number of emplacement -style *returns* of hunting practices and food security elements (also on the expense of loss of herding time/focus). He determined that they contained "ancestral" elements of the hunting culture, even though gaps and losses had happened. We think the Krasnoyarskii krai situation is not directly applicable in Southern Sakha but remains an important comparative case.

## 5. Conclusions—Nimat: Key Elements of Potential of a Renewed Emplacement

Increasingly from the early Soviet period and overwhelmingly so from 1970s onwards, the state-sponsored intrusion into Southern Sakha-Yakutia [13] has wrecked the intact nature of the taiga ecosystems of the Evenki with the construction of mines, pipelines [14,19,27,43,48] calls these 'critical infrastructures' of the state] and hydrostations, as well as road and railway lines that have altered permanently the status of the ecosystems. Wild reindeer are one of the most potent symbols of loss of natural ecosystems this large process has caused. As Varfolomeeva [49] demostrates, even the "softer solutions" such as tourism and road constructions for non-industrial needs will have consequences for the local traditional-Indigenous cultural context in Siberia. Lavrillier [30] reviewed the options of supporting culture of the region using nomadic schools, to varying degress of success. There are no easy answers.

Locally, Evenki have responded in a number of ways which can be summarised as a mix of access, avoidance, withdrawal, confrontation and ultimately acceptance [37]. This multifaceted transformation of the Iyengra Evenki cannot be summarised in one article, but we propose an alternative development for the region, building on the Evenki cultural concept—*nimat*. We do not pretend that this conceptual plan could be implemented at once in the present-day conditions. We also acknowledge its difficulties in the present socio-political realities. Rather, we draw on a cultural grounding of the emplacement potential Iyengra still has.

Nimat—here, *sharing*, is essential for survival in Iyengra. In 2005 in a reindeer brigade [4] a number of herders exclaimed during the oral history work that focused on *nimat*: *"All get an equal share! Nobody is left out! We share amongst the people of the same tent, every bone shared. We always do it."*

The Evenki taiga of Southern Sakha-Yakutia is now an altered ecosystem and society containing both intact and wrecked components. By accepting the continuity of externally induced displacement, the Evenki culture will become integrated in the new post-traditional era. However, alternatively, systematic advancement of endemic modes of living contain an imagined 'rebirth' (see the cultural roots and structural potential amongst Evenki in [2]) of the taiga that is shared by many, if not all, Evenki of Iyengra. The three critical points of a successful emplacement defined by Mustonen and Lehtinen [36] are still present in Iyengra—(1) some remaining core ecosystems, (2) knowledge of living cultural codes (and social-ecological networks) of engagement with the landscapes and (3) lastly, at least some agency to realize what this form of *nimat* could undertake.

This alternative is rich in potentials rooted in past community wisdom, both spiritual and practical, that is still remembered and commemorated among the Evenki. We should not forget what occurred in 1991 with the dissolution of the Soviet power—a return to traditional ways of life in the forest, fostered both by economic necessity and leadership of Evenki themselves, such as Keptuke and Matriona Kulbertinova [1]. Mustonen [50] points to the applicability of oral histories (see on regional translation and power languages in [51]) as a basis of ecosystem restoration.

Therefore, a societal *nimat* could solve the crisis of both the reindeer and the people [2]. We outline some of the umbrella components of an emplacement process of *nimat* which would have to be met for a land-based lifeway to be guaranteed. First, rewilding and restoration of those taiga habitats and river systems that have been adversily affected by industrial land use would need to be initiated. Second, an establishment of Indigenous community conserved areas (ICCAs) in the taiga forest that would protect the life of both the wild and domestic reindeer and the Evenki is needed to guarantee certainty of nature-based Indigenous life in the forest. Thirdly, an allowance of a taiga life of interconnected, healing ecosystems—where nature is providing her *nimat*, the eternal bounty of sharing resources, food and survival, in a unique part of Eurasia that has nurtured the Evenki for so long.

**Author Contributions:** Conceptualization, V.S., T.A., K.M., methodology, T.M. and V.S.; validation, V.S., T.M. and T.A.; formal analysis, T.M., V.S., K.M.; investigation, V.S., T.M. and T.A.; resources, T.M. and K.M.; data curation, V.S., T.M. and T.A.; writing—original draft preparation, V.S., T.M., K.M. and T.A.; writing—review and editing, V.S., T.M. and T.A.; supervision, T.M.; project administration, T.M.; Evenki knowledge and language curation, T.A. All authors have read and agreed to the published version of the manuscript.

**Funding:** No special funding for the research.

**Institutional Review Board Statement:** The research has followed in every way overall research ethical guidelines of Russian Academy of Sciences and the Finnish Academy as well as ALLEA's European Code of Conduct for Reseach and the Arctic Council Ottawa Principles of Indigenous Knowledge. Institute for Humanities Research and Indigenous Studies of the North of Siberian branch of the Russian Academy of Sciences (IHRISN SB RAS), Yakutsk, Russia reviewed the research and its framing and supported it as well as Snowchange Cooperative in Finland for Finnish researchers.

**Informed Consent Statement:** Informed consent was obtained from all subjects involved in the study.

**Data Availability Statement:** Temperature data available on open Russian meteorological sources and also from Snowchange Cooperative.

**Acknowledgments:** We are thankful to our co-researchers in Iyengra for all they have shared over the years. We also remember and acknowledge those knowledge holders that have unfortunately moved on. We thank Brie van Dam at Snowchange for the proofreading and statistical analysis of weather data. We are also thankful to ELOKA Project at National Snow and Ice Data Center in Colorado, USA for the technical innovations and solutions for the Evenki Atlas.

**Conflicts of Interest:** The authors declare no conflict of interest.

## References

1. Varlamova, G. Traditional ecological knowledge of the Indigenous Peoples of the North in the Context of Northern Climate Change; Unpublished Manuscript in Russian. 2005. Available online: http://www.snowchange.org/pages/wp-content/uploads/2021/11/varlamova.pdf (accessed on 15 September 2021).
2. McNeil, L. Recurrence of Bear Restoration Symbolism: Minusinsk Evenki and Basin-Plateau Ute. *J. Cogn. Cult.* **2008**, *8*, 71–98. [CrossRef]
3. Mertens, K. *Mobility and Economy of the Evenkis in Eastern Siberia*; Boise State University: Boise, OH, USA, 2015.
4. Wure'ertu. Evenki Migrations in Early Times and Their Relationship with Rivers. Études mongoles et sibériennes, centrasiatiques et tibétaines, Volume 49. 2018. Available online: http://journals.openedition.org/emscat/3196.2018 (accessed on 20 September 2021).
5. Shubin, A. *On the Question of Ethnographic Features in the Material Culture of the Evenks of the North of Buryatia ( ooo oaooo aao oCa )*; Ethnographic Collection: Academy of Sciences of the USSR: Siberia, Russia; Buryat, Russia; Institute of Social Sciences: Ulan-Ude, Russia, 1974; Issue 6.
6. Atlas, E. Online Atlas of the Evenki of Iyengra, Version 2.0. ELOKA and Snowchange. 2020. Available online: www.evenki-atlas.org (accessed on 15 September 2021).

7. Lavrillier, A.; Dumont, A.; Brandišauskas, D. Human-Nature Relationships in the Tungus Societies of Siberia and Northeast China, Études Mongoles et Sibériennes, Centrasiatiques et Tibétaines, Volume 49. 2018. Available online: http://journals.openedition.org/emscat/3088 (accessed on 15 September 2021).
8. Fryer, P.; Lehtinen, A. Iz'vatas and the diaspora space of humans and non-humans in the Russian North. *Acta Boreal.* **2013**, *30*, 21–38. [CrossRef]
9. Lavrillier, A. The creation and persistence of cultural landscapes among the Siberian Evenkis: Two conceptions of "sacred" space. In *Landscape and Culture in Northern Eurasia*; Jordan, P., Ed.; Routledge: London, UK, 2011; pp. 215–231.
10. Lavrillier, A.; Gabyushev, S. *An Arctic Indigenous Knowledge System of Landscape, Climate and Human Interactions: Evenki Reindeer Herders and Hunters*; SEC Publications: New York, NY, USA, 2018. Available online: http://www.siberian-studies.org/publications/PDF/lavgab.pdf (accessed on 15 September 2021).
11. Brandišauskas, D. Sensory Perception of Rock Art in East Siberia and the Far East-Soviet Archeological "Discoveries" and Indigenous Evenkis. *Sibirica* **2020**, *19*, 50–76. [CrossRef]
12. Araseyenin, B. *Sakha Historical-Cultural Atlas*; Institute of Humanities: Yakutsk, Russia, 2007. (In Russian)
13. Newell, J. *Russian Far East—A Reference Guide for Conservation and Development*; Daniel & Daniel Publishers: McKinleyville, CA, USA, 2004.
14. Yakovleva, N. Oil pipeline construction in Eastern Siberia: Implications for Indigenous people. *Geoforum* **2011**, *42*, 708–719. [CrossRef]
15. Lavrillier, A. S'orienter avec les rivières chez les Évenks du Sud-Est sibérien [from French—Rivers as a spatial, social and ritual system of orientation among the Evenks of southeastern Siberia]. *Études Mongoles et Sibériennes, Centrasiatiques et Tibétaines* **2006**, *36–37*, 95–138. [CrossRef]
16. Syulbe, B. *(aaC)*; ooa : , Russia, 2004.
17. Habeck, J.O. How to turn a reindeer pasture into an oil well, and vice versa: Transfer of land, compensation and reclamation in the Komi Republic. In *People and Land. Pathways to Reform in Post-Soviet Siberia*; Kasten, E., Ed.; Dietrich Reimer Verlag: Berlin, Germany, 2004; pp. 125–147.
18. Karjalainen, T.; Habeck, J.O. When "the environment" comes to visit: Local environmental knowledge in the Far North of Russia. *Environ. Values* **2004**, *13*, 167–186. [CrossRef]
19. Sidortsov, R.; Ivanona, A.; Stammler, F. Localizing governance of systemic risks: A case study of the power of Siberia pipeline in Russia. *Energy Res. Soc. Sci.* **2016**, *16*, 54–68. [CrossRef]
20. Krivoshapkin, A.A.; Mordosov, I. Condition of the Wild Reindeer Population Numbers in Sakha-Yakutia. 2008. Available online: https://cyberleninka.ru/article/n/sostoyanie-chislennosti-lesnyh-populyatsiy-dikogo-severnogo-olenya-rangifer-tarandus-linneaus-1758-yakutii (accessed on 15 September 2021).
21. Safronov, V. Population, Natural Habitat, Migration of Tundra Wild Reindeer in Yakutia. 2007. Available online: https://cyberleninka.ru/article/n/chislennost-areal-i-migratsii-tundrovogo-dikogo-severnogo-olenya-v-yakutii/viewer (accessed on 15 September 2021).
22. Shubin, A. *Evenks of the Baikal Region ( aa)*; Belig Publishing House: Ulan-Ude, Russia, 2001.
23. Shubin, A. *Evenki ()*; Publishing House of JSC "Republican Printing House": Ulan-Ude, Russia, 2007.
24. Vorob'ev, D. Contemporary Beliefs of Northern Wild Deer Hunters—(The Case of the Chirinda Evenki). *Anthropol. Archeol. Eurasia* **2014**, *52*, 34–58. [CrossRef]
25. Myreeva, A. *o- oa[Evenki-Russian Dictionary]*; Nauka Publishing: Novosibirsk, Russia, 2004.
26. Frainer, A.; Mustonen, T.; Hugu, S.; Andreeva, T.; Arttijeff, E.-M.; Arttijeff, I.-S.; Brizoela, F.; Coelho-De-Souza, G.; Printes, R.B.; Prokhorova, E.; et al. Opinion: Cultural and linguistic diversities are underappreciated pillars of biodiversity. *Proc. Natl. Acad. Sci. USA* **2020**, *117*, 26539–26543. [CrossRef]
27. Mustonen, T. Karhun Väen Ajast-Aikojen Avartuva Avara. Tutkimus Kolmen Euraasialaisen Luontaistalousyhteisön Paikallisesta Tiedosta Pohjoisen Ilmastonmuutoksen Kehyksessä [Peoples of the Bear: An Inquiry into the Local Knowledge of three Eurasian Subsistence Communities in the Context of Northern Climate Change]. Unpublished. Ph.D. Thesis, University of Joensuu, Joensuu, Finland, 2009.
28. Lehtinen, A.; Mustonen, T. Arctic Earthviews—Cyclic Passing of Knowledge among the Indigenous Communities of the Eurasian North. *Siberica* **2013**, *12*, 39–55. [CrossRef]
29. Mustonen, T.; Lehtinen, A.A. Lived displacement among the Evenki of Iyengra. *Int. J. Crit. Indig. Stud.* **2020**, *13*, 16–44. [CrossRef]
30. Lavrillier, A. *A nomadic School Among Evenk Reindeer Herders in Siberia: Questions and Solutions in a Book Sustaining Indigenous Knowledge: Learning Tools and Community Initiatives for Preserving Endangered Languages and Local Cultural Heritage*; Kasten, E., de Graaf, T., Eds.; Kulturstiftung Sibirien: Fürstenberg/Havel, Germany, 2013; pp. 105–127.
31. Lavrillier, A.; Gabyshev, S.; Rojo, M. The sable for Evenk reindeer herders in Southeastern Siberia: Interplaying drivers of changes on biodiversity and ecosystem services—Climate change, worldwide market economy and extractive industries. In *Indigenous and Local Knowledge of Biodiversity and Ecosystem Services in Europe and Central Asia*; Roué, M., Molnár, Z., Eds.; UNESCO: Paris, France, 2016.
32. Mamontova, N. From "Clan" to Speech Community Administrative Reforms, Territory, and Language as Factors of Identity Development among the Ilimpii Evenki in the Twentieth Century. *Sibirica* **2016**, *15*, 40–72. [CrossRef]

33. Huusko, S. The Representation of the Evenkis and the Evenki Culture by a Local Community Museum. *Sibirica* **2020**, *19*, 77–98. [CrossRef]
34. Safonova, T.; Sántha, I. Evenki Hunter-Gathering Style and Cultural Contact. In *A Book Hunter-Gatherers and their Neighbors in Asia, Africa, and South America*; Kazunobu, I., Ed.; Central European University, Hungarian Academy of Science: Budapest, Hungary, 2016.
35. Montonen, M. *Suomen Peura*; WSOY: Helsinki, Finland, 1974.
36. Mustonen, T.; Lehtinen, A.A. Endemic renewal by an altered boreal river: Community emplacement. *Clim. Dev.* **2021**. [CrossRef]
37. Huntington, H.P.; Begossi, A.; Fox Gearheard, S.; Kersey, B.; Loring, P.A.; Mustonen, T.; Vave, R. 2017. How small communities respond to environmental change: Patterns from tropical to polar ecosystems. *Ecol. Soc.* **2017**, *22*, 9. [CrossRef]
38. Slezkine, Y. *Arctic Mirrors. Ithica*; Cornell University Press: Ithaca, NY, USA, 1994.
39. Argunov, A. Population Dynamics and Usage Dynamics of the Northern Wild Reindeer Resources in Yakutia (AA CC СПВА CCB CB B //Co o a oaoa;№ 3. 2017. Available online: http://science-education.ru/ru/article/view?id=26526 (accessed on 25 July 2021).
40. Pika, A. *Neotraditionalism in the Russian North*; CCI: Edmonton, AB, Canada, 1999.
41. Sirina, A. *Katanga Evenkis in the 20th Century and the Ordering of their Life-World*; CCI Press: Edmonton, AB, Canada, 2006.
42. Davydov, V. Strategies of Using the Space and Autonomy Regimes: The Relations of Evenkis and the State in the Northern Baikal. *Etnografia* **2018**, 2. [CrossRef]
43. Porteous, D. Topocide: The annihilation of place. In *Qualitative Methods in Human Geography*; Eyles, J., Smith, D.M., Eds.; Polity Press: Oxford, UK, 1988; pp. 75–93.
44. Brandišauskas, D. *Leaving Footprints in the Taiga: Luck, Spirits and Ambivalence among the Siberian Orochen Reindeer Herders and Hunters*; Berghahn: New York, NY, USA, 2017.
45. Istomin, K. Roads versus Rivers—Two Systems of Spatial Structuring in Northern Russia and Their Effects on Local Inhabitants. *Sibirica* **2020**, *19*, 1–26. [CrossRef]
46. Evenki "a". Anonymous Oral History from Iyengra, Recorded. May 2020. Primary materials available from Snowchange. Available online: www.snowchange.org (accessed on 15 September 2021).
47. Klokov, K. Reindeer herders' Communities of the Siberian Taiga in changing social contexts. *Sibirica* **2016**, *15*, 81–101. [CrossRef]
48. Mustonen, T. Re-reading critical infrastructure—A view from the Indigenous communities of the Russian Arctic. In *Russian Critical Infrastructures: Vulnerabilities and Policies*; Pynnöniemi, K., Ed.; The Finnish Institute of International Affairs (FIIA Report 35): Helsinki, Finland, 2012; pp. 79–101.
49. Varfolomeeva, A. Lines in the Sacred Landscape-The Entanglement of Roads, Resources, and Informal Practices in Buriatiia. *Sibirica* **2020**, *19*, 2. [CrossRef]
50. Mustonen, T. Oral histories as a baseline of landscape restoration. *Fennia* **2013**, *191*, 76–91. Available online: https://fennia.journal.fi/article/view/7637 (accessed on 15 September 2021). [CrossRef]
51. Slepcov, P.A. (Ed.) *o- oa [Sakha-Russian Dictionary]*; Izdatel'stvo "Sovetskaja Snciklopedija": Moscow, Russia, 1972.

sustainability

MDPI

*Article*

# Climate Change, Farming, and Gardening in Alaska: Cultivating Opportunities

**Nancy Fresco [1,*], Alec Bennett [2], Peter Bieniek [1] and Carolyn Rosner [1]**

[1] International Arctic Research Center, University of Alaska Fairbanks, Fairbanks, AK 99775, USA; pbieniek@alaska.edu (P.B.); crosner@alaska.edu (C.R.)
[2] School of Management, University of Alaska Fairbanks, Fairbanks, AK 99775, USA; apbennett@alaska.edu
* Correspondence: nlfresco@alaska.edu; Tel.: +1-907-474-2405

**Abstract:** Ongoing climate change and associated food security concerns are pressing issues globally, and are of particular concern in the far north where warming is accelerated and markets are remote. The objective of this research was to model current and projected climate conditions pertinent to gardeners and farmers in Alaska. Research commenced with information-sharing between local agriculturalists and climate modelers to determine primary questions, available data, and effective strategies. Four variables were selected: summer season length, growing degree days, temperature of the coldest winter day, and plant hardiness zone. In addition, peonies were selected as a case study. Each variable was modeled using regional projected climate data downscaled using the delta method, followed by extraction of key variables (e.g., mean coldest winter day for a given decade). An online interface was developed to allow diverse users to access, manipulate, view, download, and understand the data. Interpretive text and a summary of the case study explained all of the methods and outcomes. The results showed marked projected increases in summer season length and growing degree days coupled with seasonal shifts and warmer winter temperatures, suggesting that agriculture in Alaska is undergoing and will continue to undergo profound change. This presents opportunities and challenges for farmers and gardeners.

**Keywords:** climate change; agriculture; Alaska; growing degree days; seasonality; plant hardiness zone

**Citation:** Fresco, N.; Bennett, A.; Bieniek, P.; Rosner, C. Climate Change, Farming, and Gardening in Alaska: Cultivating Opportunities. *Sustainability* **2021**, *13*, 12713. https://doi.org/10.3390/su132212713

Academic Editors: Stephanie Pfirman, Gail Fondahl, Grete K. Hovelsrud and Tero Mustonen

Received: 30 September 2021
Accepted: 8 November 2021
Published: 17 November 2021



## 1. Introduction

The relationship between agricultural production and climate change is of particular interest and pertinence in Alaska for several reasons. These include the accelerated pace of climate change in Alaska, the state's current low agricultural food production and associated vulnerability to supply disruptions, remoteness, the lack of diversity in Alaska's highly oil-dependent economy, and the potential for agricultural expansion.

Alaska's high-latitude setting places it at the front lines of environmental change [1,2]. Due in large part to polar amplification [3], the climate is warming in the far north at as much as three times the rate of other regions of the world [4].

Food security is an issue of particular concern in Alaska, in part because the state is remote from the contiguous United States and other agricultural regions; of all the agriculturally-produced food consumed in the state, only five percent is locally grown [5]. Alaska also has many communities which are inaccessible by road, and are thus vulnerable to interruptions in food supply [6]. Alaska has ample arable land and fresh water, and yet lags far behind northern European nations in terms of agricultural self-sufficiency, which places it at a high risk for catastrophic disruptions to supply chains [7].

Rising temperatures, altered precipitation regimes and associated shifts in growing degree days, summer season length, and the timing of spring thaw and autumn frost are among the factors that are rapidly altering natural ecosystems and agricultural opportunities [2,8]. The ability of Alaskans to predict these changes will profoundly affect their ability to adapt. The State of Alaska recognizes the scope and magnitude of these changes and has made it a priority to ensure Alaskan communities and managers incorporate anticipated change into local and regional planning. This is also a goal of the University of Alaska (UA), which seeks to apply and advance its expertise in climate science and landscape ecology to better understand the manner by which these changes affect ecosystems, food webs and human populations.

Thus far, adaptation to climate change in the agricultural sector has been slow, and studies suggest that such adaptation does not occur until farmers perceive the importance and immediacy of climate change [9]. Farmers' perceptions of climate change have been identified as an important factor for adaptation to take place [10–12] as it triggers the necessary changes that are needed for action, in addition to other factors [13].

Currently, agriculture in Alaska is climate-limited. Sparrow et al. [14] note that primary limitations include low heat energy, short growing seasons, and cold winters that prevent survival of perennial crops. Considerable research (e.g., [15–17]) has assessed how to overcome some of these limitations, particularly in the context of food security.

The scientific consensus suggests that climate change is already altering the equation, and will continue to do so [18]. Hatch [19] found that climate projections show that future growing conditions in the Fairbanks North Star Borough may be more similar to northern prairies in the lower 48 states. Sparrow [16] found that increases in growing degree days (GDD) could cause crop production to advance northward throughout the century, with increases in yields and new varieties becoming viable. Meanwhile, for some crops, climate change may not be positive. For example, burgeoning peony markets are dependent on Alaska's relatively cool climate and late summer season.

Lader et al. [20] investigated some of the potential impacts of climate warming on northern agriculture. Their research used climate projections based on regional dynamical downscaling using the Weather Research and Forecasting (WRF) Model. They used these model outputs to assess changes in growing season length (GSL), spring planting dates, and potential occurrences of plant heat stress for five regions in Alaska. Expanding the use of these data by adding additional variables and the full spatial extent of the state, this project's goal was to provide real-world tools that stakeholders around Alaska can use to plan for and adapt to agricultural change.

Currently available tools are limited in their ability to help Alaska's farmers adjust to climate change. USDA hardiness zones are relatively fine-scale within Alaska, but are based only on extreme winter temperature; thus, they serve as a reliable metric only for plants affected and limited by winter extremes. Indicator plants, as defined by the USDA, help to capture some of the nuances of range limits. However, with ongoing climate change, both winter extremes and indicator species may shift and change. Moreover, for many species, particularly annual crops, other climate indicators are likely to provide more pertinent hardiness information. Communication with stakeholders can aid researchers in creating climate change assessments that address real-world concerns. Moreover, a meaningful assessment must take into account both positive and negative potential changes, including new opportunities and new stressors.

Peony farming serves as an excellent case study for research on climate change and agriculture in Alaska, because peonies represent a burgeoning niche market, and are a crop that is uniquely lucrative in Alaska for reasons linked directly to the climate. Peonies bloom in Alaska in July, August, and September and are available commercially nowhere else in the world during this time. Commercial peony farming has seen considerable growth in recent years. There are over 100,000 peony roots in the ground on peony farms in Alaska and farmers are continuing to add roots at over 30,000 roots per year, with gross sales of well over a million dollars. Peony growers engaged in the project expressed concern about seeing shifts toward earlier blooming times, which puts Alaska's peonies in more direct competition with other markets.

The shifts in the Arctic climate will likely produce a range of impacts on different crop species, but through the development of decision support tools, those affected have a greater ability to prepare for changes proactively. This project drew upon local knowledge and the best available climate modeling techniques to build user-friendly tools that deliver practical information to farmers, ranchers, forest landowners, and Alaska Native communities to help them to adapt to climate change. The project demonstrated that over the coming years and decades profound shifts are likely in growing season length, growing degree days, and winter temperatures, and it linked these changes with potential shifts in key crops.

## 2. Materials and Methods

The project included four major stages:

1. collaborative research and information-sharing in order to determine primary questions, available data, and effective strategies;
2. development of datasets, models, and tools, to address these primary questions;
3. interpretation and refinement of models and tools in order to maximize their utility and effectiveness; and
4. final interpretation and dissemination of results and outcomes to project partners and to the public.

In the first stage, we fully reviewed the existing literature and the potential applications for existing downscaled climate projections. We then met (in person, via video conference, and by email) with collaborative partners from around Alaska, representing family-owned community-sustained farms (Calypso Farm and Ecology Center and Spinach Creek Farm), small-scale commercial peony growers (Arctic Alaska Peonies and The Alaska Peony Growers' Association), a knowledge-sharing program between UAF and rural Alaska communities (Community Partnerships for Self-Reliance) and a Tribal conservation organization dedicated to the wise use of natural resources (Tyonek Tribal Conservation District). These partners had all indicated interest in the project prior to it being successfully funded, and had provided letters of support. Some, including the peony growers, had previously approached us with their questions, while others were networking contacts with known interest in climate change research, sustainability, and community self-reliance.

Given the collaborative, rather than top-down, nature of the research, discussions were open-ended. We discussed the development of user-friendly tools that could be developed for the project. However, we did also seek each collaborators' specific thoughts regarding the greatest climate-related factors related to successful farming, with the expectation that these answers would vary by region and expertise but would reveal key patterns.

From these interactions we determined the following priorities. Those marked by a star (1, 2, and 7) were selected for modeling, based on the availability of appropriate climate data. Those not examined in this study may be pursued in future research.

1. Growing degree days (GDD) are an important variable for determining crop viability. It is optimal to map GDD spatially, given that elevation, slope, and aspect are all key. Peonies and brassica (broccoli/cabbage family) are more successful with cooler temperatures (lower GDD), while other flowers and vegetables (e.g., squash) are more successful with higher GDD. *
2. Total season length and/or the date of the first and last frost are crucial. This affects not only crops that need a long growing season (e.g., squash) but also crops that farmers like to stagger with multiple plantings, and plants for which the timing of harvest is key in order to compete in markets. *
3. Soil temperature, particularly in spring, is very important.
4. Cold spring temperatures in general are difficult for farmers.
5. Drought or constant rain are problematic.
6. For perennials, the timing of when snow arrives is an important factor.
7. The coldest monthly mean temperature (often January) and coldest winter temperature are of interest for perennials. *

The selected questions led directly to the data analysis, modeling, and tool creation described below.

Climate station data is limited in Alaska, necessitating the use of downscaled gridded reanalysis data for both historical and future projections. The climate projections used in this project were derived from regional dynamically downscaled data produced with the Advanced Research core of the Weather Research and Forecasting Model (WRF) [21]. The model simulations were driven by multiple climate datasets for past time periods, including ERA-Interim data [22], the Geophysical Fluid Dynamics Laboratory Climate Model (GFDL), version 3 [23], and the National Center for Atmospheric Research (NCAR) Community Climate System Model, version 4. Two different sets of modeled data were used in this project, referred to as GFDL and NCAR, based on these two different Global Circulation Models, in order to represent the range and uncertainty associated with use of climate projections. While both models have been shown to be highly valid in northern latitudes [1], GFDL data tend to project greater changes in temperature, while NCAR outputs are more conservative. The full dynamical downscaling methodology and WRF configuration are described in Bieniek et al. [24] and Lader et al. [20]. Data were downscaled to 20 km spatial resolution; thus, all community-specific data described in this analysis can be understood to represent the 20 km grid cell that best represents the community locations. The projected data used the 8.5 RCP (Representative Concentration Pathway) of phase 5 of the Coupled Model Intercomparison Project (CMIP5) [25], as defined by the IPCC. The limitations imposed by spatial resolution and choice of RCP are described in the Discussion section.

Temperature and precipitation data were produced at hourly time resolution. This allowed for fine-scale identification of some of the key variables that were identified by stakeholders. Because we were aiming to highlight climate trends over long periods of time (decades or longer) rather than to accentuate model variability at the annual or sub-annual level, we used decadal means and multi-decadal means in our data visualization tools.

Separate tool interfaces within a dashboard-based website were developed for each selected variable, including one for GDD, one for season length, and two separate tools for visualizing cold conditions. All four resulting tools are available online–in separate tabs—at https://www.snap.uaf.edu/tools/gardenhelper/ (accessed on 15 September 2021) (see Supplementary Materials) with an accompanying explanatory text aimed at a wide range of stakeholders from the general public.

The first tool in the online interface was designed to provide gardeners with past, current, and projected future data estimating growing season length, as defined by the longest time period during which the temperature never drops below a selected threshold Fahrenheit degrees were used throughout the tool, based on user familiarity. This familiarity is also reflected by the fact that Fahrenheit degrees are more commonly referenced in the corresponding agricultural literature. Thus, in the tool these thresholds are defined as "hard frost, 28 °F" (−2.2 °C); "light frost, 32 °F" (0 °C); "Cold crops, 40 °F" (4.4 °C); or "Warm crops, 50 °F" (10 °C). In order to make the tool locally pertinent and accessible, we created drop-down menus to offer users a choice of hundreds of Alaska communities, each linked to the appropriate latitudinal and longitudinal location in the database; a radio-button choice of the NCAR or GFDL model, and a drop-down menu choice of temperature thresholds. The accompanying text explains these choices and interprets the outputs. Based on the literature, gardeners are offered a table with appropriate threshold values and the approximate number of days necessary to produce 24 different annual crops, including a range of popular vegetables and grains. The interpretation includes an explanation for why the data appear variable, even when averaged by decade, as well as an explanation for the use of two different GCMs.

The second component of the tool focuses on daily minimum temperatures—estimates of record-breaking cold. We created an interface such that for user-selected locations and a user-selected model, as described above, a graph is generated showing the modeled data based on the coldest temperature ever recorded or projected for a chosen location, date (e.g.,

12 February or 18 July) and time period. For the purposes of this interface, we aggregated the data into color-coded thirty-year ranges to represent climatologies: 1980–2009, 2010–2039, 2040–2069, and 2070–2099. This allows users to see the clear distinctions between past, current, near-future, and far-future projections. The accompanying text explains how to use and interpret the interface.

The third tool interface calculates GDD, a metric commonly used to estimate how much heat is available and useable to crops. There are several methods for calculating GDD. Based on the available modeled data, and in order to create a user-friendly online tool with the greatest possible flexibility and clarity, we used a method in which we took the average of the daily high and daily low temperatures, and subtracted the user-selected baseline value from that average. In other words, if a user selected a baseline of 50 °F (10 °C), and if the daily high for a particular day was 70 °F (21 °C) and the low was 60 °F (16 °C), the GDD value for that day would be ((70 °F + 60 °F)/2) − 50 °F = 15 °F or in SI units ((21 °C + 16 °C)/2) − 10 °C = 8.5 °C. As in the season length tool, we offer four possible baselines: 28 °F (−2.2 °C), 32 °F (0 °C), 40 °F (4.4 °C), and 50 °F (10 °C). Daily values are cumulatively summed across the summer season, creating graphical outputs. In order to smooth the data and create a reasonable number of future projections, data are averaged by decade. Because heat stress is rare in Alaska, we did not include upper GDD thresholds in our calculations. GDD is not a familiar concept or calculation for many stakeholders, and as such we included adequate explanation in the tool interface to allow for appropriate interpretation. This included tables of sample crops identified by their growth thresholds and necessary GDD values (ten species with a threshold of 32 °F, four at 40 °F, and six at 50 °F; no species were identified with a threshold below 32 °F, but 28 °F was included in the dropdown interface to provide continuity with the growing season length tool.)

In the fourth and final tool interface, we created maps using metrics similar to those used by the USDA to define Plant Hardiness Zones. Here, users do not need to select a location, because all maps cover the full statewide spatial domain. However, users are offered the option of downloading high-resolution individual maps for four current and future time periods, or viewing all four simultaneously. For the purposes of this interface, we aggregated the data into the same thirty-year ranges that were used for the minimum temperature tool: 1980–2009, 2010–2039, 2040–2069, and 2070–2099. However, rather than being based on absolute coldest daily temperatures, hardiness maps are based on the average annual minimum winter temperature. In order to aid user interpretation, we matched map colors and labeling schemes to those used in USDA maps.

Finally, in addition to the creation of the above tool components, we assessed the potential changes to peony crops in Alaska, based on our climate projections coupled with data available from the literature and insights on agricultural research on peonies—particularly research conducted in Alaska by Patricia Holloway and others at UAF's Georgeson Botanical Garden. This included data on dormancy, stem growth, flowering, and seasonal timing.

## 3. Results

### *3.1. Length of Growing Season*

An example of the tool outputs for the length of growing season is shown in Figure 1. Although this figure shows only a single location (Fairbanks, the location of the researchers and several stakeholders engaged in this project), threshold (32 °F) and model (GFDL), it is typical of the full range of results in several ways. First, it clearly demonstrated, even to a casual viewer, that the summer growing season is projected to get longer over time. Second, it shows that this increase is likely to occur at both ends of the season, with earlier springs and later autumns. Finally, the results demonstrate that this shift, while obvious at the scale of a century, is somewhat variable or unpredictable, even with decadal averaging.

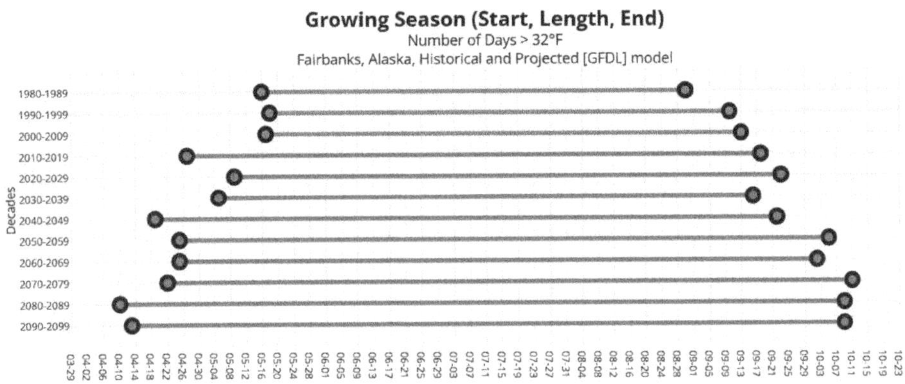


**Figure 1.** Sample of Growing Season Length tool output. Location selected was Fairbanks, temperature threshold selected was 32 °F, and model selected was GFDL.

This tool can return data that, although technically realistic (reflecting likely real-world conditions), may be confusing and not useful to end users. If users select a particularly high temperature threshold, and/or live in a very cold region, the results may appear to be short and uneven, as in Figure 2. This is because the tool finds the longest consecutive period during which the daily minimum temperature never drops below the selected temperature. This time period may be extremely short, and is unlikely to be helpful in determining when to plant crops. Users are cautioned to be sure to select thresholds that make sense for their area. In future tool iterations, feedback from users may help create visualizations that avoid this issue altogether.

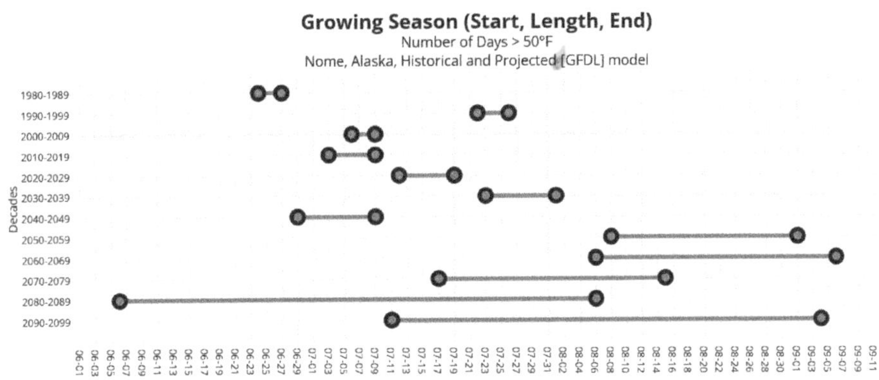


**Figure 2.** Sample of Growing Season Length tool output. Location selected was Nome, temperature threshold selected was 50 °F, and model selected was GFDL.

### 3.2. *Annual Minimum Temperature (AMT)*

Sample output from the AMT interface is shown in Figure 3. In this case, Anchorage (the largest population center in Alaska) and the NCAR model, which tends to project less extreme climate change than the GFDL model, were selected. However, outputs for other locations and for the GFDL model show similar patterns. While variability is high, as can be seen from the scattering of a few extreme values, and while there is considerable overlap between time periods, even with thirty-year time intervals, the overall pattern of projected warming is clear from time period to time period. Also of note is the fact that although warming is projected across all seasons, winter warming is likely to be much greater than summer warming. In this example, by late this century (2070–2099), very few days are expected to be below 10 °F in Anchorage, which is a stark departure from past extreme lows. In future iterations of this tool, greater contrast in dot colors may improve readability.

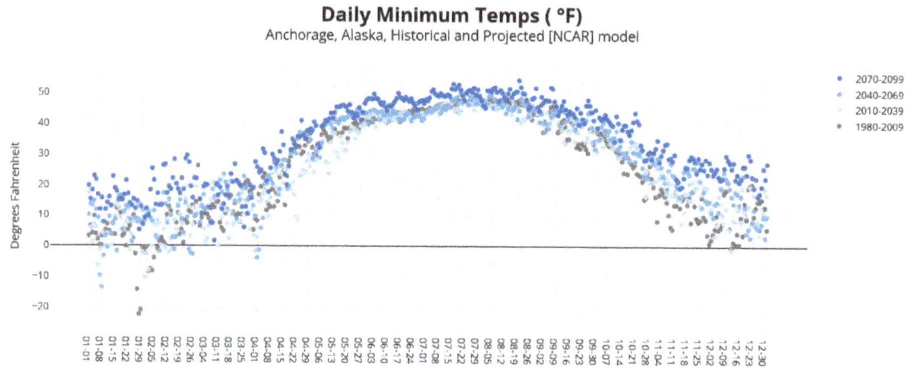


**Figure 3.** Sample of Daily Minimum Temperature tool output. Location selected was Anchorage and model selected was NCAR.

### 3.3. *Growing Degree Days (GDD)*

Sample results for the GDD tool are shown in Figure 4. These outputs are for Igiugig, a small remote village with an active community garden. Again, this example shows many features common to outputs from this tool. Heat units are added up day by day to create a cumulative total. Totals increase from 1875 °F (1042 °C) for the earliest baseline decade (1980–1989) to 4454 °F (2474 °C) for the most distant projected decade (2090–2099). The layout of the graph is designed to make the approximate magnitude of this shift clear even to users who are unfamiliar with GDD.

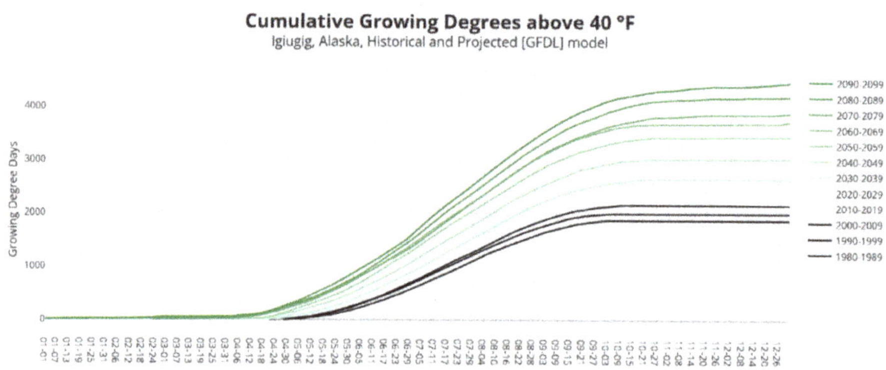


**Figure 4.** Sample of Daily Minimum Temperature tool output. Location selected was Igiugig, threshold was 40 °F and model selected was GFDL.

Plants reach particular growth stages when cumulative GDD reaches the necessary values. However, the minimum GDD necessary for growth and development varies by species and, as such, different lower thresholds are used for the calculation of GDD. Many Alaskan wild plants and cultivated crops are cold-hardy, and can take advantage of all above-freezing days, so for these species GDD can be calculated with a baseline of 0 °C (32 °F). Most crops in other regions have higher baseline temperatures, e.g., 5 °C (about

40 °F) for crops considered suitable for cool climates, such as barley and oats, or 10 °C (about 50 °F) for so-called warm-climate crops, such as corn and tomatoes.

Because many tool users may be unfamiliar with GDD and with the thresholds and total GDD needed for crop growth and maturity, we provided an explanatory text, noting that, "plants can grow when the temperature is above some minimum value, which varies by species. Many Alaska plants are cold-hardy and can grow on all above-freezing days. For these, GDD can be calculated with a baseline of 32 °F. Most crops in other regions have higher baseline temperatures, such as 40 °F for barley and oats, or 50 °F for corn and tomatoes. Choose a threshold based on what crop you plan to grow." We also provided sample tables, based on the literature. These are shown in Table 1, which also shows data on days to maturity. In the online interface, these data are shown in two separate tables associated with the two different tools, in order to avoid confusion.

**Table 1.** Common crops and their associated minimum temperature thresholds, days to maturity, and GDD.

| Baseline Temperature Threshold, °F | Species or Variety | Minimum Number of Days to Maturity | Growing Degree Days to Maturity, °F |
|---|---|---|---|
| 32 | Wheat (hard red) | 90–100 | 2800–3029 |
| 32 | Barley | 60–90 | 2316–2771 |
| 32 | Oat | 85–88 | 2701–3160 |
| 32 | Canary Seed | 95–105 | 2447–2795 |
| 32 | Flax | 85–100 | 2917–3273 |
| 32 | Canola (*B. rapa*) | 73–102 | 2280–2519 |
| 32 | Mustard (*S. juncea*) | 85–95 | 2748–2930 |
| 32 | Chickpea | N/A | 3054–3277 |
| 32 | Lentil | 85–100 | 3164–3408 |
| 32 | Sunflower | 80–120 | 3236–3581 |
| 40 | Wheat (Indiana) | N/A | 2100–2400 |
| 40 | Broccoli from starts | 46 | 1623–1702 |
| 40 | Beets | 40 | N/A |
| 40 | Brussels sprouts | 90 | N/A |
| 40 | Cabbage | 45 | 1623–1702 |
| 40 | Carrots | 60 | N/A |
| 40 | Cauliflower | 45 | 1623–1702 |
| 40 | Radish | 25 | N/A |
| 40 | Spinach | 39 | N/A |
| 40 | Kale | 25 | N/A |
| 40 | Peas | 60 | N/A |
| 50 | Sorghum | 90–120 | 1690–1944 |
| 50 | Soybeans | 100 | 1679–1992 |
| 50 | Cucumber | 60 | 682–952 |
| 50 | Sweet corn | 80 | 1134–1522 |
| 50 | Tomatoes | 60 | 1700 + |

### *3.4. Hardiness Zones*

The USDA uses Plant Hardiness Zones as the standard by which growers can determine which plants are likely to thrive at a given location. Many seed manufacturers reference these zones. Hardiness maps are based on the average annual minimum winter temperature. These zones are only a rough guide. Because they are based on winter temperatures, they are of greatest importance for perennials, such as fruit trees or peonies. The four maps shown in Figure 5 use nomenclature and color ramps similar to those used in USDA maps in order to render them more familiar to gardeners and farmers who are accustomed to the USDA zone delineations. These maps represent current estimates of hardiness zones in Alaska, plus projections of how these zones may look in three future time periods, as described in the Methods section.

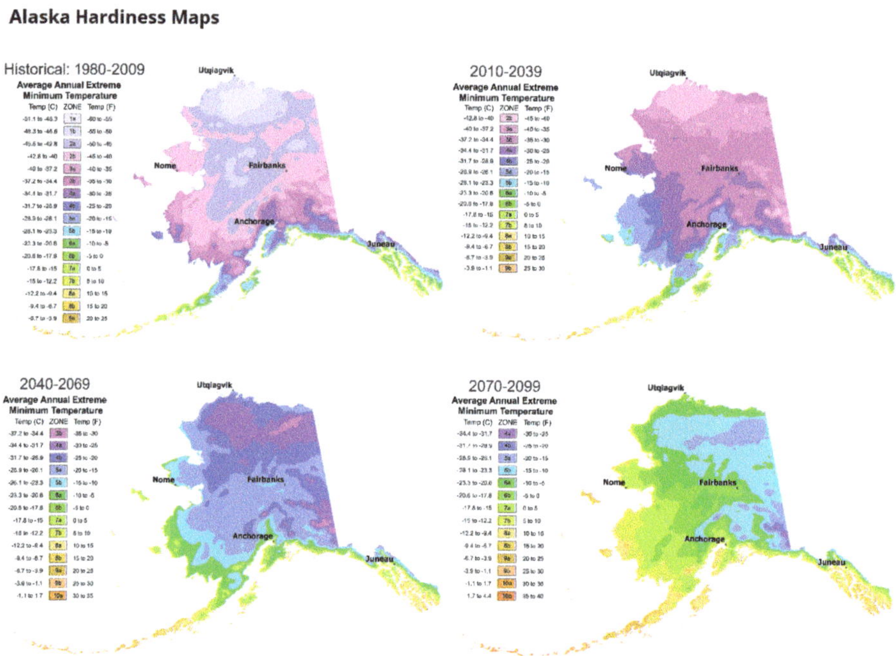


**Figure 5.** Alaska Hardiness Maps.

### *3.5. Peony Case Study*

This case study was shared directly with project partners and was summarized for the public and made available here online [26]. The summary highlights some clear continuing advantages for Alaska peony growers, including the fact that although Alaska winters are likely to remain cool enough for peony dormancy, the same may not be true for growers elsewhere. However, it also highlights challenges, particularly for growers in the parts of the state with the warmest summers, as increased spring heat spurs earlier blooming and diminishes the late-summer niche enjoyed by Alaska growers.

Previous studies show that relatively cool winter temperatures are necessary for peony roots to achieve dormancy. However, provided that temperatures are consistently below 6 °C (43 °F) for seventy days, dormancy will be achieved [27]—a condition easily met in most of Alaska. Plants break dormancy as soon as temperatures rise above freezing [28–30]. Byrne and Halevy [27] report that flowering can occur in only about 50 days in greenhouse conditions, but suggests that slower growth in cooler temperatures results in less atrophy of buds. Indeed, Kamenetsky et al. [31] found that moderate temperatures with highs of 72 °F and lows of 50 °F were best for enhancing stem length and flowering. When daily highs and lows were 82 °F and 72 °F, flowering was drastically reduced. Hall [27] similarly found that temperatures over 77 °F resulted in reduced blooms. Holloway et al. [28–30] found that flowers bloomed in all cases when cumulative GDD above a 32 °F threshold reached between 1734 and 2313. In contrast, the number of days from bud emergence to first cutting ranged from 32 in Fairbanks to 79 in much cooler Kenai.

A further case study linking tool outputs to existing or planned community gardens would offer an excellent area for future investigation.

## 4. Discussion

These tools have already been discussed, shared, and used as teaching and presentation materials within the Alaska agricultural community, particularly by project partners and participants associated with Cooperative Extension Services and/or the peony growing industry. The results of the peony case study were presented at the annual meeting of the Alaska Peony Growers' Association in 2020. Outcomes were included in a presentation in June 2021 by Dr. Glenna Gannon and Shannon Powers which focused on Variety Trials in the Matanuska Susitna region. In addition, the Fairbanks Daily Newsminer ran a feature on the tool [32].

Given the goals of this project in relation to stakeholder needs, small-scale gardens and farms, and food security, all of the results must be interpreted within the simultaneous contexts of users' ability to successfully access the information, correctly understand and interpret the information, and apply the information.

One overarching aspect of model transparency is the clear explanation of data uncertainties. With this in mind, uncertainties are clearly explained in plain language in conjunction with all outputs, including online tools and fact sheets in order to avoid misinterpretation and misapplication. Across all outputs, some uncertainties can be attributed to underlying differences in the complex atmospheric modeling used in the GCMs. By offering two models, we gave users a chance to explore a model that tends to produce more extreme results, and a model that tends to produce conservative results. Additional uncertainty stems from spatial limitations. As noted, all model outputs are at 20 km resolution. Especially in areas of complex topography, growing conditions can vary enormously across areas of this size. As such, users are reminded to consider their local microclimate. Uncertainties inherent to short-term variability in weather are inherent to agriculture. While averaging across decadal or multi-decadal time periods helps smooth data for the purposes of highlighting long-term trends, users are cautioned that short-term variability will nonetheless play a large role in year-to-year gardening and farming outcomes.

All of the tools developed during this research were intended to improve the current state of information readily and easily available to Alaska gardeners. For example, with regard to the season length tool, many seeds offer estimates of how many days the crop may take to mature. Typically, planting guides refer to "last frost" in spring and "first frost" in fall, implying daily minimum temperatures of 0 °C (32 °F). By offering additional thresholds, our tool allows for more flexibility in considering cold-hardy crops that may be harvested only when a hard frost is reached (28 °F), or more delicate crops that cannot effectively grow when temperatures are below a higher threshold. Such plants might be kept as starts in a greenhouse until a later planting date, and harvested earlier. Moreover, the results show that season length is increasing statewide. In many regions, longer frost-free seasons may make it possible to plant crops that were not previously suitable for the region.

Very little information on GDD is currently available to gardeners who do not read the scientific literature. Understanding GDD and knowing the approximate number of growing degree days that can be expected in an area, for a given baseline temperature, can help gardeners plan what to plant, and what not to plant, especially when the length of the frost-free season does not provide enough information. For example, with a baseline temperature of 50 °F and over 2000 GDD necessary for maturation, corn is not likely to be successful in most parts of Alaska, even though many varieties can mature in only 60–80 days, given enough heat. However, the results indicate that GDD is shifting rapidly and dramatically statewide, with values projected to double or even triple by the end of the century. This may prove a productive avenue for additional study and seed trials.

Both the Annual Minimum Temperature tool and the Hardiness Zone Maps offer important information for those who are interested in perennials, such as fruit trees and shrubs, which have to be hardy to survive Alaska winters. Many cannot withstand temperatures below certain thresholds—but model results make it clear that these thresholds are changing rapidly statewide. The results suggest that many perennials that were not suitable to Alaska may soon become potential crops in large areas of the state. This may prove to be an important area for further research and experimentation. However, tool users are reminded that "cold hardiness" is just one gauge of whether a crop is suitable to a particular region. Many other factors affect winter survival, such as the insulating value of snow, the moisture content of the ground, the presence or absence of permafrost, and the number of freeze–thaw cycles that occur. Future versions of this tool may include some of these factors.

Alaska's peony growers may see both gains and losses due to climate change. Winters are likely to remain cool enough for peony dormancy, while growers in other parts of the world may find challenges in this regard. This may provide some local advantage. Using the Growing Season tool to look at the 32 °F threshold can help to provide an estimate of when peonies are likely to break dormancy in the future in communities around Alaska. However, late springs and cool summer weather are better for Alaska's peony growers for two reasons: first, because such conditions promote healthier flowering, and second, because they promote later flowering, which allows Alaska to capture the late-season niche market. Given the results obtained by Holloway et al. [28–30], using the Growing Degree Days tool to plan for peony growth once buds have emerged is likely to be more effective than using the Growing Season tool. For peony farming, as late springs and cool summers become more elusive, growers may need to adapt. This may be hardest for those who already farm in regions of the state that are warmest in the summer, such as Fairbanks. For new growers who have yet to invest in land, picking cooler parts of the state may be practical if relocation is possible, or selecting cooler sites within a community—such as north-facing slopes—might aid at the local level. New storage methods for cut blooms can also extend the season for sales.

Taken together, the outcomes of this research, as well as the feedback received by tool users, point toward potential refinements in tool development as well as many new possible avenues for expansion of Alaska's agricultural potential at the local scale. Especially in the context of climate-related agricultural uncertainty, challenges in other regions and possible climate-related needs for greater local autonomy (due to disruptions in supply chains and/or reduced use of fossil fuels for transportation of crops), and the need to diversify Alaska's economy, such new opportunities for farms and gardens, may prove important areas for further study and development.

**Supplementary Materials:** The Alaska Garden Helper tool described in this article is available online at https://www.snap.uaf.edu/tools/gardenhelper/ (accessed 3 October 2021).

**Author Contributions:** Conceptualization, N.F. and A.B.; methodology, N.F. and A.B.; software, A.B.; validation, N.F., A.B., P.B., and C.R.; formal analysis, N.F. and A.B.; investigation, N.F.; resources, N.F., A.B., P.B. and C.R.; data curation, A.B. and P.B.; writing—original draft preparation, N.F.; writing—review and editing, N.F., A.B., P.B. and C.R.; visualization, A.B. and C.R.; supervision, N.F.; project administration, N.F.; funding acquisition, N.F. All authors have read and agreed to the published version of the manuscript.

**Funding:** This research was funded by the United States Geological Survey (USGS) via the Alaska Climate Adaptation Science Center. The APC was also funded by the USDA via the Alaska Climate Adaptation Science Center.

**Informed Consent Statement:** Not applicable.

**Data Availability Statement:** The data presented in this study are openly available in the Scenarios Network for Alaska and Arctic Planning (SNAP) Data Portal at http://ckan.snap.uaf.edu/dataset (accessed 3 October 2021).

**Acknowledgments:** The authors would like to acknowledge the support of the faculty and staff at UAF's International Arctic Research Center. All work on this project was built upon past efforts in model selection, model downscaling and data analysis. We would also like to thank all the agriculturalists who contributed their expertise and advice, including the Alaska Peony Growers Association, Community Partnerships for Self-Reliance, Calypso Farm and Ecology Center, Spinach Creek Farm, Arctic Alaska Peonies Co-op, and Tyonek Tribal Conservation District.

**Conflicts of Interest:** The authors declare no conflict of interest. The funders had no role in the design of the study; in the collection, analyses, or interpretation of data; in the writing of the manuscript, or in the decision to publish the results.

## References


1. Walsh, J.; Chapman, W.L.; Romanovsky, V.; Christensen, J.H.; Stendel, M. Global Climate Model Performance over Alaska and Greenland. *J. Clim.* **2008**, *21*, 6156–6174. [CrossRef]
2. Wolken, J.M.; Hollingsworth, T.N.; Rupp, T.S.; Chapin, F.S.; Trainor, S.F.; Barrett, T.M.; Sullivan, P.F.; McGuire, A.D.; Euskirchen, E.S.; Hennon, P.E.; et al. Evidence and implications of recent and projected climate change in Alaska's forest ecosystems. *Ecosphere* **2011**, *2*, 1–35. [CrossRef]
3. Bekryaev, R.V.; Polyakov, I.V.; Alexeev, V. Role of Polar Amplification in Long-Term Surface Air Temperature Variations and Modern Arctic Warming. *J. Clim.* **2010**, *23*, 3888–3906. [CrossRef]
4. AMAP. *Arctic Climate Change Update 2021: Key Trends and Impacts. Summary for Policy-Makers*; Arctic Monitoring and Assessment Programme (AMAP): Tromsø, Norway, 2021.
5. Ken, M.; Goldenberg, M.P. Building Food Security in Alaska. Crossroads Resource Center. Minneapolis. Commissioned by the Alaska Department of Health and Social Services, with collaboration from the Alaska Food Policy Council. 2014. Available online: https://www.crcworks.org/akfood.pdf (accessed on 28 September 2021).
6. USDA/NASS. 2014 Census of Horticultural Specialties. United States Department of Agriculture National Agricultural Statistics. Service. 2014. Available online: https://www.nass.usda.gov/Publications/AgCensus/2012/Online_Resources/Census_of_Horticulture_Specialties/ (accessed on 8 September 2021).
7. Stevenson, K.T.; Alessa, L.; Kliskey, A.D.; Rader, H.B.; Pantoja, A.; Clark, M. Sustainable Agriculture for Alaska and the Circumpolar North: Part I. Development and Status of Northern Agriculture and Food Security. *Arctic* **2014**, *67*, 271. [CrossRef]
8. Swenson, N.Y. Modeling Changes in the Length of the Agricultural Growing Season in Interior Alaska. Master's Thesis, University of Alaska, Fairbanks, AK, USA, 2013. Available online: http://hdl.handle.net/11122/4604 (accessed on 2 September 2021).
9. Lyle, G. Understanding the nested, multi-scale, spatial and hierarchical nature of future climate change adaptation decision making in agricultural regions: A narrative literature review. *J. Rural Stud.* **2015**, *37*, 38–49. [CrossRef]
10. Abid, M.; Schilling, J.; Scheffran, J.; Zulfiqar, F. Climate change vulnerability, adaptation and risk perceptions at farm level in Punjab, Pakistan. *Sci. Total Environ.* **2016**, *547*, 447–460. [CrossRef] [PubMed]
11. Feola, G.; Lerner, A.; Jain, M.; Montefrio, M.J.F.; Nicholas, K. Researching farmer behaviour in climate change adaptation and sustainable agriculture: Lessons learned from five case studies. *J. Rural Stud.* **2015**, *39*, 74–84. [CrossRef]
12. Wheeler, S.; Zuo, A.; Bjornlund, H. Farmers' climate change beliefs and adaptation strategies for a water scarce future in Australia. *Glob. Environ. Chang.* **2013**, *23*, 537–547. [CrossRef]
13. Mugi-Ngenga, E.W.; Mucheru-Muna, M.W.; Mugwe, J.N.; Ngetich, F.K.; Mairura, F.S.; Mugendi, D.N. Household's socio-economic factors influencing the level of adaptation to climate variability in the dry zones of Eastern Kenya. *J. Rural Stud.* **2016**, *43*, 49–60. [CrossRef]
14. Sparrow, S.; Lewis, C.E.; Juday, G.P. *Climate Change and High Latitude Agriculture*; ACIA: Fairbanks, AK, USA, 2007; Available online: http://66.160.145.48/coms/cli/uaf_stephan_sparrow.pdf (accessed on 23 February 2017).
15. Holloway, P.S. *The Challenge of Cultivating Plants in Cold Soils*; Georgeson Botanical Notes No. 12; University of Alaska Fairbanks Agricultural and Forestry Experiment Station: Fairbanks, AK, USA, 1993.
16. Van Veldhuizen, R.M.; Knight, C.W. *Performance of Agronomic Crop Varieties in Alaska 1978–2002*; AFEF Bulletin 111:110; School of Agriculture and Land Resources Management, Agricultural and Forestry Experiment Station: Fairbanks, AK, USA, 2004.
17. Caster, C. Assessing Food Security in Fairbanks, Alaska: A Survey Approach to Community Food Production. Ph.D. Dissertation, University of Alaska, Fairbanks, AL, USA, 2011.
18. Juday, G.P.; Barber, V.; Duffy, P.; Linderholm, R.T.S.; Sparrow, S.; Yarie, J. Forests, land management, and agriculture. *Arct. Clim. Impact Assess.* **2005**, 205. Available online: https://acia.amap.no/ (accessed on 2 September 2021).
19. Hatch, E. Micro-Hardiness Agriculture Zones in the North Star Borough, Alaska: Past and Future Scenarios. Ph.D. Dissertation, University of Alaska, Fairbanks, AK, USA, 2011.
20. Lader, R.; Walsh, J.E.; Bhatt, U.S.; Bieniek, P.A. Agro-Climate Projections for a Warming Alaska. *Earth Interact.* **2018**, *22*, 1–24. [CrossRef]
21. Skamarock, W.; Klemp, J.; Dudhia, J.; Gill, D.; Barker, D.; Duda, M.; Huang, X.; Wanf, W.; Powers, J. A Description of the Advanced Research WRF Version 3. 2008. Available online: http://www2.mmm.ucar.edu/wrf/users/docs/arw_v3.pdf (accessed on 13 September 2021).
22. Dee, D.P.; Uppala, S.M.; Simmons, A.J.; Berrisford, P.; Poli, P.; Kobayashi, S.; Andrae, U.; Balmaseda, M.A.; Balsamo, G.; Bauer, P.; et al. The ERA-Interim reanalysis: Configuration and performance of the data assimilation system. *Q. J. R. Meteorol. Soc.* **2011**, *137*, 553–597. [CrossRef]
23. Donner, L.J.; Wyman, B.L.; Hemler, R.S.; Horowitz, L.; Ming, Y.; Zhao, M.; Golaz, J.-C.; Ginoux, P.; Lin, S.-J.; Schwarzkopf, M.D.; et al. The Dynamical Core, Physical Parameterizations, and Basic Simulation Characteristics of the Atmospheric Component AM3 of the GFDL Global Coupled Model CM3. *J. Clim.* **2011**, *24*, 3484–3519. [CrossRef]

24. Bieniek, P.A.; Bhatt, U.S.; Walsh, J.; Rupp, T.S.; Zhang, J.; Krieger, J.R.; Lader, R. Dynamical Downscaling of ERA-Interim Temperature and Precipitation for Alaska. *J. Appl. Meteorol. Clim.* **2016**, *55*, 635–654. [CrossRef]
25. Taylor, K.E.; Stouffer, R.J.; Meehl, G.A. An Overview of CMIP5 and the Experiment Design. *Bull. Am. Meteorol. Soc.* **2012**, *93*, 485–498. [CrossRef]
26. *Peonies in a Changing Climate: A Case Study*; Alaska Climate Adaptation Science Center, University of Alaska Fairbanks: Fairbanks, AK, USA; Available online: https://uaf-snap.org/wp-content/uploads/2020/06/Peony-report.pdf (accessed on 9 September 2021).
27. Hall, A.; Catley, J.; Walton, E. The effect of forcing temperature on peony shoot and flower development. *Sci. Hortic.* **2007**, *113*, 188–195. [CrossRef]
28. Holloway, P.; Hanscom, J.; Matheke, G. *Peonies for Field Cut Flower Production. First-Year Growth*; Agricultural and Forestry Experiment Station Research Prog. Report 41; University of Alaska Fairbanks: Fairbanks, AK, USA, 2003; 4p.
29. Holloway, P.; Hanscom, J.; Matheke, G. *Peonies for Field Cut Flower Production. Second-Year Growth*; Agricultural and Forestry Experiment Station Research Prog. Report 43; University of Alaska Fairbanks: Fairbanks, AK, USA, 2004; 8p.
30. Holloway, P.; Matheke, G.; DiCristina, K. *Peony Phenology in Alaska*; School of Natural Resources and Agricultural Sciences Agricultural and Forestry Experiment Station, University of Alaska Fairbanks: Fairbanks, AK, USA, 2013.
31. Kamenetsky, R.; Barzilay, A.; Erez, A.; Halevy, A.H. Temperature requirements for floral development of herbaceous peony cv. 'Sarah Bernhardt'. *Sci. Hortic.* **2003**, *97*, 309–320. [CrossRef]
32. Riley, J. Alaska Garden Helper: A New Computer Tool for Planting in a Changing Climate. Fairbanks Daily Newsminer, 8 March 2020. Available online: https://www.newsminer.com/features/sundays/gardening/alaska-garden-helper-a-new-computer-tool-for-planting-in/article_88cc0f6a-60ee-11ea-be87-bf18cc1e499f.html (accessed on 16 September 2021).

sustainability MDPI

Essay

# Developing a Sustainable and Inclusive Northern Knowledge Ecosystem in Canada

**Gary N. Wilson**

Department of Political Science, University of Northern British Columbia, 3333 University Way, Prince George, BC V2N 4Z9, Canada; gary.wilson@unbc.ca

**Abstract:** A knowledge ecosystem is a collection of individuals and organizations who are involved in the creation, management and dissemination of knowledge, both in the form of research and lived experience and teaching. As is the case with ecosystems more generally, they thrive on variation and diversity, not only in the types of individuals and organizations involved but also in the roles that they play. For many decades, the northern knowledge ecosystem in Canada was dominated and controlled by Western scholarly approaches and researchers based in academic institutions outside the North. More recently, this research landscape has started to change, largely in response to the efforts of Indigenous peoples and northerners to realize greater self-determination and self-government. Not only have these changes led to the development of research and educational capacity in the North, but they have also changed the way that academic researchers engage in the research process. The keys to maintaining the future sustainability and health of the northern knowledge ecosystem will be encouraging diversity and balance in the research methodologies and approaches used to generate knowledge about the North and ensuring that the needs and priorities of northern and Indigenous peoples are recognized and addressed in the research process.

**Keywords:** knowledge ecosystem; north; research; education

**Citation:** Wilson, G.N. Developing a Sustainable and Inclusive Northern Knowledge Ecosystem in Canada. *Sustainability* **2021**, *13*, 9213. https://doi.org/10.3390/su13169213

Academic Editor: Tan Yigitcanlar

Received: 9 July 2021
Accepted: 13 August 2021
Published: 17 August 2021



## 1. Introduction

In March 2018, Inuit Tapiriit Kanatami (ITK), the national association for Inuit in Canada, released the National Inuit Strategy on Research, a policy document that outlines a vision for the future of research in Inuit Nunangat, the Inuit homeland in Canada [1]. It presents Inuit priorities for research conducted in their territories, including greater Inuit oversight and control over research funding and projects and the information generated by those projects, so that research benefits Inuit communities and regions. It also aims to build research capacity so that Inuit communities can lead and conduct research without having to rely on researchers based at southern post-secondary institutions and research organizations.

The National Inuit Strategy on Research is one example of how the northern knowledge ecosystem is changing and adapting to the new realities created by the efforts of Indigenous peoples to realize greater self-determination. In the circumpolar north, as in many other regions of the world, the knowledge ecosystem, which consists of the individuals and organizations who are involved in all aspects of knowledge production and dissemination, has been dominated by academic, state and corporate organizations and actors. Historically, research was conducted without the consent or input of northern and Indigenous peoples, often in ways that were inconsistent with their values and harmful and disruptive to their ways of life. Although much has changed over the last several decades, as Natan Obed, the President of ITK reminds us, "In this era of reconciliation, research governance bodies, procedures, and practices must be transformed to respect Inuit self-determination in research" [2].

The changes that are reshaping the northern research landscape are driven by both internal and external forces but are ultimately the product of the efforts of northerners and, in particular, Indigenous northerners to exercise greater control over research conducted in the North. These trends are not only happening in Canada, but across the circumpolar north. Many in the academic community, especially the younger generation of academic researchers, support such changes and have taken steps in their own research to collaborate more extensively with northern and Indigenous communities in the research process through the co-production of research and community-based participatory research. Others have gone further by aligning their research with the cause of social justice and substantive political and economic change.

Rather than present a defined body of research, the purpose of this article is to reflect on the current and changing landscape of the northern knowledge ecosystem in Canada. In addition to outlining, in a very general manner, the main elements of this knowledge ecosystem, the article will consider the various roles that academic researchers and educators can play in ensuring the sustainability and inclusiveness of this ecosystem in the future. The northern knowledge ecosystem is complex and draws sustenance and inspiration from a variety of sources. As is the case with any ecosystem, each actor, large and small, contributes in some way to the sustainability of the whole. Sometimes, these contributions go unnoticed or are underappreciated or misunderstood, but they are, nonetheless, important to the ecosystem's future viability.

As a non-Indigenous scholar at a small, northern post-secondary institution, my perspectives on the northern knowledge ecosystem in Canada have been shaped by my experiences living and working in the Canadian provincial north and travelling throughout the Canadian Arctic and circumpolar north. They have also been informed by my research on Indigenous self-governance and politics in the Canadian and circumpolar north and by my involvement in several northern research and education associations. The views represented in this article, however, are my own and do not necessarily represent the opinions and perspectives of these associations.

This article is divided into two parts. Part one begins by considering the concept of a knowledge ecosystem and then reviews the contours of the northern knowledge ecosystem in Canada. Part two examines the opportunities and challenges confronting the northern knowledge ecosystem and speculates about the ways in which it could evolve in the future.

## 2. Northern Knowledge Ecosystem in Canada

Thomson defines a knowledge ecosystem as "the complex and many-faceted system of people, institutions, organizations, technologies and processes by which knowledge is created, interpreted, distributed, absorbed and utilized" [3]. The concept has been used in a variety of academic disciplines to describe the collection of individuals and organizations who are involved in the creation, management and dissemination of knowledge, both in the form of research and lived experience and teaching. As is the case with ecosystems more generally, knowledge ecosystems thrive on variation and diversity, not only in the types of individuals and organizations involved but also in the roles that they play in ensuring the sustainability of the ecosystem as a whole. As anthropogenic climate change has clearly demonstrated, the overwhelming dominance of one actor or force can have dire consequences for ecosystem sustainability. The most sustainable knowledge ecosystems draw strength from a variety of knowledge and perspectives that coexist in an open and transparent forum.

Knowledge ecosystems exist in many different contexts and settings but tend to be dominated by post-secondary institutions and independent research organizations in the public, nongovernmental and private sectors. In the past, universities and other post-secondary institutions have focused primarily on curiosity-driven research and teaching, but more recently, they have been actively involved in a "third mission" that mobilizes their considerable resources to address the pressing social and economic challenges facing society [4]. Regional development has become a common theme of academic research and teaching, especially in remote and peripheralized regions that typically lack knowledge infrastructure and research capacity [5]. In this and other respects, the experiences of northern Canada and the circumpolar north are similar and relevant to those of other remote regions such as northern Australia and Amazonia [6].

The northern knowledge ecosystem in Canada and other parts of the circumpolar north is evolving in ways that suggest it is becoming more sustainable and diverse. For many decades, research and knowledge about the North was dominated and controlled by Western scholarly approaches and researchers based at academic institutions outside the North. This research was disseminated mainly through academic publications and reports that were inaccessible and not very useful to northern and Indigenous communities. Northern and Indigenous knowledge about the North, although known and used by people living in the North, rarely informed wider discussions about northern conditions and priorities, and government policies designed to address northern issues. Over the past several decades, however, the research landscape has started to change. Universities and colleges have been established in the North, providing opportunities for northerners to access post-secondary education without leaving the North. This trend has been supported and supplemented by online and remote programming; although the challenges of accessing the Internet in many remote, northern communities has limited the impact of such programming. In addition to opening the North to the world, northern post-secondary institutions focus their attention on the pressing issues and needs of the northern regions and communities they serve. For example, they were among the first post-secondary institutions to establish research and academic partnerships with northern and Indigenous communities and, in doing so, started to challenge the existing orthodoxies around research approaches and methodologies.

In certain respects, Canada has lagged behind other countries in the circumpolar north when it comes to nurturing the development of post-secondary institutions in the North. Countries such as Russia and Norway have invested heavily in educational infrastructure and programming in northern regions, establishing world-class institutions of higher learning in the Arctic. The educational landscape in Canada, however, has evolved over the last several decades, starting with the establishment of universities in the provincial norths (for example, the University of Northern British Columbia, University College of the North in Manitoba and Université du Québec en Abitibi-Témiscamingue) and then expanding to the territorial north. The recent transition of Yukon College to Yukon University represents the first of a number of expected developments in post-secondary education in the Canadian territorial north [7]. The establishment of Yukon University was done in full consultation and partnership with Indigenous peoples in Yukon and was a response to a longstanding demand from Yukon First Nations, dating back to the early 1970s, for a university in the territory [7]. The post-secondary landscape in northern Canada is complemented and diversified by other educational institutions and opportunities such as the Dechinta "bush university", a land-based educational initiative that is run by the Dechinta Centre for Research and Learning and recently received CAD 13 million in federal funding over five years [8].

In Canada, the gradual expansion of post-secondary education in the North has also led to the establishment of northern research institutes whose primary role is to oversee research being conducted in particular region or territory by academic institutions and organizations and to encourage northern-led research projects. An example is the Aurora Research Institute in the Northwest Territories. In some cases, these institutes have developed research ethics principles and protocols, similar to those at southern post-secondary institutions, to monitor the research process and provide avenues for northern and Indigenous involvement throughout. Northern focused institutes are also based in "southern" institutions. These institutes facilitate the mobilization of significant research capacity at large post-secondary institutions in the interests of northern research. Some examples include Centre d'Études Nordiques at Université Laval in Québec City, UAlberta North at the University of Alberta in Edmonton and the Arctic Institute of North America at the University of Calgary. Some Indigenous communities and regions have established their own research institutes and ethics processes to provide oversight of and approval for the research being conducted in their territories. Examples from Inuit Nunangat include the Nunatsiavut Research Centre, the Nunavik Research Centre and Inuit Qaujisarvingat (the Inuit Knowledge Centre), which is connected to ITK. For Indigenous governments and organizations, this represents an important and necessary step towards building critical research capacity and infrastructure, focusing attention on issues and topics that are important and meaningful to Indigenous communities, ensuring Indigenous control over the data generated by research and, ultimately, contributing to the self-determination of Indigenous peoples [2].

While researchers at post-secondary institutions outside the North still dominate research agendas and funding competitions, they are increasingly doing so in partnership with northern and Indigenous organizations and communities. Co-production and community-based participatory research have become regular and, in some cases, expected methodological approaches to conducting research in the North. Many academic researchers, especially early-career scholars, view research as a partnership with Indigenous rights-holders and northern stakeholders and invest time and energy in relationship-building prior to starting the research. Such investments often lead to research projects and agendas that respond to northern and Indigenous priorities.

Post-secondary institutions and other research organizations based outside the North have also supported northern development through research and pedagogical programming designed to strengthen the capacity of northern and Indigenous communities. One of many examples is the Community-Based Teacher Education Program, a partnership between the Faculty of Education at Memorial University, the Labrador Institute, the Nunatsiavut Government and other local organizations. The aim of the program is to focus on primary and elementary teacher education in a northern and Indigenous context in order to enhance Nunatsiavut's educational capacity [9]. In addition to reaching out to northern regions and communities, post-secondary institutions have also started to address longstanding and entrenched systems of colonization by reforming their internal structures and processes and engaging in processes of Indigenization as part of a broader effort to create a more inclusive academic environment that respects and values Indigenous ways of knowing and being [10].

The northern knowledge ecosystem is supported by a variety of organizations and agencies, including governments, philanthropic organizations and academic associations, which provide funding and other resources for research projects and initiatives. In Canada, a considerable amount of funding for northern research is channeled through arms-length government bodies such as the Tri-Council agencies: Social Sciences and Humanities Research Council (SSHRC); Natural Sciences and Engineering Research Council (NSERC); the Canadian Institutes of Health Research (CIHR). In the past, these agencies have been accused of perpetuating exploitative and colonial research relationships; for example, individuals or organizations that are not based at a recognized post-secondary institution were not allowed lead research projects as Principal Investigators (PIs). In a recent statement, however, the SSHRC said that it is "committed to supporting and promoting research by and with Indigenous peoples. This commitment emphasizes the importance of Indigenous perspectives and knowledge systems to increase and expand our knowledge and understanding about human thought and behaviour in the past and present, as well as the future" [11]. This commitment is supported by other initiatives such as the Tri-Council Policy Statement: Ethical Conduct for Research Involving Humans—TCPS2 (2018) [12].

The Tri-Council agencies' funding mandate is broad and extends beyond the North. Other government agencies and programs, however, focus specifically on supporting northern research and knowledge production and dissemination. Polar Knowledge Canada, which falls under the jurisdiction of Crown–Indigenous Relations and Northern Affairs Canada, "is responsible for advancing Canada's knowledge of the Arctic, strengthening Canadian leadership in polar science and technology, and promoting the development and distribution of knowledge of other circumpolar regions, including Antarctica" [13]. It operates the Canadian High Arctic Research Station (CHARS), in Cambridge Bay, Nunavut, an important part of Canada's Arctic research infrastructure. (Other important northern and Arctic research stations include the Churchill Northern Studies Centre in northern Manitoba and the Labrador Institute Research Station in Newfoundland and Labrador). Polar Knowledge Canada also houses the Northern Scientific Training Program (NSTP), a research funding program for students doing fieldwork in the North.

There are a number of Canadian academic associations focused on northern and Arctic research and education, including the Association of Canadian Universities for Northern Studies (ACUNS) and ArcticNet. Although the specific mandates and foci of these associations differ, they share a commitment to promoting research and knowledge creation in, about and with the North and play a critical role in connecting the various actors within the northern knowledge ecosystem. For over 40 years, ACUNS has served as a network for post-secondary institutions engaged in northern research and education. It has supported the next generation of northern researchers through its administration of grants and other funding opportunities for students and early career scholars and engaged in outreach to northern and Indigenous organizations to learn about and promote their research priorities [14,15]. It recently unveiled a new five-year strategic plan that is designed to align its goals more closely to the changing northern knowledge landscape [16]. Research and education in the North are further supported by philanthropic and charitable organizations, such as the Walter and Duncan Gordon Foundation, the W. Garfield Weston Foundation and, most recently the Mastercard Foundation. In addition to targeting northern research in their funding portfolios, their activities also support the development of northern research infrastructure and encourage collaborations between organizations, decision makers, rights-holders and stakeholders. These organizations often work in partnership with academic associations, post-secondary institutions and government agencies.

We should also remember that the northern knowledge ecosystem in Canada does not exist in geographic isolation; it is connected to circumpolar and international networks of institutions, associations and organizations involved in northern research and scholarship. One of the clearest examples of such a network is the University of the Arctic (UArctic), a consortium of over 200 post-secondary institutions, research institutes and other organizations. The UArctic promotes education and research in an about the North by building and strengthening "collective resources and infrastructures that enable member institutions to better serve their constituents and regions" [17].

As with any ecosystem, the northern knowledge ecosystem in Canada is dominated by large entities which leave a significant footprint on the landscape. It is important to remember, however, that all of these entities are comprised of individuals who perform multiple roles and contribute different forms of knowledge. This includes local knowledge holders who have experience living and working in the North, as well as individuals who have more formal educational and research training. I want to focus on one group of individuals—post-secondary educators—as a means of dispelling some myths about their place within the northern knowledge ecosystem. In the past, educators, especially those based in the South, have been derided as agents of colonization or as esoteric "experts" holed up in their ivory towers and far removed from the issues and challenges facing northerners. Neither of these myths are true anymore. In fact, the vast majority of educators are dedicated to supporting and helping the North and its inhabitants. With the growth of northern post-secondary institutions and research organizations, some are residents of the North and are active in the life of the communities where they live. Those who live outside the North often spend significant amounts of time and effort building relationships with northern communities, conducting research that is beneficial to the North and volunteering for the organizations and associations discussed earlier, whose mandates are to promote research and education in the North and improve the lives of northerners.

One of the most important roles played by post-secondary educators and, indeed, one that is often overlooked is to inform post-secondary students and the broader public about the issues confronting the North. The vast majority of Canadians have never travelled to, let alone having lived or worked in, the North and have little or no knowledge about the challenges and opportunities facing northern and Indigenous communities. The concerns of the North tend to get lost in the political and popular discourse, which is often dominated by the issues that are important in larger urban centers in the South. Post-secondary educators whose research focuses on the North are involved in educating students about the North and why it is important. It may seem like a very small and inconsequential act compared to the important work being done in the North by organizations and governments but raising awareness about the North among students, a group of people who will be the country's future political leaders, businesspeople, activists and citizens, will pay dividends in the years ahead. The impact of post-secondary educators also extends beyond academia. They are often called upon by the media to provide analysis of issues and events that are affecting the North. In doing so, they are able to reach a much wider audience of people who otherwise would not be aware of what is happening in the North. By engaging in public discourse about the North, post-secondary educators contribute to a broader and much-needed national dialogue about the changes that are taking place in this important region and the implications for Canada and Canadians [18].

## 3. The Evolving Northern Knowledge Ecosystem: Drivers, Challenges and Opportunities

The changes we are seeing in the Canadian North with regards to research and education are being driven by a number of developments. Perhaps the most significant of these is connected to the efforts of Indigenous peoples to realize greater self-determination. At its heart, self-determination is about exercising control over one's destiny. Usually, we think of it in political or economic terms, but it could also easily be characterized in terms of regaining control over knowledge and knowledge production. International treatises such as the United Declaration on the Rights of Indigenous Peoples (UNDRIP) and, in Canada, the recommendations and findings of the Royal Commission on Aboriginal Peoples and the Truth and Reconciliation Commission's Calls to Action provide the moral and legal impetus for changes in the relationship between Indigenous peoples and the State and broader society that will allow Indigenous peoples to self-determine. At the level of research and education, they provide general guidance on the appropriate behaviour and conduct of non-Indigenous researchers working with Indigenous communities. They have also informed other parts of the northern knowledge ecosystem by initiating changes in the ways that governments, organizations, and funders develop and administer their research support programs.

Over the past five decades, northern Indigenous peoples have negotiated and signed comprehensive land claims agreements (treaties) that have provided them with resources and the authority to exercise greater control over activities taking place in their territories. Although these agreements do not allow northern Indigenous peoples to exercise full sovereignty, they have been able to establish governance institutions that allow for greater regional and local autonomy. In doing so, many have established research infrastructure to oversee, encourage and conduct research that is important to their communities and regions. Researchers are now not only required to gain local approval for research projects well in advance of the commencement of the research, but they are also strongly encouraged to co-create research projects in close consultation with local communities and organizations. This new institutional framework is often supported and encouraged by research ethics boards at post-secondary institutions and by the major funding agencies as a precondition of their approval of research projects [19]. As a result of these changes, we have seen a profound shift in attitudes on the part of academic researchers from being in control of the research process from start to finish, to co-producing research with local partners and conducting research that meets the needs of communities.

The changes taking place in the northern knowledge ecosystem raise some interesting questions about its future purpose and direction. Since colonization, the research process has been controlled by outsiders to the North and curiosity-driven research based on Western methodological approaches and critical inquiry has been at the centre of academic endeavours. Although this research has produced valuable insights on the North, its peoples and environments, in some cases, it has been conducted in ways that have been disrespectful and even harmful to northerners. The changes outlined above have started to ensure that such research practices do not continue; however, we also need to be careful that the pendulum does not swing too far to the other end of the spectrum, where research is being conducted purely to serve the needs and expectations of a particular group or organization. As Alcantara, Lalonde and Wilson have argued:

Academic researchers walk a fine line between getting to know the communities and regions they study (and supporting those communities and regions by trying to understand and reflect on their issues and problems) and maintaining a sufficient distance from those communities so that they can preserve a level of autonomy and reflection that is important to producing diverse and useful knowledge [20].

Balancing community-based and community-driven research with the academic freedom to pursue curiosity-driven research will be a difficult and increasingly contentious task in the future [19]. There is both room and a need for both types of research. Community-based research provides an important local perspective and involvement in the research process and works well for projects that aim to address or bring attention to local needs or issues. It is often imbued with a sense of social justice that seeks to shed light on and address the historical injustices suffered by Indigenous and northern communities at the hands of the settler state and its agents. Other forms of research, however, might be better suited to western-based methodological approaches; for example, comparative studies that are truly circumpolar in scope and compare and contrast developments in different regions of the Arctic, thereby yielding important information and insights for both communities and researchers. Such research can reveal important observations on a range of topics affecting northern communities and peoples, as well as innovative solutions to similar problems that have been experienced in other jurisdictions. In the future, we are likely to see research that employs elements of both western and Indigenous approaches. However, to do so, "some type of 'translation bridge' is needed to narrow epistemological gaps and to recognise and respect the distinctiveness, context, and origins of [Indigenous] knowledge when used alongside Western science" [19].

The northern knowledge ecosystem is home to researchers from a variety of disciplines, including those in the natural and physical sciences, social sciences, health sciences and humanities. For many years, the natural and physical sciences have dominated research in the North, both in terms of the number of research projects and the amount of research funding. In certain respects, this is understandable, given the fact that research projects in these disciplines usually require more expensive equipment and travel to remote locations far away from communities. The imbalance in focus and funding, however, has been recognized by academic associations such as ACUNS, as well as northern organizations such as ITK which are trying to build research capacity and use research as a means to address some of the pressing socioeconomic issues facing people living in northern communities [1,14,15]. As the National Inuit Strategy on Research has noted: "The current investments in Inuit Nunangat research reflect a biological-physical science research bias that diminishes the prominence and attention given to other Inuit research priorities, such as health and social science" [1].

Addressing the myriad socioeconomic challenges facing northern and Indigenous communities will require insights from a variety of experts, including people working for non-governmental and community organizations in the North, governments at all levels and academics trained in the social and health sciences and humanities. With the increasing involvement of Indigenous and northern communities and organizations in the research process, there will not only be a need to refocus research on the pressing social, economic and political issues facing the North, but also to break down the silos that exist between the different disciplines. In the academic world, we are already seeing this shift taking place, with many research projects now incorporating interdisciplinary approaches that include the natural scientists and social scientists. However, as is the case with the involvement of Indigenous peoples in the research process, it is critical that such collaboration occurs from the outset of the project, rather than as an afterthought once the project is designed and underway.

## 4. Conclusions

The northern knowledge ecosystem in Canada is an incredibly complex and organic entity that has evolved over time to include many different actors. It consists of a variety of organizations and individuals, often with different agendas and perspectives on the research process, but with a general interest in promoting the creation and dissemination of knowledge about the North. While it has been historically dominated by southern-based academics and academic institutions, in the last couple of decades, this knowledge ecosystem has become more diverse and complicated, largely due to the political and social changes taking place in the North and the resulting demands from northern and Indigenous organizations and peoples for greater involvement in and control over research and education. Generally speaking, academics have responded to these calls by modifying their research methodologies to recognize northern and Indigenous demands for more inclusive and respectful research. Some may feel that the pace of change is too slow or not sweeping enough, but for those who have watched the development of the northern knowledge ecosystem over the last decade or so, it is clear that important changes have occurred and that a new equilibrium is emerging which values and integrates non-Western perspectives, alongside and in partnership with Western research approaches and emphasizes interdisciplinary and cross disciplinary research. Such diversity is key to the future sustainability of the northern knowledge ecosystem, but more work needs to be done to build the "translation bridge" between different research approaches. The academic community must continue to work with northern, and Indigenous organizations and peoples to ensure that their needs and priorities are reflected in the research process and outcomes.

**Funding:** This research received no external funding.

**Conflicts of Interest:** The author declares no conflict of interest.

## References

1. Inuit Tapiriit Kanatami. National Inuit Strategy on Research. Available online: www.itk.ca/wp-content/uploads/2020/10/ITK-National-Inuit-Strategy-on-Research.pdf (accessed on 6 May 2021).
2. Obed, N. *Unlocking the Potential of Inuit Nunangat Research*; Northern Public Affairs: Ottawa, ON, Canada, 2018; pp. 57–59.
3. Thomson, A.J. How should we manage knowledge ecosystems? Using adaptive knowledge management. In *Sustainable Forestry: From Monitoring and Modelling to Knowledge Management and Policy Science*; Reynolds, K.M., Thomson, A.J., Köhl, M., Shannon, M.A., Ray, D., Rennolls, K., Eds.; CABI: Wallingford, UK, 2007; pp. 461–479.
4. Campagnucci, L.; Spigarelli, F. The Third Mission of the university: A systematic literature review on the potentials and constraints. *Technol. Forecast. Soc. Chang.* **2020**, *161*, 1–30. [CrossRef]
5. Pinheiro, R.; Benneworth, P.; Jones, G. *Universities and Regional Development: A Critical Assessment of Tensions and Contradictions*; Routledge: London, UK, 2012.
6. Coates, K. The Discovery of the North: Towards a Conceptual Framework for the Study of Northern/Remote Regions. *North. Rev.* **1994**, *12/13*, 15–43.
7. Russell, R. The Shiny New University in Canada's North (University Affairs, 17 July 2020). Available online: https://www.universityaffairs.ca/features/feature-article/the-shiny-new-university-in-canadas-north (accessed on 30 June 2021).
8. Zingel, A. Dechinta 'Bush University' Eyes Expansion with 5 Years of Federal Funding (CBC News, 28 March 2019). Available online: https://www.cbc.ca/news/canada/north/dechinta-federal-funding-1.5074493 (accessed on 30 June 2019).
9. Memorial University. Bachelor of Education (Primary/Elementary) Nunatsiavut/Inuktitut Option. Available online: https://www.mun.ca/educ/programs/undergraduate/primary_elementary/ibed.php (accessed on 10 August 2021).
10. Newhouse, D. The Meaning of Indigenization in Our Universities (CAUT Bulletin June 2016). Available online: https://bulletin-archives.caut.ca/bulletin/articles/2016/06/the-meaning-of-indigenization-in-our-universities (accessed on 10 August 2021).
11. Social Sciences and Humanities Research Council. Indigenous Research Statement of Principles. Available online: https://www.sshrc-crsh.gc.ca/about-au_sujet/policies-politiques/statements-enonces/indigenous_research-recherche_autochtone-eng (accessed on 7 May 2021).
12. Government of Canada. Tri-Council Policy Statement: Ethical Conduct for Research Involving Humans—TCPS2 (2018). 2020. Available online: https://ethics.gc.ca/eng/policy-politique_tcps2-eptc2_2018.html (accessed on 5 July 2021).
13. Polar Knowledge Canada. Available online: https://www.canada.ca/en/polar-knowledge.html (accessed on 21 May 2021).

14. Association of Canadian Universities for Northern Studies. Research Excellence in Yukon: Increasing Capacity and benefits to Yukoners in the Social Sciences, Humanities and Health Sciences. Available online: https://acuns.ca/wp-content/uploads/2013/09/YUKON-ENGLISH-WEB.pdf (accessed on 8 July 2021).
15. Association of Canadian Universities for Northern Studies. Research Excellence in the Northwest Territories: Holistic, Relevant and Ethical Research in the Social Sciences, Humanities and Health Sciences. Available online: https://acuns.ca/wp-content/uploads/2018/07/2018-NWT_English-Web.pdf (accessed on 8 July 2021).
16. Association of Canadian Universities for Northern Studies. Available online: https://acuns.ca/homepage/why-we-exist/ (accessed on 6 July 2021).
17. The University of the Arctic. Available online: https://www.uarctic.org/about-uarctic/ (accessed on 10 August 2021).
18. Abele, F.; Prince, M.J. Aboriginal Governance and Canadian Federalism: A To-Do List for Canada. In *New Trends in Canadian Federalism*; Rocher, F., Smith, M., Eds.; Broadview Press: Peterborough, ON, Canada, 2003; pp. 135–165.
19. Felt, L.F.; Natcher, D. Ethical foundations and principles for collaborative research with Inuit and their governments. *Études Inuit Stud.* **2011**, *35*, 107–126. [CrossRef]
20. Alcantara, C.; Lalonde, D.; Wilson, G.N. Indigenous Research and Academic Freedom: A View from Political Scientists. *Int. Indig. Policy J.* **2017**, *8*, 1–19. [CrossRef]

*sustainability*

MDPI

*Perspective*

# Shaping Arctic's Tomorrow through Indigenous Knowledge Engagement and Knowledge Co-Production

**Tatiana Degai [1,2,3,*], Andrey N. Petrov [2,4], Renuka Badhe [5], Parnuna P. Egede Dahl [6,7,8], Nina Döring [9], Stephan Dudeck [10], Thora M. Herrmann [11,12], Andrei Golovnev [13], Liza Mack [14], Elle Merete Omma [15], Gunn-Britt Retter [15], Gertrude Saxinger [16], Annette J. M. Scheepstra [17], Chief Vyachelav Shadrin [18,19], Norma Shorty [20] and Colleen Strawhacker [21]**

1 Department of Anthropology, University of Victoria, Victoria, BC V8P 5C2, Canada
2 Department of Geography, ARCTICenter, University of Northern Iowa, Cedar Falls, IA 50614, USA; andrey.petrov@uni.edu
3 Council of Itelmens of Kamchatka "Tkhsanom", 683000 Petropavlovsk-Kamchatski, Russia
4 International Arctic Social Sciences Association (IASSA), Nord University, Universitetsalleen 11, 8049 Bodø, Norway
5 Gaffelstraat, 2584SJ Den Haag, The Netherlands; renuka@cantab.net
6 Department of Arctic Social Science & Economics, Institute of Social Science, Economics & Journalism, University of Greenland, 3905 Nuussuaq, Greenland; parnuna@plan.aau.dk
7 Department of Planning, Technical Faculty of IT and Design, Danish Centre for Environmental Assessment, Aalborg University, 9000 Aalborg, Denmark
8 Inuit Circumpolar Council-Greenland, 3900 Nuuk, Greenland
9 Institute for Advanced Sustainability Studies, 14467 Potsdam, Germany; nina.doering@iass-potsdam.de
10 Centre for Arctic Social Studies, European University at Saint Petersburg, 191187 Saint Petersburg, Russia; sdudeck@ulapland.fi
11 Helmholtz Centre for Environmental Research-UFZ, 04318 Leipzig, Germany; thora.herrmann@idiv.de
12 German Centre for Integrative Biodiversity Research (iDiv) Halle-Jena-Leipzig, 04103 Leipzig, Germany
13 Peter the Great Museum of Anthropology and Ethnography (the Kunstkamera) of the Russian Academy of Sciences, 199034 Saint Petersburg, Russia; andrei_golovnev@bk.ru
14 Aleut International Association, Anchorage, AK 99503, USA; liza.mack@aleut-international.org
15 Saami Council, 9735 Karasjok, Norway; elle.merete@saamicouncil.net (E.M.O.); gbr@saamicouncil.net (G.-B.R.)
16 Department of Political Science, Uni Vienna (AT) , Austrian Polar Research Institute APRI, 1090 Wien, Austria; gertrude.saxinger@univie.ac.at
17 Arctic Centre, University of Groningen, 9700 AB Groningen, The Netherlands; a.j.m.scheepstra@rug.nl
18 Yukaghir Council of Elders, 677007 Yakutsk, Russia; odul.shadrin@gmail.com
19 The Institute for Humanities Research and Indigenous Studies of the North, Russian Academy of Sciences, 677007 Yakutsk, Russia
20 Arctic Athabaskan Council, Whitehorse, YT Y1A 5X9, Canada; 1012secondstreet@gmail.com
21 Office of Polar Programs, National Science Foundation, 2415 Eisenhower Ave, Alexandria, VA 22314, USA; colstraw@nsf.gov
* Correspondence: tatianadegai@uvic.ca

**Citation:** Degai, T.; Petrov, A.N.; Badhe, R.; Egede Dahl, P.P.; Döring, N.; Dudeck, S.; Hermann, T.M.; Golovnev, A.; Mack, L.; Omma, E.M.; et al. Shaping Arctic's Tomorrow through Indigenous Knowledge Engagement and Knowledge Co-Production. *Sustainability* **2022**, *14*, 1331. https://doi.org/10.3390/su14031331

Academic Editors: Stephanie Pfirman, Gail Fondahl, Grete K. Hovelsrud and Tero Mustonen

Received: 21 September 2021
Accepted: 7 December 2021
Published: 25 January 2022



**Abstract:** This perspective presents a statement of the 10th International Congress of Arctic Social Sciences Indigenous Knowledge and knowledge co-production panel and discussion group, 20 July 2021. The statement is designed to serve as a characterization of the state-of-the-art and guidance for further advancement of Indigenous Knowledge and knowledge co-production in the Arctic. It identifies existing challenges and provides specific recommendations for researchers, Indigenous communities, and funding agencies on meaningful recognition and engagement of Indigenous Knowledge systems.

**Keywords:** Indigenous Knowledge; co-production; Arctic; Indigenous Peoples

## 1. Introduction

Indigenous Knowledge and knowledge co-production are central for both research and policymaking in the Arctic, now and in the future. The International Arctic Social Sciences Association (IASSA), a professional society that brings together social scientists, humanities and Indigenous scholars, has long elevated the Indigenous Knowledge systems in its agenda. In 2017 IASSA adopted a statement on the Indigenous Knowledge and in 2021 it revised its Principles and Guidelines for Conducting Ethical Research in the Arctic to ensure productive and equitable engagement of Indigenous Knowledge, unconditional adherence to principles of Indigenous data and knowledge sovereignty and commitment to decolonizing research through knowledge co-production. IASSA members continued this work, and these efforts culminated in developing a new vision for Indigenous Knowledge engagement and co-production in the Arctic that is discussed below.

## 2. Indigenous Knowledge Is Key to Understanding Natural and Social Systems in the Arctic

On 20 June 2021, the International Congress of Arctic Social Sciences (ICASS X) hosted a plenary and other sessions devoted to the Indigenous Knowledge and research in the Arctic. The panelists and presenters have developed the following statement.

The Indigenous Peoples are the original Arctic researchers who hold unique knowledge, grounded in multigenerational experiences, of land and environment. This knowledge is time tested and implies deep understanding of the Arctic environment, socioeconomic systems, and human-environment relations. Indigenous Knowledge provides a foundation for individual and collective well-being of past, present, and future generations of Arctic Indigenous Peoples. Indigenous Knowledge systems have their own ontologies, epistemologies, and methodologies, and possess internal validation principles and processes based on reciprocity and respect. Indigenous Knowledge is key to accurate interpretation of dynamics in the natural and social systems in the Arctic. Science and policy that are not inclusive of the Indigenous Knowledge cannot be considered adequate to address the Arctic Peoples' needs. A major advancement in Arctic science will be achieved through Indigenization of Arctic research.

While working with Indigenous communities, one has to be mindful of the systemic trauma they have experienced in their history, and allow time, and channel resources so that these communities can heal and reconcile with their land, histories and languages that were disrupted due to colonization.

## 3. Co-Production Must Become a Priority

In order to ensure the vitality of Indigenous Knowledge systems and to improve the quality and relevance of Arctic research, collaborative efforts across disciplines under the guidance of the Indigenous Knowledge holders must become a priority, and knowledge co-production must be seen as central for Arctic research. Co-production must be based on ethical, equitable, meaningful and mutually beneficial engagement of knowledge systems that is embedded in Indigenous rights, recognition of Indigenous Knowledge sovereignty and ownership. Co-production must be recognized, promoted and supported by funding agencies, academic institutions, and researchers regardless of their discipline, area of research and affiliation.

Co-production should imply co-identification of research needs, co-creation of research ideas, co-design of research questions, co-definition of research objectives, co-development of research programs, co-authorship of research results, co-implementation of research projects and co-evaluation of research outcomes. Co-production must ensure that Indigenous and non-Indigenous research partners share a common vision of what these, and other terms, mean in the research process. In addition to being based on co-production, Arctic research must also make room for Indigenous Peoples' knowledge systems to stand on their own without being validated by research partnerships with non-Indigenous scholars.

Finally, co-production should generate practical results important for Indigenous communities. It is important to acknowledge that these processes take time.

## 4. Practical Steps to Be Taken Now

To take steps towards achieving the above goals, we recommend the following: Recognize and respect Indigenous Knowledge in understanding natural and social systems in the Arctic including the importance of data sovereignty, intellectual property and ownership by Indigenous rights holders.

- Support the Indigenous Peoples to identify, define, research and act upon their own research priorities and methodologies, for example by providing financial, organizational and institutional capacities.
- Enable and encourage development of equitable relationship and understanding between the Indigenous Peoples and researchers necessary to co-create meaningful, relevant research guided by Indigenous Knowledge and societal values.
- Focus on reciprocal, mutually enriching capacity building between researchers and Indigenous communities:
  - this includes building capacity among researchers to recognize Indigenous rights and fully engage Indigenous Knowledge in Arctic research.
  - engaging Indigenous youth and elders to have an active role in knowledge production.
- Encourage research institutions and funding agencies to support and enable meaningful collaboration at all stages of research projects in the social and natural sciences and humanities to meet the expectations of knowledge co-production.
- Urge funding agencies to provide research-planning (seed) funding, flexible funding, and long-term funding options to researchers and Indigenous organizations to establish, build, and maintain relationships with Indigenous communities and to lay the foundation for knowledge co-production prior to actual research.
- Recommend that research institutions and funding agencies engage and support Indigenous Peoples to evaluate the research before, during and after the research to ensure that the research is progressing in a way that meets their needs.
- Prioritize collaboration, co-creation, Indigenous-led projects and capacity building research initiatives in funding calls and in the project selection process.

Finally, recognizing sustained commitment and extensive work that the International Arctic Social Sciences Association (IASSA) has done to advance the role of the Indigenous Knowledge in Arctic research and promote knowledge co-production, it is important to continue by focusing on:

- facilitating the equitable and ethical application of Indigenous Knowledge and engagement of Arctic Indigenous communities by developing guidance to the international research community in all aspects of Arctic science and research.
- working on creation of intellectual space for Indigenous Knowledge holders at international fora.
- further engaging Indigenous Knowledge holders in IASSA.
- developing internal IASSA strategies, structures and resources to establish a support system of Indigenous Knowledge holders within the IASSA, e.g., an Indigenous Knowledge Working Group.

## 5. Conclusions

Diverse Indigenous Knowledge systems in the Arctic are critical for ensuring the well-being of Arctic communities and ecosystems. They are central for accurate interpretation of the natural and social dynamics in the Arctic. Equitable engagement and co-production are the primary mechanisms for decolonizing and Indigenizing Arctic research and policy-making that will secure a sustainable Arctic tomorrow. Attaining these goals will take a collective effort and individual commitment. The authors and IASSA are determined to continue this work. Ultimately, we call on individual researchers to ask themselves: what can I do to make this happen?

**Author Contributions:** Conceptualization, all; formal analysis, all; writing—original draft preparation, T.D. and A.N.P.; writing—review and editing, all; project administration, T.D.; funding acquisition, A.N.P. All authors have read and agreed to the published version of the manuscript.

**Funding:** This research was funded by the US National Science Foundation, grant number 2001989.

**Institutional Review Board Statement:** Not applicable.

**Informed Consent Statement:** Not applicable.

**Data Availability Statement:** Not applicable.

**Acknowledgments:** This paper was informed by the research conducted on traditional lands that continue to be stewarded by the Indigenous Peoples of the Arctic. The authors are thankful to participants of the ICASS X plenary and other sessions devoted to the Indigenous Knowledge in the Arctic and anonymous reviewers for their valuable comments and suggestions. We are also grateful to Sustainability Journal and Special Issue editors for waving the publication fee for this contribution thus demonstrating commitment to enhancing equity in access to publishing for diverse Arctic knowledge holders.

**Conflicts of Interest:** The authors declare no conflict of interest.

sustainability

MDPI

Article

# Northern Research Policy Contributions to Canadian Arctic Sustainability

Alison D. Perrin [1,*], Gita Ljubicic [1,2] and Aynslie Ogden [3]

[1] Geography and Environmental Studies, Carleton University, Ottawa, ON K1S 5B6, Canada; gita.ljubicic@mcmaster.ca
[2] School of Earth, Environment & Society, McMaster University, Hamilton, ON L8S 4L8, Canada
[3] Polar Knowledge Canada, Cambridge Bay, NU X0B 0C0, Canada; aynslie.ogden@polar-polaire.gc.ca
* Correspondence: alison.perrin@carleton.ca

**Abstract:** Academic research plays a key role in developing understanding of sustainability issues in the Canadian Arctic, yet northern organizations and governments struggle to find research that is relevant, respectful of local interests, and that builds local capacity. Northern science and research policies communicate expectations for how research should be prioritized, planned, conducted, and disseminated. They discuss northern leadership of research and outline the diverse roles that northerners and northern organizations could fill in research programs and projects. Many of these documents are founded on the need for research to improve environmental, economic, and social sustainability in the Canadian North and provide insight into how academia can support a northern-led Arctic sustainability research agenda. The goal of this study is to examine northern research-policy documents to identify commonalities amongst the goals and priorities of northern organizations and their shared expectations for research in northern Canada. The objectives are to understand how organizations expect researchers to engage in and conduct research, how research programs can align with northern science policy objectives, and how academic research can support policy and decision-making related to sustainability. Through a quantitative content analysis combined with a qualitative thematic analysis, this comprehensive review examines research policy, strategy, guidance, and program documents produced by northern and northern-focused governments and Indigenous organizations. Relationships, partnership, and communication are the foundations of relevant and applicable research, requiring both resources and time for local and partner participation. Our analysis shows that researchers should consider potential policy applications for sustainability research early on in the development of research projects, ensuring that relevant local and policy partners are involved in designing the project and communicating results.

**Keywords:** northern research policy; research governance; research priorities; Arctic; sustainability; Yukon; Northwest Territories; Nunavut; Nunavik; Nunatsiavut

**Citation:** Perrin, A.D.; Ljubicic, G.; Ogden, A. Northern Research Policy Contributions to Canadian Arctic Sustainability. *Sustainability* **2021**, *13*, 12035. https://doi.org/10.3390/su132112035

Academic Editors: Stephanie Pfirman, Gail Fondahl, Grete K. Hovelsrud and Tero Mustonen

Received: 2 September 2021
Accepted: 27 October 2021
Published: 31 October 2021

**Publisher's Note:** MDPI stays neutral with regard to jurisdictional claims in published maps and institutional affiliations.



## 1. Introduction

Northern governments and organizations have identified the need for relevant research, accessible results, and engagement with northern communities as they increasingly focus on integrating evidence into decision-making [1–4]. Yet, northern organizations and governments struggle to find research that is relevant and is developed in a way that is respectful of local interests and builds local capacity [4–6]. Given the pressing importance of sustainability issues in the Canadian North, academic research can play a key role by contributing information to support sustainability transformations in the region [7–9]. However, there are gaps in understanding the relationship between sustainability research and practice, including how research can support sustainability policy and decision-making [7].

Northern experiences of research have been widely documented on a local scale, but not brought together as a whole in a way that enables a broad, transformative vision for northern research [4–6,10,11]. There have been discussions within the scientific literature about improving how northern research is conducted in terms of individual projects [10–14]. However, the guidance and expectations coming from northern governments and organizations are generally not reflected within the academic literature; rather, they are captured in policy documents and grey literature. Systematic and realist reviews mainly capture themes and trends from available scientific literature based on literature-database search protocols. This policy review seeks to contribute to a better understanding of northern Indigenous, organization, and government perspectives on research policy through a unique systematic approach to grey literature review.

There is a wide variety of organizations developing policies or providing guidance on research in northern Canada. There are strategy, guidance, and program documents produced by northern and northern-focused governments, Indigenous, and non-governmental organizations that discuss research policy. While there is a general discussion of research in these documents that applies to both Indigenous and non-Indigenous northerners, there is also an emphasis on research with Indigenous northerners, which is a result of both the colonial history of research and the importance of Indigenous rights across the North. Northern organizations often play a role in mediating how researchers engage with northerners, and their research-policy documents outline how they would like those relationships to unfold. Research-policy documents discuss northern leadership of research and provide insight into the diverse roles that northerners and northern organizations fill in research programs and projects. Through an analysis of these documents, we explore how organizations define and guide research in the Canadian North and the anticipated contributions of research to northern sustainability.

The goal of this study is to examine northern research-policy documents to identify commonalities amongst the goals and priorities of northern organizations and their shared expectations for research in northern Canada. The objectives are to understand how organizations expect researchers to engage in and conduct research and how research programs can align with northern science policy objectives. A related objective is to identify pathways for academic research to support policy and decision-making related to sustainability. This analysis examines the governance structures, policies, priorities, and organizations that guide, regulate, and interact with northern research. We identify themes related to principles, priorities, and guidelines, as well as the different steps in the lifecycle of a research project and the roles different actors play in research. It is clear that sustainability, particularly in terms of climate change and resource development, is a priority for northern regions. Research is identified as an important means of supporting sustainability transformations. Through this analysis, we provide insight into how academia can contribute to a northern-led sustainability research agenda that supports evidence-informed policy.

## 2. Background

Over the past sixty years, the Canadian North has been the focus of several national and international scientific programs and numerous government science strategies. Calls for a more coordinated response across northern jurisdictions are motivated by the impacts of resource development, sovereignty issues, and more recently, the impacts of increasingly rapid environmental changes [15–19]. Since the 1970s, efforts have been made by academia to articulate the need for a single unifying northern science policy that guides investments in northern research [15,16,19]. Serious gaps in Canadian Arctic research were identified in the 1990s [20], and from that realization, investments were made, including the International Polar Year and a northern research chair program. These were opportunities to build up northern research, invest in students, and develop robust research programs in the North.

Historically, northerners have felt left out of research decision-making in their region. Northerners, particularly Indigenous northerners, did not trust researchers due to a long history of poor or no communication, misunderstanding, and extractive research [6,10,21,22]. This distrust of research is directly related to the colonial legacy in the North, including loss of control over decision-making, land, and other aspects of their lives [4,23]. Poor communication and uneven power relationships have existed between researchers and communities, which contributes to ongoing distrust of the contemporary discussion of research and colonialism in the North [4,10,11,24–27]. The relationship between Indigenous northerners and research has been fraught with issues, for example, research data and results not being shared with participating communities, sensitive data being published without consultation, and researchers not attributing Indigenous or local knowledge to knowledge holders [4,22,23,28]. As Indigenous northerners have asserted governance and leadership in the North, this tenuous relationship has begun to change, but there are still uneven power dynamics.

The settlement of land-claim agreements and resulting self-government, co-management, and other expressions of Indigenous control over resource and land management, have influenced how northerners, in general, and Indigenous northerners, in particular, interact with research. This evolution in governance is reflected in an emerging dialogue about self-determination in research and a desire for control over both decision-making around research and guiding how research is conducted [2,4,6,10,29,30]. Indigenous research is intrinsically linked with governance, particularly control of the research agenda, whereby "Indigenous research . . . is in itself an enactment of governance" [22]. Indigenous northerners have used research to challenge colonial policy, defend land rights, and contest resource development, often employing western academics and methodologies with Indigenous values and knowledge [22]. Although there are gaps in opportunities for northern Indigenous scholarship [7], Indigenous northerners are asserting their role as leaders in northern research [13,31–36]. There is a desire in the North for greater access to the necessary resources and facilities to conduct and manage research locally [4,6,10,17,37]. This includes support for research capacity for governments, co-management bodies, and institutes of self-government as they pioneer new forms of governance in the North [22,36]. Through the International Polar Year program, northern organizations saw opportunities to change how research was being done, to build partnerships to further their own research priorities, and to build northern research capacity through training and infrastructure [14,38–43].

### 2.1. Geographic Context

While there are many different ways to delineate the Canadian North [2,4,44,45], the geographic focus of this policy analysis includes five regions in northern Canada: Yukon, the Northwest Territories (NWT), Nunavut, Nunavik (the Inuit land-claim region in northern Québec), and Nunatsiavut (the Inuit land-claim region in northern Labrador) (Figure 1). We acknowledge that this excludes the northern provincial regions of British Columbia, Alberta, Saskatchewan, Manitoba, Ontario, Québec, and Labrador. Areas of the provincial North often face similar challenges in relation to sustainability issues. However artificial, political boundaries tend to frame policy jurisdiction, and thus, for the purpose of this study, the three territories, along with Nunavik and Nunatsiavut, were selected as our focus because they are most consistently included in pan-northern policies and publications.

The five regions of focus are all large in land area but have small populations, often with a majority clustered in the administrative capital while the rest of the population is spread throughout smaller communities [44]. In Nunavut, Nunavik, and Nunatsiavut, communities are mostly coastal and serviced by air and ship. In Yukon and the NWT, there is a mix of access, including year-round road, winter road, boat, and air. The three territories operate like provinces, with elected governments overseeing territorial programs and services, although some control lies with the federal government. Federal control of lands and resources was devolved to the Government of Yukon in 2003, giving the Yukon similar powers to the provinces, and to the Government of the Northwest Territories in 2014 [45–47]. Nunavut, which separated from the Northwest Territories to become a territory in 1999, is currently in devolution negotiations with the federal government to bring control of crown lands and resources to the territorial government [48]. Within the three territories, there are Inuit, First Nation and Métis land-claim and self-government agreements. Nunavik and Nunatsiavut are Inuit land-claim regions within the provinces of Québec and Newfoundland and Labrador, respectively.

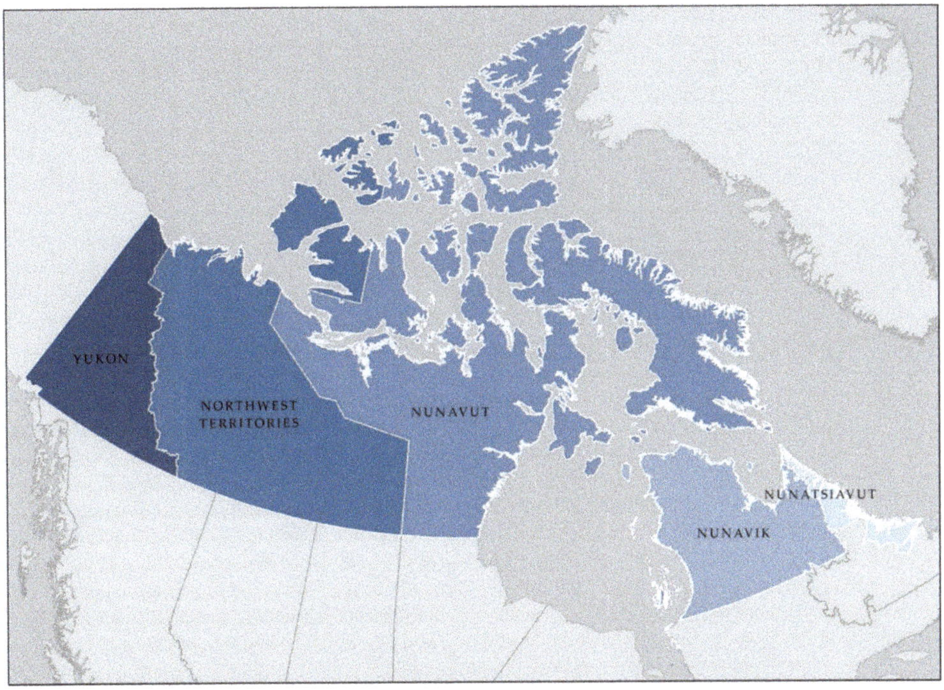


**Figure 1.** Map showing regions of northern Canada included in study.

Although the five regions of focus have distinct differences in terms of governance, cultures, and landscapes, they also experience similar challenges. Since the late nineteenth to mid-twentieth century, colonial expansion, resource development, and exploration were major sources of change across the northern regions in Canada. Colonization has had serious impacts on northern Indigenous peoples and their ways of life, including moving nomadic people into permanent settlements, forced relocations, land dispossession, the removal of children to attend residential schools, the impacts of physical and sexual abuse, intergenerational trauma, the loss of traditional cultures and languages, and the imposition of western governance systems [27,49–52]. However, since the 1960s and 1970s, Indigenous land-claim and self-government agreements have been driving innovations in Indigenous governance to achieve self-determination [45,49,53–57]. Indigenous governance is enacted in different ways across the North, and there are numerous different types of organizations that arose out of land-claim processes, including Indigenous governments, corporations, co-management boards, and institutes of public governance.

*2.2. Organizational Context*

Governance and responsibility over research in the North are spread amongst territorial and provincial governments, Inuit, First Nations, and Métis organizations and governments, federal agencies, academic institutes, non-governmental organizations, and the private sector. All these different actors can play roles in setting policy, providing guidance on how research should be done, distributing funding, and/or conducting research [2,4,6,16,19]. Government, academic, non-profit, and consultant research have all made considerable contributions to northern scholarship. Some research is being conducted by governments themselves or through consultants on behalf of governments or other organizations. This research is usually done to answer a specific question and under direct control of the contracting organization, and while some of it is tracked through research licensing or public reports, much of it is not publicly available. Policy documents, however, are a way of communicating needs, priorities, or expectations to those engaged in research that is not directly controlled by the organization. Such policy documents are particularly aimed at academic research, although they may be relevant to all types of researchers.

Historically, northern regions have relied on federal (or provincial) funding for science activities; however, they have recently started taking a more active role in shaping northern science by outlining expectations for research that is relevant, respectful, and solution-oriented [4,6,58]. The three territories, as well as Nunatsiavut, have research licensing or permitting processes that are designed to ensure that researchers consult with relevant rights- and stakeholders and that research does not cause environmental or societal harm. Research licenses provide an opportunity for northern governments, communities, and researchers to negotiate relationships and influence decisions around research [10].

Across all five regions, there are northern academic and research institutes that play a role in advancing northern science through their growing capacity to conduct and direct research [6,17]. Along with the territorial and regional research centres, there are numerous research stations across northern regions operated by universities, federal departments, and other organizations that engage in conducting research or hosting visiting researchers, including the Government of Canada's Canadian High Arctic Research Station (CHARS). The Canadian Network of Northern Research Operators (CNNRO) represents a membership of 31 operators and 10 associate members that range from community-based organizations to government departments and university-run research stations (e.g., Kluane Lake Research Station, Polar Environment Atmospheric Research Laboratory, Centre d'études nordiques) [59]. Across the North, there are non-governmental organizations (NGOs), often community-based, that are involved in conducting research (e.g., Ittaq Heritage and Research Centre, Institute for Circumpolar Health Research, Yukon Wildlife Preserve, and the recently closed Arctic Institute of Community-Based Research). There are also national and international NGOs that maintain northern research offices, including the Wildlife Conservation Society and the World Wildlife Fund. There are also a few research networks that are currently or formerly active in northern research and that greatly influence the research context. ArcticNet is a Network Centre of Excellence (NCE) focused on Arctic research in Canada that has recently initiated a funding program to support northern and Inuit-led research. The Canadian Mountain Network (CMN) is another NCE that is active in northern mountainous regions and offers several funding streams based on different modes of research, including Indigenous-led research.

#### 2.2.1. Yukon

In Yukon, there are fourteen First Nations, eleven of which have settled land claims and finalized self-government agreements [60]. There are also several transboundary Indigenous groups from NWT and British Columbia with traditional territories and land claims in the Yukon. Some Yukon First Nations have research or Indigenous knowledge protocols and policies (e.g., Vuntut Gwitchin First Nation), and the Council of Yukon First Nations (CYFN) also has a number of research-policy documents. The Government of Yukon released their Science Strategy in 2016 and the Yukon Science Policy in 1986, both of which set out goals for developing research capacity in the Yukon [3,61]. The Science Strategy and online Compendium of Research and Monitoring are products of YG's Office of the Science Advisor, housed in the Executive Council Office [3]. The Yukon Scientist and Explorer's Act outlines requirements for research licenses, which are administered by the Government of Yukon's Department of Tourism and Culture [62]. The Act covers the physical and social sciences, with the exception of archaeological research, which goes through a separate permitting process. The Act applies to anyone "who enters the Yukon to undertake scientific research" [63], while researchers based in the Yukon are not required to apply for a license. There are also several other permitting or permissions processes that may be required in addition to a research license for research with wildlife, in parks or protected areas, or on Yukon First Nations' settlement lands. The Yukon is home to Yukon University, which recently transformed from Yukon College in 2020, and hosts campuses in most Yukon communities. The YukonU Research Centre, based in Whitehorse, is part of Yukon University and is focused on conducting research in the Yukon and across the North. There are also southern Canadian universities with Yukon-based headquarters, including both the University of Alberta North (a Yukon University partner) and University of Calgary's Arctic Institute of North America, which runs the Kluane Lake Research Station.

#### 2.2.2. Northwest Territories

Negotiation of Indigenous land claims and self-government agreements is ongoing in the NWT, with many communities or regions having finalized their negotiations. Within the NWT, there are Inuit, First Nations, and Métis rightsholders. The Dene Nation is a coordinating organization that brings together First Nation governments in NWT, while the Inuvialuit Regional Corporation (IRC) represents Inuvialuit. Several Indigenous governments and land-claim organizations in NWT have research or Indigenous knowledge protocols and policies, and some have research offices within their governments (e.g., the Dedats'eetsaa: Tłı̨chǫ Research & Training Institute [64]). The Government of the Northwest Territories (GNWT) released a knowledge agenda in 2017 and a related action plan, which are coordinated and led by the Senior Science Advisor, housed within the Department of Environment and Natural Resources [1,65]. The knowledge agenda builds off an earlier science strategy with the same broad research priorities [66]. GNWT has a research partnership with Wilfrid Laurier University, which has offices in Yellowknife and an active research program in the territory. Aurora College is the main postsecondary institution in NWT, with campuses in several communities. The Aurora Research Institute (ARI) is part of Aurora College and hosts research centres in Inuvik, Fort Smith, and Yellowknife. ARI is mandated by the NWT Scientists Act to license and coordinate research in the NWT [67]. They operate a research ethics board (REB), and any researcher applying to conduct research with humans in NWT that is not going through a research-ethics approval process with their home institution is required to go through the REB process at ARI. The Dedats'eetsaa: Tłı̨chǫ Research & Training Institute hosts Hotìı ts'eeda, a Strategy for Patient-Oriented Research (SPOR) Support Unit funded by the Canadian Institutes for Health Research (CIHR) and governed by NWT Indigenous governments.

#### 2.2.3. Nunavut

Nunavut became a territory as part of a land-claim process, and Inuit rights are recognized in the public government's mandate and through Inuit institutions of public governance [45,68]. The Government of Nunavut (GN) represents all Nunavummiut (residents of Nunavut), while Nunavut Tunngavik Incorporated (NTI) represents Inuit beneficiaries and is responsible for land-claim implementation and managing Inuit-owned lands. Nunavut's Scientists Act requires that researchers from all disciplines obtain a scientific license unless they require an archaeological permit issued under the Nunavut Act [69]. In addition to the research license, there are also separate permitting processes, depending on the type of research (e.g., wildlife, water, parks). When conducting research on Inuit-owned lands, a separate approval process is required through the relevant regional Inuit associations in addition to the research license [68]. The Nunavut Research Institute (NRI) administers the scientific research license in Nunavut, operates several research stations, supports visiting researchers, and conducts research in the region. The NRI is part of Nunavut Arctic College, Nunavut's postsecondary institute that has campuses across the territory and has recently partnered with Memorial University to provide degrees. Nunavut-based NGOs like the Aqquimavvik Society and the Qaujigiartiit Health Research Centre are actively involved in conducting research, as well as providing guidance on research ethics and methodologies.

#### 2.2.4. Nunavik

Nunavik is a land-claim area in the northern part of Québec and was the first of the four Inuit regions to sign a modern land-claim agreement as part of the James Bay and Northern Québec Agreement (JBNQA) in 1975. The Kativik Regional Government is a product of the JBNQA and is responsible for providing public services to the region, while Makivik Corporation is the land-claim organization established to represent Inuit beneficiaries [70]. Nunavik does not have a research license, but Makivik Corporation is currently leading the development of a regulatory system for research in the region. Makivik operates the Nunavik Research Centre based in Kuujjuaq, which collaborates with northern organizations and southern academic institutions to respond to research questions that directly link to community priorities and to support policy development by Makivik Corporation [71]. Université Laval also has two research stations in the region as part of its Centre d'études nordiques.

#### 2.2.5. Nunatsiavut

Nunatsiavut is an Inuit land-claim and self-governing region within Newfoundland and Labrador. The Inuit of northern Labrador negotiated the Labrador Inuit land-claim agreement and self-government at the same time, resulting in the Nunatsiavut Government (NG) [70,72]. The NG is actively engaged in conducting research and regulates research through an advisory committee that reviews all proposed projects [73,74]. The Nunatsiavut Government operates the Nunatsiavut Research Centre in Nain, which hosts field and lab-based research conducted by NG staff, collaborators, and visiting researchers [73]. Memorial University is engaged in research partnerships in the region, particularly through the Labrador Institute and the School of Arctic and Subarctic Studies based in Happy Valley-Goose Bay.

#### 2.2.6. Government of Canada

Even with devolution, the federal government still plays a key role in northern policy and programming, including through science, health, Indigenous services, and economic development. Science, resource development, and sovereignty issues have preoccupied federal government interest in the North for many years, leading to a range of political strategies and initiatives [45,75]. The current elected Government of Canada has developed an Arctic Policy Framework (APF), which replaces the previous federal government's policy document, Canada's Northern Strategy, and the Statement on Canada's Arctic Foreign Policy [48,76,77]. The APF provides overarching direction to the Government of Canada on priorities, activities, and investments in the Arctic, with a vision to 2030 [77]. Polar Knowledge Canada (POLAR), established to replace the Canadian Polar Commission, is the department that represents and coordinates the Government of Canada's Arctic science interests. They also facilitate a networked approach to northern research by bringing together other northern research centres and organizations. POLAR operates CHARS, provides research-project funding, and administers support for student research through the Northern Scientific Training Program [78]. The CHARS campus in Cambridge Bay, Nunavut is home to POLAR staff and visiting federal and academic researchers. There are numerous other federal government departments involved in administering, conducting, or supporting Arctic science, including Environment and Climate Change Canada, Health Canada, Natural Resources Canada (Polar Continental Self Program), and Crown-Indigenous Relations and Northern Affairs Canada (Northern Contaminants Program and Climate Change Preparedness in the North).

#### 2.2.7. National Indigenous Organizations

Inuit Tapiriit Kanatami (ITK) is the national organization that represents Inuit in Canada and works closely with the four Inuit regions that make up Inuit Nunangat (Inuit homeland within Canada): Nunavut, Nunavik, Nunatsiavut, and the Inuvialuit Settlement Region (in NWT) [79]. ITK's focus is on advocacy, research, outreach, and policy guidance on issues affecting Inuit, with a particular focus on Inuit relations with the Government of Canada [80]. ITK is engaged in science policy by advocating for better research practices, relationships, and outcomes within Inuit Nunangat. Research is one of ITK's current priority areas, and they have developed a National Inuit Strategy on Research (NISR) [4], among other research-guidance documents [49,81–84]. Self-determination in research is the backbone of the NISR, and the four Inuit regions have been directly involved in its development and implementation [4]. Through their own research programs, they support research in key topic areas, including health, education, and climate change. ITK has also played an important role advising ArcticNet, where they were instrumental in establishing Inuit research advisors. The Assembly of First Nations (AFN) represents First Nations in Canada nationally. They are involved in political advocacy but also play a coordinating role for First Nations on a number of issues. The First Nations Information Governance Centre (FNIGC) is an organization that promotes data governance in relation to First Nations, specifically the ownership, control, access, and possession (OCAP) principles. Although no longer operating, the National Aboriginal Health Organization (NAHO) also produced several documents promoting data governance for Indigenous Canadians. The NAHO had First Nations (FNC-NAHO), Métis (MC-NAHO), and Inuit centres (IC-NAHO) that addressed issues particular to each Indigenous group.

### 2.3. Sustainability Context

Arctic sustainability is a growing research area that aligns with many of the priorities articulated by northern governments and organizations. Yet, despite the numerous contributions this research area has already made, there are still big gaps in our understanding of sustainability in the Arctic [85]. A broadly accepted definition of sustainability includes meeting the needs of today without compromising the future [8,9,86,87]. However, there is no clear definition of what that means specifically for the North; it can have different meanings depending on the context of the discussion and the perspective of those involved [9,88]. Sustainability frameworks include several dimensions of sustainability, including sociocultural, economic, environmental, and institutional [8,9,89]. Indigenous sustainability frameworks incorporate holistic approaches to sustainability, emphasizing cultural values that are not necessarily inherent in western scientific-sustainability paradigms [88,90].

Research that supports sustainability is a core goal shared by diverse organizations, despite differences in some of their specific sustainability priorities [1,66,91,92]. In northern research, sustainability is often linked with wicked problems characterized by long-term and complex interactions, such as climate change, resource development, health and well-being, impacts of colonialism and residential schools, and economic development [7,86]. Whereas these broad priorities are shared across the North, communities and regions have their own specific challenges and priorities, such as waste and water management, infrastructure, food or energy security, safe and healthy communities, housing, and economic diversity [1,8,86]. Although there is general agreement that environmental, sociocultural, and economic sustainability are important end goals for research, there are still knowledge gaps around how to achieve these transformations.

Northern governments and organizations have declared their interest in moving towards a science-policy framework that is informed by northern priorities and produces research that supports policy development, decision-making, and local action for environmental, sociocultural, and economic sustainability. The policy documents included in this analysis emphasize the importance of research that aligns with local needs, benefits local communities and regions, and produces research outcomes that improve the lives of northerners or affect broader change. In northern Canada, scientific knowledge is not the only type of evidence used in decision-making. "openness to alternative epistemologies creates a space for envisioning a future that is different from the present" [93]. Putting Indigenous knowledge at the forefront of policy development promotes resiliency in northern communities and prepares them to adapt to change. Indigenous knowledge is valued by the territorial governments as equal to scientific knowledge and has an equivalent role in informing evidence-based policy [1–3]. Sustainability transformations require action and policy-relevant research attuned to local needs and values. This policy analysis provides insight into the contributions of research to sustainability and how that can be improved to ensure that there are valuable outcomes from research in the North.

## 3. Methodology

Document analysis is a systematic method for reviewing and evaluating documents using qualitative interpretive analysis [94–96]. The document analysis undertaken in this project includes a focused review and thematic analysis of the policy documents typically found within grey literature. This policy review and analysis method was critical to gain insight into the context, values, and origins of northern research policies [94,97]. To clarify terminology, a policy is a decision, a commitment to a course of action, or a guideline for action that is based on a combination of values, beliefs, evidence, and political ideology [98,99]. Policy for science refers to all decision-making related to science or the systematic pursuit of knowledge, including research [99]. Policy documents provide insight into the priorities, values, and belief systems that guide decision-making within a government or organization, as well as identifying practical issues that are current and emergent [97,98]. The policy documents included in this study have a variety of uses, including communicating information on research in the North and the regulatory context, as well as providing insight into the goals, values, and motivations of the organizations that influence northern research. Therefore, documents were examined with attention to intention and credibility, including the purpose, the author, and the influence over actions or decisions [97]. For many northern-based organizations, science-policy documents are used to advocate for their research expectations and priorities, while in some cases, they communicate where an organization is focusing their research resources or how they approach their own research.

### *3.1. Sourcing Documents and Inclusion Criteria*

We conducted a thorough online search using a common search engine and searching websites of known governments, Indigenous organizations, and non-governmental organizations (e.g., territorial governments, ITK, Dene Nation, CYFN). Documents were selected based on their relevance to the overall study goals and approach and according to three main inclusion criteria (Figure 2) [94,100–103]. The initial search identified 128 relevant documents; however, 18 were removed after reviewing the full text. Therefore, 110 documents were included for in-depth analysis. Most of the documents were in a standalone format (e.g., report, guide, pamphlet, discussion paper), but webpages were also recognized and included as policies when they discussed substantial policy information or guidance [104,105].

We aimed to be as comprehensive as possible, including all documents that met the inclusion criteria and were available digitally (see Supplementary Materials Table S1 for a list of documents, including author organizations and websites). We also recognize that there are limitations to the study based on our goals and approach, which focus on policies for research, as opposed to policies developed from research, and we therefore did not include research-program results [97]. We did not examine research funding programs and forms, which change regularly and are often updated or removed from the host website. We also did not include application forms for research licenses or protocols, which, in some cases, were the only documentation available from an organization (e.g., Nunatsiavut Government, Vuntut Gwitchin Government). For the federal and territorial governments, we focused on government-wide or corporate policies. These governments are actively conducting research in the North and may have additional departmental-level science policies that were not included within the scope of this study. There are potentially relevant policies being enacted by northern organizations that are either internal and not shared publicly or not documented in a written format and were therefore not part of this analysis. There are northern organizations and many First Nation governments that are actively involved in influencing, conducting, and guiding research in their region that are not represented in the study because no relevant policy documents were found online. Some of those organizations have information online about their involvement with research, but without extensive policy information, their webpages were not included in the study (e.g., Dechinta Centre for Research and Learning, Dedats'eetsaa: Tłı̨chǫ Research and Training Institute).

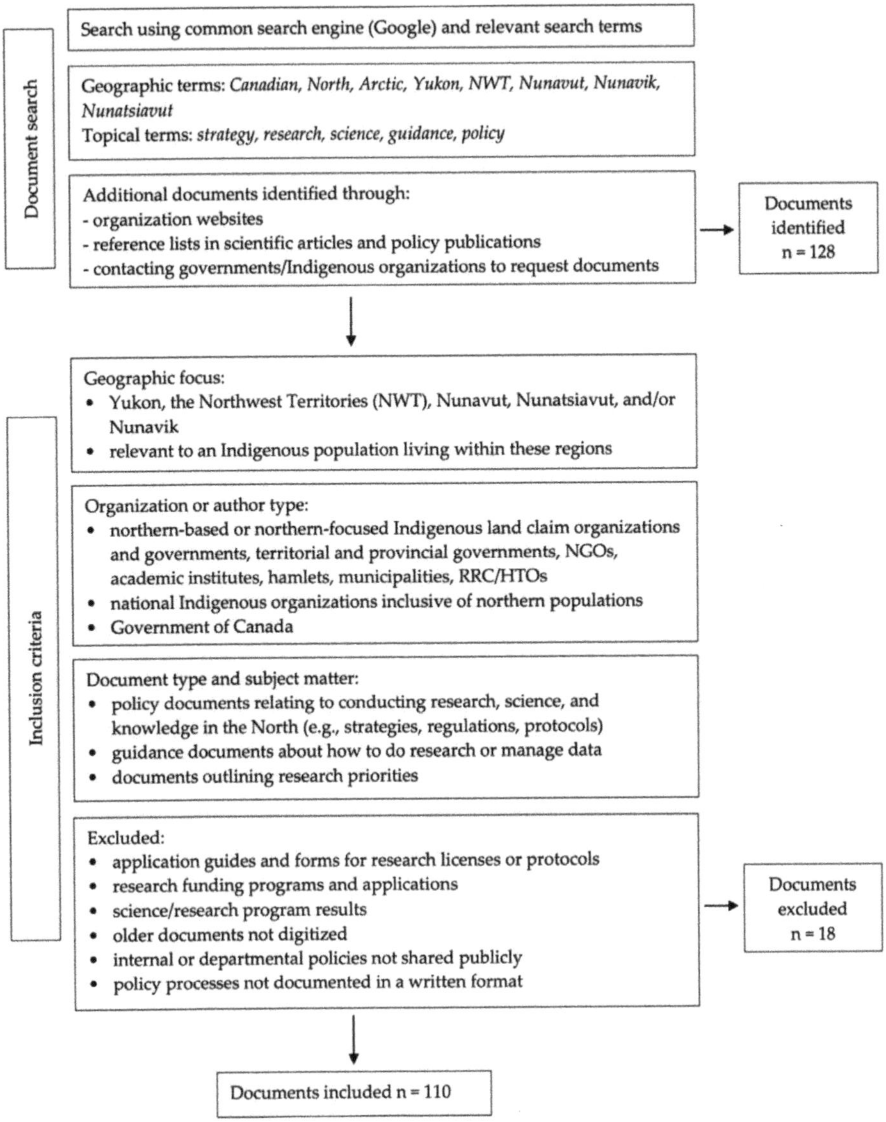


**Figure 2.** Document-sourcing and screening process.

### 3.2. Coding

In the process of reviewing relevant documents, descriptive information related to each document was recorded in an Excel spreadsheet to track and categorize documents, as well as to assist with initial analysis. These include author type, geographic scale (regions it is relevant to), publication focus and type (policy, research priorities, or research guidance), discipline focus, relevance to or focus on Indigenous peoples, mention of Indigenous knowledge and local knowledge, and focus on local or regional protocol or research license. Descriptive statistics were produced in Excel to provide insight into the types of documents that exist and their geographical and topical coverage. These characteristics were then entered as document attributes in NVivo to be used in the thematic analysis.

The coding process combined a deductive and inductive approach. An initial coding scheme was developed based on a literature review focused on research issues and approaches relevant to the North, and new codes were created as they emerged during the coding process. Research-planning and guidance documents were used to identify the different steps in the lifecycle of a research project and the types of actors that are involved in northern research. A comprehensive list of codes (Appendix A Table A1) guided the coding process, including codes relating to: (i) elements of the research project lifecycle (e.g., research design and approval, conducting research, sharing results); (ii) concepts relevant to research practice (e.g., partnership, community engagement, participation, capacity building); and (iii) roles that different actors play in northern research (e.g., funder, regulator, student). Within the groups of codes, there are parent codes that represent general or broad themes and subcodes that are more specific. Codes were applied when concepts were named explicitly but also through interpretation when a concept was discussed without using key terms. We created a clear description of each of the codes to help reduce error and bias in the coding process. Documents were coded in NVivo software, which supports document analysis by enabling the comparison of document codes and attributes within a large number of documents. Using NVivo, we explored relationships between codes and attributes and compared themes across regions and author types using crosstab queries. We compared the three types of codes to get a better understanding of how key concepts should be applied to the steps of a research project and which actors should be involved.

Coding for Sustainability

Sustainability was identified early in the coding process as a driving issue for northern research. It was added to the coding framework and tracked throughout the documents, providing insight into research priorities. Sustainability was coded by searching for explicit mentions of the word sustainability, as well as coding for discussions of issues or priorities related to complex changing environmental, sociocultural, or economic conditions. Environmental sustainability issues included impacts of climate change, resource development, and shipping. Sociocultural sustainability was identified in relation to issues like community well-being, health disparities in the North, food and water security, and cultural concerns like language revitalization. Economic sustainability was discussed in relation to economic diversification, tourism, institutional research capacity, and funding stability.

### *3.3. Description of Documents*

The analysis focused on several types of documents, including strategies, action plans, discussion papers, reports, guides, statutes, information brochures, and webpages (Figure 3). Some of these documents communicate internal policies, such as how an organization will support and fund research or how they will incorporate science into decision-making [1,3]. Other documents communicate guidance or expectations to external stakeholders, for example, by discussing concerns with current research practices and expectations for how research should be conducted. Some research-guidance documents outline internal policies on how an organization approaches their own research [106,107], while others provide guidance to external researchers or organizations in conducting research [108–110]. Most of the documents were produced from 2000 onwards (91%). This temporal scope reflects the limited availability of digital documents prior to that time period, the project focus on contemporary science policy, and the more recent emphasis by various organizations on articulating science and research policy.

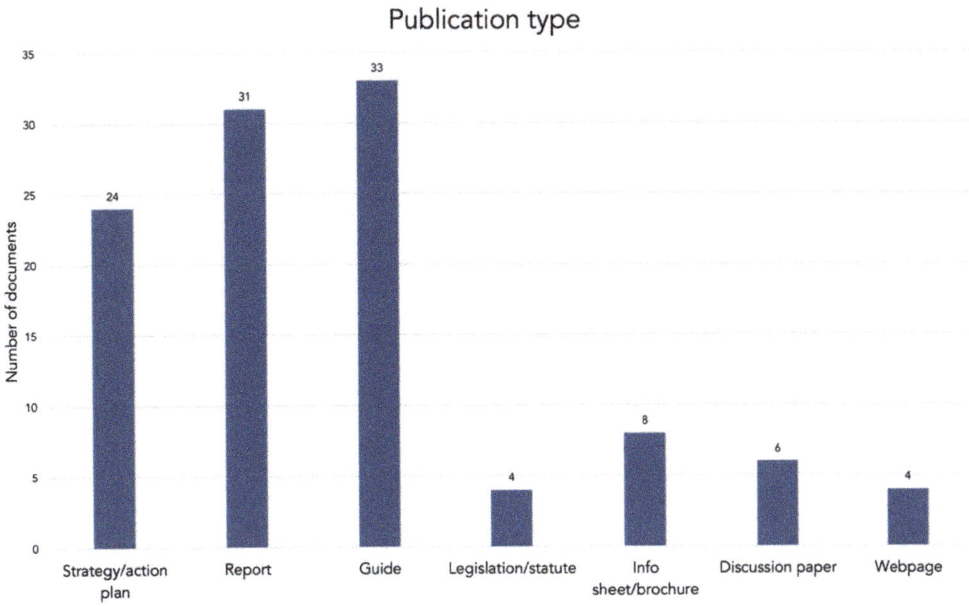


**Figure 3.** Documents analyzed according to publication type, $n$ = 110.

Most of the documents (88, or 80% of the total) focused on research in general, although there were also some with a discipline-specific focus, including social sciences (4%), humanities (2%), health research (16%), physical and natural sciences (3%), and Indigenous research (meaning research using Indigenous research methodologies) (6%). Some of the policy documents communicate a protocol or provide guidance on how to apply a protocol, such as Indigenous knowledge (or Traditional knowledge) protocols and policies and regional research licensing protocols. Documents produced by the Canadian Arctic Research Licensing Initiative (CARLI), which reviewed the licensing and permitting processes in the three territories during IPY, were also included in this category. Although they are not regulatory documents themselves, they provide information on regulatory processes. Protocols and associated documents were communicated through different types of publications, including legislation, guides, and information sheets (Figure 3). Within the 110 documents reviewed, 19 were about a protocol at the regional or pan-Arctic scale, while 6 were focused on a local protocol targeted at a community or a subregion.

All documents were tracked according to author type (Figure 4), and any documents produced by a consultant on behalf of another organization were included in the main author type of the organization that initiated the document (e.g., NGO documents cited under the individual author's name). Most of the documents we reviewed were produced by the Government of Canada (27%), including by POLAR, the Canadian Polar Commission, and federal advisory boards (e.g., Science Council of Canada). National Indigenous organizations produced 19% of documents, including organizations that represent and advocate for Inuit, First Nations, and Métis on the national or pan-northern scale (e.g., ITK, AFN). Regional and local Indigenous governments produced 17% of documents, including Indigenous governments, Indigenous land-claim and co-management organizations, and other northern Indigenous organizations (e.g., Sambaa K'e Dene Band, NTI, Gwich'in Renewable Resources Board (GRRB), CYFN). Documents produced by the territorial governments (13%) include policies produced by the governments of the Yukon, NWT, and Nunavut. Northern NGOs (13%) include regional and community-based non-profit organizations that are involved in research (e.g., Arctic Institute of Community-Based Research, Qaujigiartiit Health Research Centre). Northern academic institutes (6%) include the northern colleges, universities, and research institutes (e.g., Aurora Research Institute, Yukon Research Centre). Southern NGOs (5%) include documents produced by associations focused on northern research (e.g., Association of Canadian Universities for Northern Studies, International Arctic Social Sciences Association). Southern academic institutes (2%) include documents produced by research networks led by southern universities (e.g., Resources and Sustainable Development in the Arctic (ReSDA), Yukon Initiating Group of the Canadian Mountain Network). The one provincial government document (1%) was produced by the Gouvernement du Québec and relevant to Nunavik. Two groups that were identified as potential author types but did not produce policy documents included in this study include renewable resource councils/hunter and trapper organizations (RRC/HTOs) and communities/hamlets.

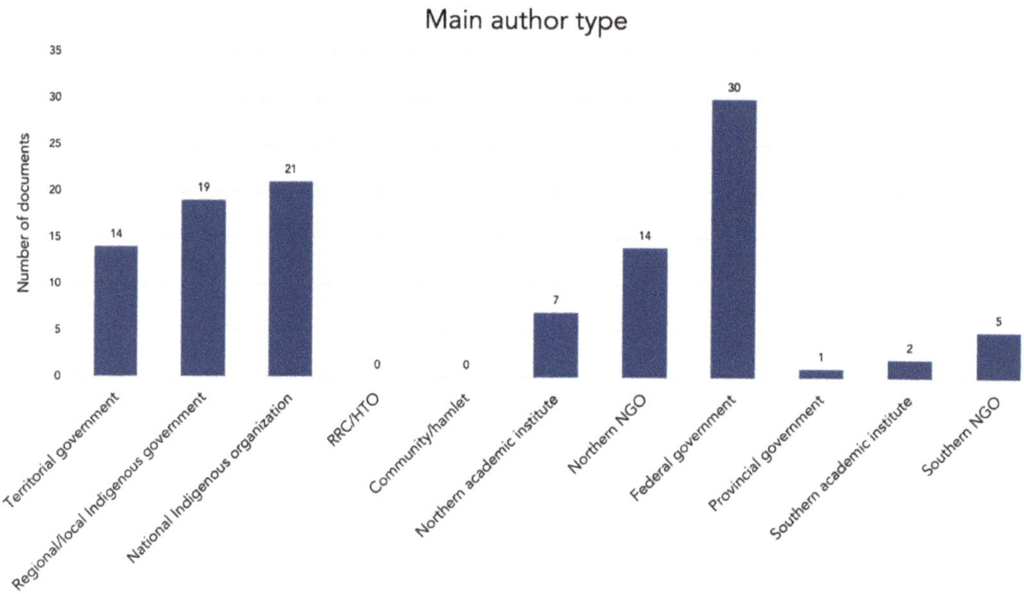


**Figure 4.** Documents analyzed according to author type, $n = 110$ *. * Number of authors is larger than the total number of documents as 3 documents are co-authored.

In terms of geographic coverage, half of the policy documents analyzed have a pan-northern focus, with many others having a regional focus (Table 1). Far fewer focus on the subregional level, with only one focused on the community level. Furthermore, the documents analyzed are relevant across the five northern regions (Figure 5). Of the 53 pan-northern documents, 27 were focused on all five regions and were mainly produced by the Government of Canada, southern NGOs, or southern academic networks. However, some of the pan-northern documents focused on a subset of regions, usually either at the pan-territorial scale (the three territories) or Inuit Nunangat scale (the four Inuit land-claim regions). The only region-specific document identified in this study for Nunavik was the Québec research and innovation strategy 2017–2022 [111], and none was identified for Nunatsiavut. All other documents coded to those two regions are pan-northern, either focused on all five regions or across Inuit Nunangat. Although the Inuvialuit Settlement Region (ISR) is geographically in the NWT and the Yukon, settled Inuvialuit communities are all in the NWT, so relevant documents were coded to the NWT. There are two exceptions: the documents produced by the Wildlife Management Advisory Council (North Slope), a co-management board related to the Inuvialuit Final Agreement, are specific to the Yukon North Slope region of the ISR but relevant to the Inuvialuit population in NWT and were therefore coded to both the Yukon and NWT.

**Table 1.** Documents according to regional-scale coverage, $n = 110$.

| Regional Scale | Number of Documents | Description |
|---|---|---|
| Pan-northern | 53 | Relevant to 2 or more regions |
| Regional | 43 | Relevant to a region (Yukon, NWT, Nunavut, Nunavik, or Nunatsiavut) |
| Sub-regional | 13 | Relevant to a subregion or traditional territory (e.g., ISR) |
| Community | 1 | Relevant to a single community |

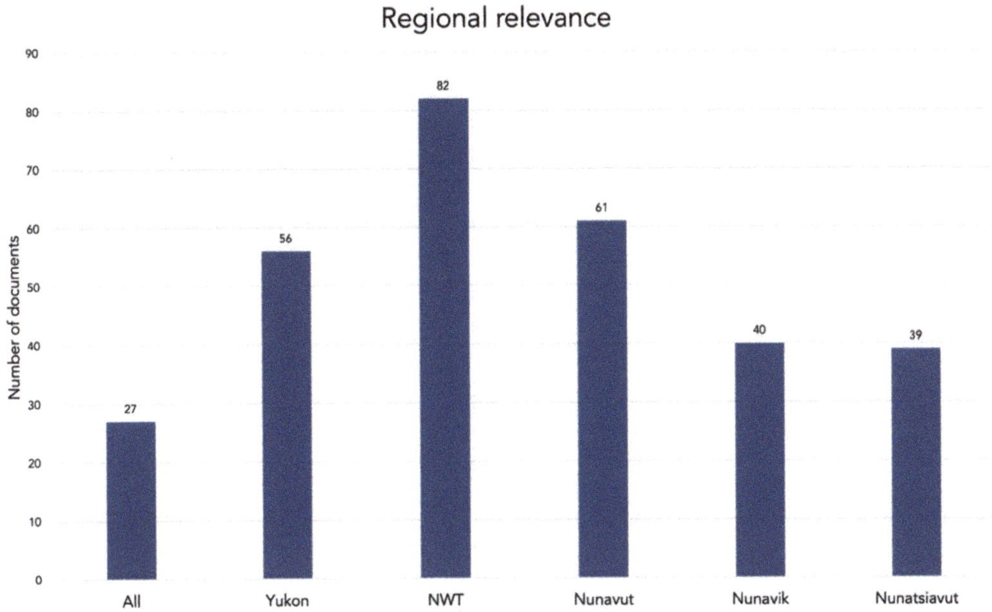


**Figure 5.** Documents analyzed according to regional relevance, $n = 110$.

## 4. Results

The coding process and thematic analysis identified key themes discussed throughout the documents, although diverse lenses were applied to these concepts. The most commonly identified codes include partnership, Indigenous knowledge, research capacity, communicating research, governance capacity, consultation engagement, informing policy decisions, outcomes of research, research benefits, and training (Figure 6). Partnership was emphasized most (in 86 documents), followed by Indigenous knowledge (76), research capacity (72), and communicating research (69), while the remaining codes were evenly distributed (66–64). Several codes are related to the impact that research has (informing policy decisions, outcomes of research, research benefits) or the context in which research is conducted (research capacity, governance capacity), while others focus on approaches to research (partnership, communicating research, and consultation engagement). Two of the codes were related to the steps in a research project lifecycle (informing policy decisions, communicating research), but none was from codes related to roles in research. Many of the documents reviewed discussed the role of researchers generally, without specifying specific types of researchers. There were also numerous mentions of Elders, students, local researchers, and local coordinators, yet the occurrence was much lower than codes from the other two categories.

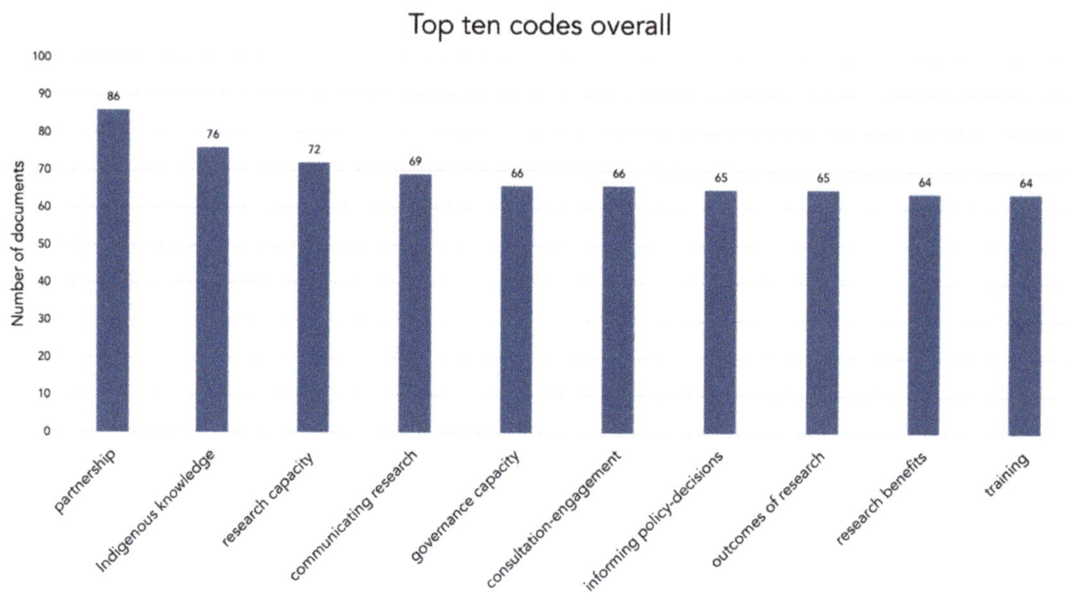


**Figure 6.** Top ten codes used overall by number of documents they are referenced in.

The top ten codes were then cross-referenced with document attributes to explore how they were prioritized by region (Figure 7) and author type (Figure 8). The number of coded documents was compared as a percentage of the total number of documents per region and author type.

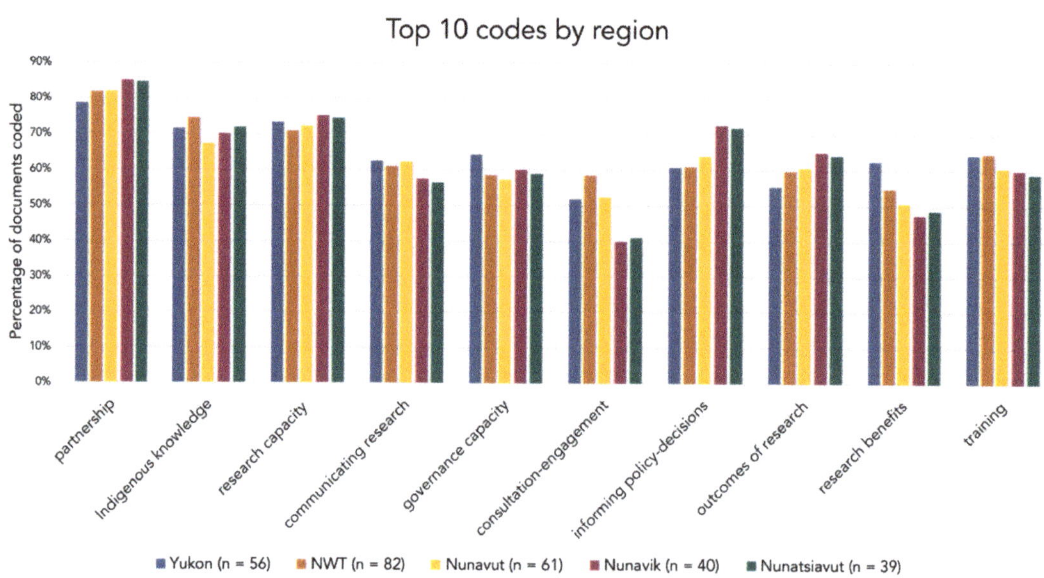


**Figure 7.** Top ten codes analyzed by region.

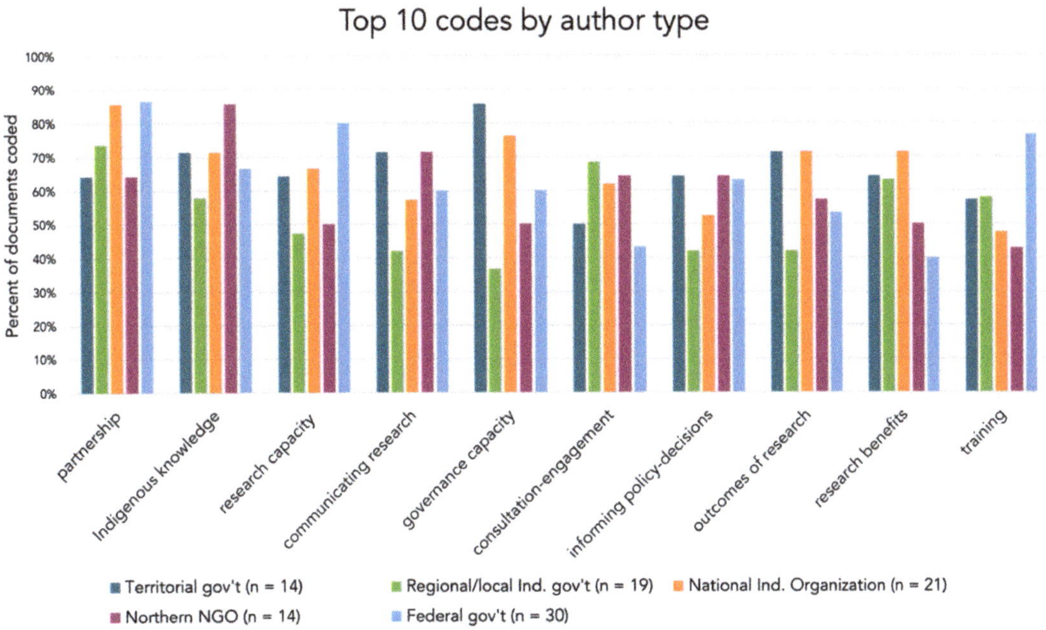


**Figure 8.** Top ten codes analyzed by author type.

There is not a lot of variation amongst the top ten codes overall across the five regions (Figure 7). There is a slightly greater importance put on governance capacity and research benefits in documents relevant to the Yukon, on consultation engagement and Indigenous knowledge in documents relevant to NWT, and on informing policy decisions and outcomes of research in documents relevant to Nunavik and Nunatsiavut (Figure 7). We analyzed the documents related to each region separately to identify the most-used codes in each region, and seven of the top ten codes ranked consistently high in all regions, including partnership, Indigenous knowledge, research capacity, informing policy decisions, governance capacity, and communicating research. Other themes with regional importance were northern participation (Yukon, NWT, and Nunavut), and research agreements (Nunavik and Nunatsiavut).

There was more variability in the thematic coding when the top ten overall codes were cross-referenced by the five most common author types (Figure 8). Indigenous knowledge and communicating research were more commonly coded in documents produced by northern NGOs (Figure 8). Many of these organizations centre Indigenous knowledge or, at the very least, emphasize its importance in their research approach. Communicating research is an important role for northern NGOs (Figure 8), many of which focus on research with communities and Indigenous northerners, often promoting participatory methodologies that require ongoing communication throughout the research project.

Governance capacity was emphasized by territorial governments and national Indigenous organizations as a key focus for research in the North (Figure 9). For northern governments, this directly relates to their role in legislating research in their own regions and advocating for relevant priorities in federal research programs. For national Indigenous organizations, this was discussed in relation to individual research projects and data management [4,112]. They also identified some of the challenges that communities or Indigenous organizations face when engaging in processes that govern research, such as advising on funding programs, participating in research committees, or advocating for priorities in federal research programs. Whereas larger organizations may have the capacity to engage in these processes, supports are required for many Indigenous organizations or community members to participate, including funding for positions, training for necessary skills, or the opportunity to take part in relevant boards and committees [4]. National Indigenous organizations also focused on how research is done, with an emphasis on partnership and research benefits (Figure 8). The federal government emphasized partnership, research capacity, and training, often in relation to their role supporting other organizations to participate in research.

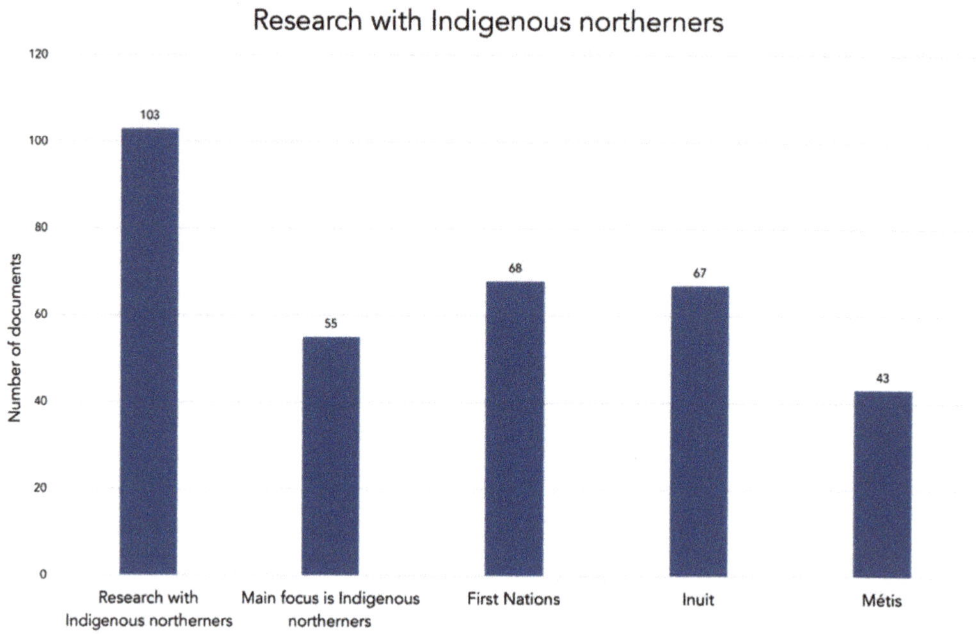


**Figure 9.** Documents that include discussion of research with Indigenous northerners.

The major themes identified through the coding analysis are explained and explored in detail in the following section. The main theme of partnership is explored in Section 4.1 as a foundational element of research relationships. Indigenous knowledge and how Indigenous northerners are included in these documents is the focus of Section 4.2. Research and governance capacity are interrelated and are explored together in Section 4.3, along with training. The outcomes of research, including research benefits, are discussed together in Section 4.4. Informing policy decisions was a major theme that required a separate analysis from the other outcomes of research and is discussed on its own in Section 4.5. Communicating research and consultation engagement are discussed together as related themes in Section 4.6.

### *4.1. Partnership as the Foundation for Research Relationships*

Partnership is identified as a key component of research, although it is rarely defined or contextualized within the policy documents we reviewed. Generally, the concept of partnership is presented in relation to researchers and their relationships with other stakeholders, particularly in terms of developing strong researcher-community partnerships. However, there is also a discussion of the role partnerships can play in governing research, including relationships between governments, other organizations, and communities in setting expectations for research, guiding priorities and topics, and developing northern research programs. In the analyzed documents, the theme of partnership is highly related to the steps in the research-project lifecycle that are involved in developing relevant, useable research, such as identifying priorities and goals for the project, informing policies and decisions, and communicating research.

National Indigenous organizations like ITK, AFN, and the Métis and First Nations centres of the NAHO, emphasize the importance of partnership in research. These organizations advocate at the national scale for Inuit, First Nations, and Métis to be equal partners in research. "In order to truly benefit from research focusing on our people, Inuit must be included as equal partners with researchers at every step of the process" [113]. Partnerships are a means of achieving research goals where capacity is limited for researchers, communities, and northern organizations. Trust, respect, and accountability are identified as integral to a research partnership, and northern participation in research projects was also identified as a core priority. Research relationships can be challenging, but as described by the MC-NAHO [114], "Support and commitment from both sides are needed for quality research ... a mutual commitment between researchers and communities to resolve these issues ... will ensure that the process is respectful."

In an era of Indigenous self-determination in research, Indigenous partners have expectations for how research projects will unfold. Research agreements are recommended as a tool to facilitate partnerships and ensure that community and partner needs are considered, along with the needs of the researcher. However, formal agreements were not discussed as frequently as the importance of building good relationships. A research agreement does not create a relationship; it is a way of formalizing a pre-existing relationship by clearly outlining expectations of all involved. Although a research agreement may not be a place to outline all aspects of the relationship, it can include a process for dealing with challenges and negotiating conflicts if they arise during the project. It is also an opportunity to document plans for storing, using, and sharing data, which has been a contentious issue between Indigenous communities and researchers in the past [112,115]. In some cases, research agreements are required, particularly when communities have a protocol for working with Indigenous knowledge [116–118].

Partnerships are a key factor affecting both governance and research capacity in the North. For the Government of Canada, partnership and governance capacity arose as themes in terms of Canada's relationship with other circumpolar countries, particularly the United States [76]. However, in recent documents produced by the federal government, more focus is placed on the relationship they have with northern governments and ITK. For northern governments and ITK, partnership is an essential component of their capacity to govern or influence the northern research agenda. Each government or organization only operates within its specific sphere of influence (e.g., as a regulator, funder, research agency, or partner) and must look to other governments for those areas where they do not have direct influence. This means that organizations can outline expectations and priorities for research but need to work with others to achieve and promote them. As an example, ITK [4] promotes Inuit self-determination in research by working with partners like governments, universities, research institutes, and academics to enhance the outcomes of research for Inuit. Specifically in relation to the Government of Canada, "the development of an Inuit Nunangat research policy is necessary to coordinate research initiatives among the more than 10 federal departments and agencies that carry out Inuit Nunangat research, and to formalize guidelines for advancing Inuit governance in research" [4].

This approach to partnership is necessary, yet it is also fraught with challenges. In her work on the consultation process for the development of the Arctic Policy Framework, Mary Simon heard numerous concerns about a lack of accountability by the Government of Canada to these relationships:

> "I encountered in my discussions a profound sense of disillusionment, and sometimes distrust, related to agreements with the Government of Canada ... The term co-development of policies with Canada was looked upon with suspicion. My overall impression was that there was a longstanding disconnect between the aspirational intentions and commitments of Ministers, and the paternalistic,

at times obstructionist, approach by the bureaucracy to the implementation of these ideas." [119]

Within the policy documents produced by Government of Canada, there is a changing discourse in relation to partnership that highlights changes in how they approach partnerships, including the recognition of community and regional priorities [48,120].

The contributions of partnerships to the research capacity of northern individuals, communities, and organizations are well documented; however, they can benefit researchers as well. As northerners gain comfort with research and find new ways to contribute to projects, there is more interest in seeking out research opportunities for their communities. As outlined by the GNWT, "The creation and sharing of knowledge is an important legacy for any research project; however, capacity–to plan, initiate and participate in research–is also an important legacy. If researchers are able to share their approaches, ideas and successes the stage will be set for more positive community involvement in research and monitoring in the future." [121]. Local organizations can provide key supports to researchers through partnership if research goals are relevant to the community [99,107]. Reciprocity in research relationships is a fundamental principle for research with Indigenous communities [110,122], and the ability to meet the needs of both the researcher and community is emphasized as part of a strong partnership.

### *4.2. Indigenous Knowledge in Relation to Research*

Most of the reviewed documents recognize Indigenous northerners as a key population for consideration in relation to research, with 103 documents (94%) at least mentioning or discussing research with Indigenous northerners (Figure 9). Many of these documents had a pan-Canadian focus and were relevant to all northern Indigenous populations, as well as non-Indigenous northerners, while others were regionally or locally specific. A few documents are not specific to the North but focus on research with First Nations or Métis on a national scale, and therefore have relevance for First Nations and Métis living in northern Canada. These include documents produced by AFN, FNIGC, the FNC-NAHO, and the MC-NAHO. Pan-northern documents that had relevance to all five regions and did not specify a specific Indigenous group were coded to First Nations, Métis, and Inuit. For regional documents, those relevant to the Yukon were coded to First Nations; those relevant to Northwest Territories were coded to First Nations, Métis, and Inuit; and those relevant to Nunavut, Nunavik, and Nunatsiavut were coded to Inuit. Overall, 68 of the 110 documents (62%) were relevant to First Nations, 67 (61%) to Inuit, and 43 (39%) to Métis (Figure 9).

Of the documents that mention or discuss research with Indigenous populations, 55 (50%) have research with Indigenous northerners or Indigenous peoples as the main focus of the document (Figure 9). These include Indigenous knowledge protocols and guidance documents for research with Indigenous communities [108,116,118]. These also include strategies, priorities, and research policies produced by Indigenous governments and organizations [117,123]. Indigenous-focused documents were more likely to be produced by a national Indigenous organization, regional or local Indigenous government or land-claim organization, or a northern NGO (Figure 10). Regionally, a greater number of these documents were relevant to the NWT (Figure 10), but that reflects the diversity of First Nation, Métis, and Inuit populations in the territory. Among the three Inuit regions, it is interesting to note the higher number of Indigenous-focused documents in Nunavut. This may reflect the increased capacity in Nunavut to engage in research governance, where there are more local organizations producing research policies.

About 80% of all documents (88) included a discussion of Indigenous knowledge as a key source of understanding for northern environments and society, but only 25% (27) provided specific guidance on how to include or engage with Indigenous knowledge (Figure 11). These documents showed a wide range in their depth of discussion on Indigenous knowledge. In some cases, this is limited to a small mention of the importance of including Indigenous knowledge, while others included an in-depth discussion of Indigenous knowledge in research or guidance on how to collect Indigenous knowledge.

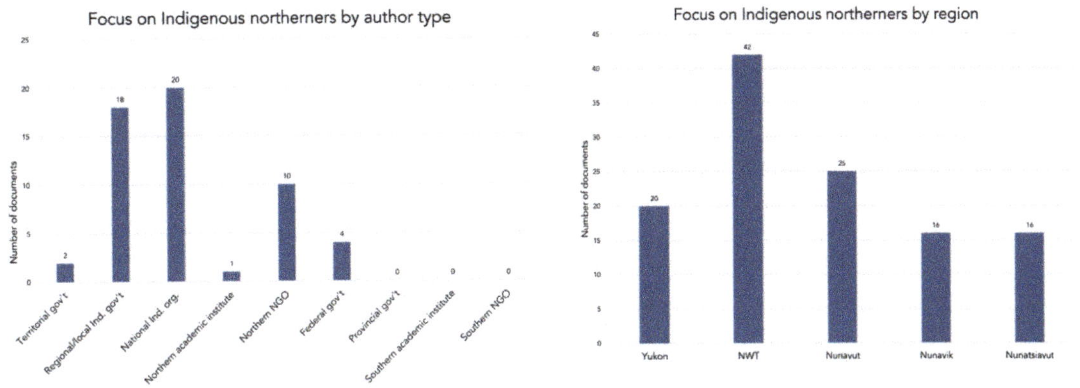


(**a**) (**b**)

**Figure 10.** Inclusion of Indigenous northerners in document focus: (**a**) by author type; (**b**) by region.

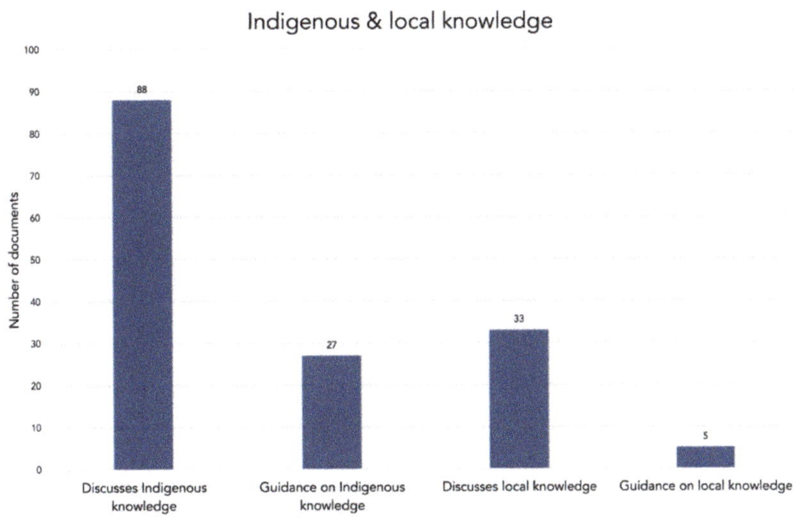


**Figure 11.** Documents with a main focus on Indigenous knowledge and local knowledge.

The 27 documents that provide guidance on Indigenous knowledge are produced by several different types of authors, including mainly regional and local Indigenous governments or land-claim organizations but also national Indigenous organizations and northern NGOs (Figure 12a). There are double the number of documents relevant to the Northwest Territories that provide guidance on Indigenous knowledge compared with the other regions (Figure 12b), again, reflective of the diversity of NWT's Indigenous population.

As well as being used as a document-level attribute to track how many documents referenced Indigenous knowledge, it was also one of the most applied codes in the thematic analysis. The code was only applied where Indigenous knowledge is discussed in depth, often with associated methodological and ethical considerations. Many northern organizations recognize that Indigenous knowledge and science come from different worldviews and should be considered in different ways. The language around Indigenous knowledge changes over time and varies between different organizations. Organizations that come from a western science tradition, particularly federal government departments, tend to discuss Indigenous knowledge as a data source that supports western science. In more recent publications, particularly those by Indigenous organizations and northern governments, the emphasis is placed on respecting Indigenous knowledge as a standalone knowledge system, as opposed to a source of data to be integrated with science [124]. Indigenous organizations emphasize that Indigenous knowledge is a knowledge system and worldview.

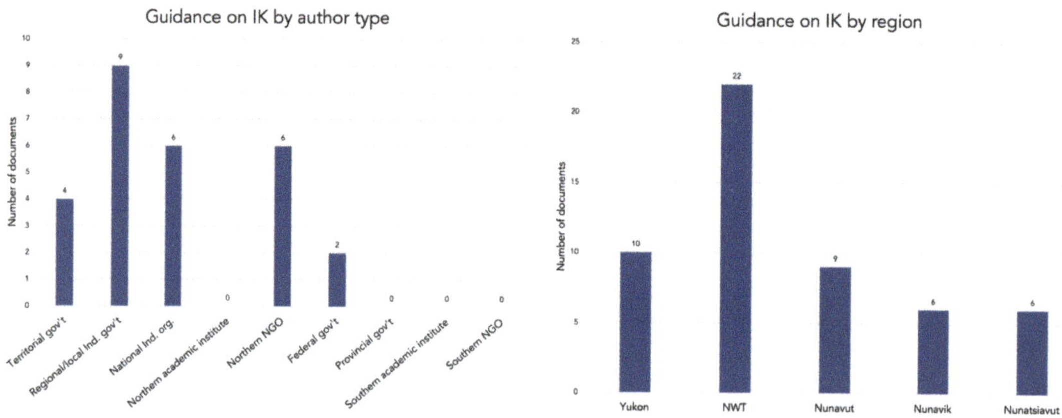


(a) (b)

**Figure 12.** Guidance on Indigenous knowledge: (**a**) by author type; (**b**) by region.

Researchers often approach Indigenous knowledge as another form of topical knowledge that can inform research results and outputs. However, Indigenous knowledge should inform the start of the project, including how research is done, the principles and values that are incorporated in research design, and the relationship between the researcher and the community [124,125]. Some northern and Indigenous organizations advise researchers to recognize and include the interests and priorities of Indigenous northerners in the development of the project and to focus on a respectful, inclusive approach with Indigenous communities, as opposed to prioritizing the integration of Indigenous knowledge in scientific research [104,125]. As an example, the youth of Nunavut that are part of Ikaarvik: Barriers to Bridges highlight that Inuit Qaujimajatuqangit (IQ) is more than just knowledge. "IQ ... incorporates knowledge, customs and values. It is a way of life. It is as much about how we interact with one another, our attitudes and behaviours, as it is about what we know." [125]. This can also extend to ensuring that Indigenous interests and priorities are reflected in broader research policies, for example, in funding programs [4].

Indigenous knowledge was frequently cross-coded with informing policy decisions, indicating the value Indigenous knowledge can bring to the decision-making process. While there was some discussion of bridging Indigenous knowledge and science in designing and conducting research, bridging knowledge sources also comes up as a strategy for informing policy. This can be accomplished by bringing the two knowledge sources together at the decision-making level, as opposed to in an individual research project [104]. Concerns have been expressed that Indigenous knowledge collected by researchers may not result in practical applications, so approaches that ensure that Indigenous knowledge is available for use by Indigenous northerners would ensure that Indigenous knowledge is available for their policy needs [118]. For this to happen successfully, Indigenous knowledge needs to be valued as an equal source of evidence with scientific knowledge. Despite noted challenges in bringing together Indigenous knowledge and science, the Government of Canada has framed Canada as a leader and innovator in bridging Indigenous knowledge and science for evidence-based policy [77].

Data sharing and data ownership were widely used codes also cross-coded with Indigenous knowledge. This highlights the importance of OCAP and other data management frameworks for protecting the rights of Indigenous peoples to own and access their data and to control how it is used and communicated. As the Guidelines for Research with Yukon First Nations points out, "researchers have the ability to construct legitimate arguments for or against ideas, theories or practices. Researchers are collectors of information and producers of meaning, which can be used for, or against Yukon First Nations interests" [108]. There are numerous references to Indigenous knowledge or data being shared without permission or used for applications that were not approved. "Research has been damaging in the past in instances where genetic material is used, sensitive information is published and confidential cultural information is shared inappropriately" [126]. Northern research has been described as extractive, where data is used by researchers but not always returned to the community. "Communities have seen countless researchers come and go, taking samples and data to build their careers and leaving communities feeling used." [127]. In their guide to OCAP, the FNC-NAHO [112] highlights a few ways that Indigenous peoples have been wronged by research, as well as providing options for First Nations communities to apply OCAP principles to support self-determination.

Many organizations note the importance of Indigenous knowledge contributing to our understanding of key social, economic, and environmental sustainability topics in the North, including climate change, wildlife ecology, and well-being. In some cases, for community uptake of research, it is essential that Indigenous knowledge be considered in developing the project, particularly in identifying relevant research questions. "Inuit occasionally dismiss scientific studies (especially those on harvested wildlife species) as unnecessary and irrelevant where they believe the studies will provide knowledge that Inuit already possess" [83]. Indigenous knowledge was discussed as a key source of information for developing policy in relation to land and wildlife management, resource development, and environmental assessment. There is also interest in better understanding methodologies for incorporating Indigenous knowledge into management and planning processes. Sociocultural sustainability topics like language, traditions, and cultural practice are pan-northern priorities and are intrinsically linked with Indigenous knowledge. Language revitalization is a concern across the North, and while it can be the focus of research, it can also be integrated into research through use of Indigenous languages wherever possible.

### 4.3. Research and Governance Capacity

Within this thematic analysis, governance capacity refers to the leadership, oversight, and promotion of research, whereas research capacity is about having the capacity, whether it is tools, resources, infrastructure, expertise, personnel, skills, or other requirements, to engage in the practice of research, including conducting research, managing data, and implementing results in programs and policies [110,128]. Capacity building is not always a north-south relationship; it can happen within a region or across the North [30]. Capacity building is a partnership that requires efforts from both community and research partners and ultimately benefits both [110]. Research is often identified as a sector that requires funding and support to sustain ongoing activities, coordination, employment, and training opportunities for northerners and students. While the federal government has been a primary support for building local capacity by investing in colleges, research institutes, and infrastructure, there are still barriers to local research capacity and a need for support at the institutional, community, and individual scale in the North [48,127]. Funding, partnerships, support for human resources, technology, information and data management, assistance for proposal writing, and research-specific training can all contribute to both institutional and community research capacity [65,112]. Capacity to conduct research includes the need for adequate and sustainable research funding at the community level and for northern organizations and institutions to coordinate, contribute, and participate in research projects, networks, and partnerships, including infrastructure, equipment, or training as required [38,104,115].

Capacity to conduct research and be involved in research is also intertwined with capacity to govern research. Where there is institutional capacity to engage in the oversight of research in the region and in research partnerships, there are more research opportunities for community members and organizations [38]. Involving northerners in research can contribute to both capacity to do research and capacity to use research; participating in research helps to develop skills, but it also helps build local understanding of an issue, which can help with applying the results [115]. Research that builds on existing capacity or community strengths is respectful of Indigenous values and contributes to meeting community needs [65,110]. Data governance was not only identified as a key concern, but it was also an area where Indigenous organizations have made big strides in asserting sovereignty by implementing OCAP or other data-ownership principles. Data management and ownership require the institutional capacity to apply the principles of OCAP [112,128]. Questioning the capacity of Indigenous and northern organizations when they assume control of research or data, instead of supporting their efforts, can interfere with northern leadership in research [112].

International partnerships also play a role in Canada's research capacity, particularly regarding emerging international priorities like climate change. In 2002, it was identified that Canadians were falling behind in research capacity, specifically in funding. Canadian researchers were reliant on their international partnerships to participate in international committees and projects [20]. International partnerships are still an important part of advancing northern scholarship and knowledge of northern systems, allowing Canadian researchers to share knowledge from other regions with northern Canada [77]. International collaboration can happen through research projects but also through formal partnerships, committees, or science associations [77]. Canada's participation in these international arenas depends on the support of the Canadian government [20]. IPY and the northern research chairs were successful examples of national funding to support northern research and foster international collaboration. POLAR and CHARS are examples of how the federal government is continuing to support this capacity. [48,76,77,129,130].

Governance capacity in northern research affects organizations at the local, regional, and national scale. For smaller organizations, this means identifying their limits in terms of support and partnership potential and prioritizing where to expend their resources [104]. Local governance usually includes organizational or advisory-group oversight of research. That can include reviewing research proposals, monitoring projects, and negotiating research relationships, as well as advocating for local priorities and interests [4]. Not all communities or local organizations have the capacity to provide this kind of oversight or have equivalent experience working with researchers [83,115]. Whether or not a community has a formal governance structure for research should not change how researchers approach the community. They should still consult with community leaders and organizations, providing similar opportunities for engagement. They may need to consider providing resources for the community to engage in the project.

At the regional level, territorial governments play multiple roles in governing research: leading and conducting research, developing partnerships, advocating for northern research priorities, influencing other organizations, and regulating research [1,3,65]. The ability to regulate research through licensing or permitting is a key component of governance capacity for the northern regions. However, to be effective, this system requires the active participation of community reviewers. More support for communities to engage in this would improve the licensing process, whether that is through community research advisors or other personnel [38,131]. The capacity of communities to review licensing applications can be affected by staff turnover and the loss of institutional knowledge.

Governance capacity is directly related to research sovereignty, or self-determination in research. When organizations have the capacity to engage in the oversight of research, they can directly influence the research agenda. As compared with funding for training, which supports the development of individual research capacity, support for organizations to engage in the oversight and governance of research can affect whether research meets local and regional needs [38]. Local research capacity is also intertwined with governance capacity, as funding for research capacity in the North relies on the advocacy of northern organizations. Governance capacity includes formal policies and strategies that commit funding for the participation of northerners and northern organizations in research oversight, without which research capacity can be vulnerable [20]. Governance capacity at the organizational level includes the ability to engage in research agreements and data-management agreements, which can require human resources, policy support, and data infrastructure [128]. This capacity ensures that community and regional-scale organizations can advocate for their research priorities and expectations and can implement OCAP and other data-management principles.

### *4.4. Outcomes of Research*

A common expectation across different policy documents is that research should be relevant and have beneficial outcomes for the land, animals, and people in the North and provide specific contributions to the communities and regions where it takes place. Outcomes of research can include tangible benefits but can also include how local residents, organizations, and decision-makers can use the information produced or their experience with the research project.

> "For Inuit, knowledge can only be described as such if it is used to improve the lives of others. If one has knowledge but does not share it or use it for the common good, then it is seen to have no value. In this light, all research must result in direct application to improving the lives of the people who contributed to the knowledge development. This understanding is critical to the design of all research being carried out with Inuit populations, but should be essential in all research anywhere." [106]

Researchers are recognized as experts who possess specialized skills and knowledge and who can help provide information for sound decision-making. Northern organizations and governments value research for the contribution it can make to local issues and northern society, particularly regarding environmental and sociocultural sustainability [108]. Research from all disciplines can contribute to decision-making; however, health and social science research in particular is expected to contribute to sustainability transformations [4,7,132]. Despite being identified as key priorities, health and social sciences, as well as humanities, have generally been underrepresented in northern research [4,7,29,30,48,133]. In Canada's Arctic and Northern Policy Framework, the Government of Canada [48] has recognized that need and committed to supporting more social science research.

Many northerners regard scientific research as a valuable tool to protect public well-being, generate wealth, and to advance knowledge for the benefit of communities and society at large. At the local scale, research can benefit communities by providing employment, training, equipment, contributions to local economy, and honoraria for participants [30,134,135].

> "Research provides much-needed capacity transfer between communities and researchers, opportunities for Indigenous peoples to address issues of local priority, and jobs that put food on the table for many families, and that can become pathways to educational and knowledge development opportunities and stable employment." [136]

However, there are also concerns that research can have adverse impacts on communities and the natural environment. Regulatory processes are in place across the North, including a research licensing process in the three territories and a research approval process in Nunatsiavut, to identify and mitigate potential negative impacts. A commitment to ethical approaches and communication with communities while designing research projects facilitates the pathway to positive outcomes and creates opportunities for northerners to engage with research projects.

### 4.5. *Informing Policies and Decisions*

The role of research in informing policies and decisions was a key theme that arose in the document analysis. Having scientific knowledge available for use in decision-making processes is a key priority, particularly in documents produced by territorial and federal governments, ITK, and regional or subregional Indigenous organizations (e.g., the IRC and CYFN). The code informing policy decisions was applied to two different aspects of research informing policy: (i) as a key step in the lifecycle of a research project (the act of sharing results to affect decisions or policies); and (ii) as an outcome of research (policy has been informed by research). With respect to policy as an outcome of research, there is a desire for research to contribute to social, health, and environmental solutions [2]. While research and science are recognized as important contributors to sustainability decision-making in the North, this is usually discussed in the documents as something that will happen in the future, as opposed to a current reality.

The prevalence of northern policy and decision-making as a theme in the analyzed documents highlights the importance of useable science in the North, which is also a key issue noted in the literature [6,137,138]. While all different types of organizations discuss the contributions research can make to policy and decision-making, the importance of research informing policy is particularly emphasized by Indigenous organizations that are involved in policymaking (e.g., ITK) and governments. However, little advice is given as to how this can be accomplished. There seems to be a common assumption that if research is based on local priorities, there will be uptake by policymakers. Yet, there is not always a causal link between relevant research and action or decision-making. Policy development incorporates a variety of information sources beyond academic research and science, and those can conflict with research results. Policymakers may require data at a different scale or format or may have identified different data gaps than researchers or communities. Participatory research methods are identified as a pathway to impact policy through inclusion of relevant decision-makers in the process. By including decision-makers early in the project, ideally in developing the project goals and research question, the hope is that research will be relevant to their needs and easily accessed.

The documents highlight the disparity between priorities at different scales, leaving gaps in the knowledge that is available and valued for decision-making.

> "The next step in the evolution of scientific practice in the Arctic is linking community-driven Arctic research priorities with national policy development to ensure scientific investments benefit communities and answer key questions facing the Arctic. I firmly believe that the foundation of effective decision-making is good information. In the Arctic, that means being committed to placing equal value on Indigenous knowledge and western science." [119]

There is potential for tension between community needs and policy needs; however, open dialogue early in the project provides the opportunity to address these issues. Effective communication between research policymakers, funders, researchers, and research users can encourage the sharing of results for decision-making.

### 4.6. *Communication and Engagement*

Sharing knowledge is a dynamic process with knowledge flowing in multiple directions between researchers, knowledge holders, policymakers, community members, practitioners, and boundary organizations [4,122,139]. Relationships are the foundation of research in the North, and communication is part of building and maintaining relationships [83,110]. Consultation and engagement relate directly with the themes of communicating research and building research partnerships. Ensuring that research is relevant to the community or partner organization requires consulting with them on priorities and goals for the project. Consultation is not only encouraged; to a certain extent, it is mandated through permitting and licensing processes. Across the North, there are requirements for consulting with rightsholders regarding research; however, there are also examples of researchers not complying or ignoring those requirements. Local protocols, like Indigenous knowledge protocols [116,140], also set expectations for engagement and communication. There is more of an obligation put on social science and health researchers to ensure their research is relevant, particularly as they are more likely to require local participation in their research [110]. However, research priorities for many northern organizations include natural and physical science topics that are relevant to community decision-making and can benefit from local consultation. For example, the GRRB has a list of priority wildlife topics that they recommend researchers engage with them on [123,141]. Researchers in all disciplines are also encouraged and expected to engage with the community before defining their research question to explore how the researcher's goals might align with the community's interests and needs. Early engagement and communication with communities will help researchers avoid initiating projects that the community is not interested in [127].

Although the onus is on the researcher to initiate communication with the community or relevant organizations, ongoing communication throughout a project requires commitment from all partners [114]. As an example, if the community clearly communicates their research expectations during preliminary engagement, there is less room for misunderstanding and conflict, and the project will be more likely to fulfill the community's needs [142]. Clear and ongoing communication was highly related to several of the steps in the lifecycle of a research project, including planning for a new project, identifying priorities and goals, defining questions, and reporting to communities and partners. For the Aqquimavvik Society, communicating research to the whole community starts while identifying goals and developing a shared understanding of the research project, which ultimately translates to informing local decision-making.

> "This is also a time when all of this information is shared and promoted across the community so that there is a collective awareness of the issues and the process being proposed and that by sharing the background information, every community member is then able to consider the topic through the lens of personal experiences and ideas. In this way, the process of research is shared beyond those who are selected informants. This is an approach which seeks to raise critical consciousness across the community and build critical mass through engagement around the issues. It also sets the stage for meaningful knowledge translations of the data results." [106]

Clear communication using plain language and translation where needed is important for establishing an agreement and ensuring that everyone understands what will happen throughout the process.

Communities or local organizations can play a role in developing communication products, particularly in putting results in the broader context. An example is the Inuit Health Survey and the role NTI played in developing relevant communication products [27]. In cases where community members or organizations have directly contributed to the project, there may be an expectation that any reporting of results requires the consultation of local leadership and participants. "Sometimes researchers have published without consulting the community, resulting in negative consequences from publications where communities had no opportunity to correct misinformation or to challenge interpretations" [122]. Expectations for community input or control over the communication process might vary from project to project, depending on the relevance and importance of the topic for the community and the level of community involvement. Of utmost importance is identifying how and when results will be communicated and who has control over those decisions [112,128]. Community reporting is a key communication milestone and step in the research process that is reiterated throughout the documents analyzed.

In terms of when communication is essential within the research process, communication was highly related with the code identifying goals and priorities, which is a critical step for northern participation. When a community or partner organization has influence over research goals, the project is more likely to produce information that is useful locally. The next step in ensuring that results are used is sharing the data and results with interested communities and organizations in a format that is accessible and applicable. These two steps are essential for achieving research outcomes at the local scale. Recommendations for returning results to communities include sharing them through posters, community radio, pamphlets, and hosting a community presentation or open house [83,143]. Depending on the type of information being produced, there may also be a need to present key findings to local leadership for their feedback before publishing. Local governments or partner organizations may want a copy of the data and a report summarizing the project and key findings, particularly if there is relevance for policy. For example, the Gwich'in Tribal Council requests that researchers conducting Indigenous knowledge research present findings to the Chief and Council, the Renewable Resource Council and the Designated Gwich'in Organization while also encouraging researchers to present their findings to community members [144].

## 5. Discussion

Discussions of sustainability are intrinsically linked with how organizations communicate their desired outcomes of research. In discussing the need for and value of research, all different types of organizations conceptualize research as a contributor to environmental, sociocultural, and economic sustainability in the North. Both broad and focused sustainability priorities are identified across the policy documents included in this study. Although physical and natural science topics are a high priority in the North, there are opportunities for all disciplines to contribute to sustainability research. While northern research guidelines and policies are relevant to all research happening in the North, they are critical for sustainability research because of the importance of these issues for northerners. Sustainability research is often interdisciplinary, bringing together researchers from different methodological backgrounds and highlighting disciplinary divides in research approaches. Northern policies address some of those gaps by providing interdisciplinary guidance. The following section discusses key insights into how to align research with policy objectives to ensure research outcomes are relevant to the questions that northern governments and organizations are dealing with. We also bring together the expectations that northern organizations have for researchers as they engage in sustainability research and identify key avenues for academic research to support sustainability decision-making that is inclusive of northern values.

### *5.1. Aligning Research with Northern Policy Objectives*

A key objective in conducting this research was to identify how research programs can align with northern science-policy objectives. With the pressure of growing sustainability issues in the North, finding ways to bridge research with policy is of utmost importance; however, there is a gap in identifying how that knowledge transfer occurs. Using research results to inform policy is usually identified as a step that happens at the end of a research project once the results are compiled and analyzed. However, by identifying potential users of research or policymakers at the beginning of the project and including them in key steps along the way, there is an opportunity to ensure the project can meet policy and program needs. Sustainability issues are often complex and multifaceted and can implicate several different organizations. Part of developing relevant projects is understanding the local context and local policy needs. Therefore, it is important to identify key groups in the community and region, such as organizations and decision-making bodies involved in the issue of concern. Engaging stakeholders that will be involved in incorporating results into decisions or programs during early stages of project development can promote knowledge transfer later in the project. This goes beyond tailoring communication products and includes designing a research question and methodology that are suitable for arriving at data and results that are relevant to policy. While this approach ensures policy priorities are considered at the outset, it needs to be done in a way that is respectful of the leadership and participation of local and Indigenous partners.

Some northern organizations, including the IRC, have recognized that if they want knowledge available to inform their decision-making and programs, they need to play an active role in setting the research agenda for their region [30]. While Arctic research plays an important role in addressing global and circumpolar sustainability issues, it should also support northern and Indigenous self-determination. This involves more than just providing knowledge to support local sustainability but also considerations around local consultation, ethical protocols, and accountability. For northern Indigenous governments and organizations, improving relationships and partnerships is intrinsically linked to their capacity to govern and influence the research agenda and process and, ultimately, self-determination in research [4,5,36,112]. There is strong support within both the academic and policy literature for sustainability research that is relevant and based on community priorities and goals [13,14]. Research that responds to community needs, respects local perspectives of sustainability, and incorporates community strengths can contribute to developing both policy and local research capacity. For researchers, ensuring that research is relevant to the community or partner organization requires consulting with them on priorities and goals for the project. Consultation is encouraged already in northern regions, and to a certain extent, it is mandated through permitting and licensing processes, but it mainly focuses on individual research projects and does not typically extend to policy relevance and governance organizations. Local protocols like Indigenous knowledge protocols set expectations for engagement and communication in research. However, such protocols must also be extended to developing respectful and meaningful policy that considers evidence based on Indigenous ways of knowing, values, and oral traditions.

Respecting northern leadership and self-determination requires careful consideration of how to include northern voices throughout research decision-making. There was a noticeable lack of documents at a community scale or subregional scale focusing on research policy broadly. Instead, those that did exist were focused on Indigenous knowledge, specific methodologies, or guidance on community-based research. This could reflect a lack of resources within those organizations and a need to focus on their immediate experiences with researchers. This puts an onus on the larger organizations that are producing policy documents to engage meaningfully with northern rightsholders and communities and effectively represent their input when putting forth broad northern research policies. It also indicates that there may be a need for resources to support smaller organizations and communities to engage in policy-setting exercises or to develop and communicate their sustainability research priorities to regional organizations and governments. The GRRB provides an example of how this can be done through their community consultations. On their website, they communicate both community research priorities that arise out of those consultations, as well as GRRB organizational research priorities. This process provides a voice for community members while also respecting that the GRRB has a specific mandate and research interests [104,123,141].

### 5.2. Expectations for Engagement in and Conduct of Research

One of the objectives of this study was to understand how organizations expect researchers to engage in and conduct research. The process of developing and conducting research can be as important as the knowledge that is produced [10,13,125]. There are already ongoing conversations in both policy and scientific literature about the importance of participatory methodologies in northern research [14,145,146], and there is guidance available on how to build relationships, engage in partnerships, and coproduce knowledge with northern communities [11,13,57,147–152]. These developments are not unique to the Canadian North, as similar conversations are happening in other jurisdictions, where policies and guidelines are being developed to address many of the same challenges [105,153–156]. Northern policy documents support the foundational concepts of relationship, partnership, and communication that underpin the development of relevant and applicable research.

Although research relationships can be challenged by conflicting values, different worldviews, and competing needs and priorities, these challenges can be overcome by ongoing communication, shared goals, ethical approaches grounded in local needs, and formal research agreements [110,122,128]. The voices and perspectives of northerners are not necessarily reflected in the scientific or policy literature. However, participatory approaches and the coproduction of knowledge are approaches that have shown success in bringing these northern voices in. When these approaches are not suitable and other methodologies are being employed, an emphasis on relationships and communication with local partners can still result in useable research to inform sustainability decision-making. There is no single approach to participation that can be applied to all research projects; rather, there is a continuum of participation that can be implemented according to local needs [83,146]. Identifying where a project appropriately fits on that continuum comes from conversations with communities and local partners and requires dedication of appropriate time and resources for community and partner participation in project development. Local organizations may need financial or in-kind support for personnel to engage with researchers or resources to participate in priority-setting exercises.

Distrust of researchers is the result of past relationships where researchers were not accountable or did not provide tangible results [1,4,27,127,157]. Social license to conduct research is important in the northern context and comes from open communication and follow-through on commitments. Fulfilling the needs of both the community and the researcher can be a source of tension and is often negotiated through an ongoing dialogue. However, there may be research questions that are not addressed in some communities or regions because there is no social license to move forward. Yet, the North is not homogenous, and interest in different topics and methods may vary amongst communities and regions.

Expectations for research conduct are generally upheld by ethical-review processes or other regulatory processes; however, these can conflict with local and Indigenous protocols that value community (collective) consent, as well as the individual consent of participants [122,128]. Northern legislators and regulators support the community-consent process through the review of applications by local and regional organizations; however, without any capacity to enforce compliance, the process is not always accountable to communities [4,122]. Accountability mechanisms are often lacking and can undermine research relationships, programs, and ethical processes.

There is a long history of communities not receiving the results of research or of receiving a copy of a scientific publication that is not understandable or useful. When the community partners play a role in guiding and leading communications, there is more likely to be trust in and uptake of the research results. Communicating research is an iterative process that happens throughout a project, particularly if the goal is to influence policy. Communication with potential research users can create pathways for implementing research results in decision-making. Partner organizations with established relationships in the region may be better suited to communicating results than the researcher, effectively playing the role of a boundary organization by bridging science and policy through two-way communication [158]. This is particularly relevant when communicating sensitive information or Indigenous knowledge, which may require contextualization by community members or partners.

### 5.3. *Supporting Sustainability Decision-Making*

The final objective was to identify avenues for academic research to support policy and decision-making related to sustainability. Northern research is expected to directly impact local decision-making, supporting sustainability transformations in resource development, wildlife management, healthy communities, cultural revitalization, and economic development, among other issues [1,2,148]. Incorporating northern voices, context, and values in developing priorities and goals related to sustainability research is essential to approaching sustainability in a way that is relevant and potentially transformational [86,149]. Sustainability transformation requires local context and input; coproduction and participatory methodologies can support the incorporation of local perceptions of sustainability [90,149]. An example of this is Inuit Qaujimajatuqangit, which has sustainability principles embedded within Inuit societal values, particularly the principle of avatimik kamattiarniq, which "encourages sustainable social/environmental stewardship" [124]. Although the term coproduction was not widely used in the analyzed policy documents, many of the principles of knowledge coproduction were discussed throughout, highlighting that while northern organizations may be slower to adopt the term than the research community, the process itself is important. Self-government and self-determination can support the capacity for Indigenous northerners to engage in sustainability research and to apply the results within their own decision-making organizations [8,150].

Within the broad sustainability priorities discussed throughout the policy documents, variability exists between regions and communities regarding specific research priorities and questions. Overwhelmingly, environmental sustainability is discussed throughout the analyzed documents as a key concern for northern regions and as an expected outcome of northern research. Climate change and resource development are the two major threats discussed in relation to northern environmental sustainability, and climate change is identified as "arguably the single greatest challenge facing the Arctic and its residents" [57]. Social and economic sustainability priorities in the North also include issues like language revitalization and preservation, cultural heritage, mental health and suicide prevention, impacts of residential schools, healthy lifestyles, economic opportunities, and the impacts of tourism and other economies [1,57,66,77,91,113,151,159]. While some sustainability topics that have global significance may not appear to be a local priority, through thoughtful communication and consultation, they may garner local interest. There are countless sustainability research gaps with local relevance and scientific significance.

Research on climate change is noted as a community, regional, and circumpolar issue, yet it is discussed in different ways by different author types. Although climate change, broadly, is a shared priority, there are divergent priorities between different types of organizations. For territorial governments, Indigenous governments, and ITK, climate-change research provides information that can improve local decision-making. The Government of Canada has obligations to both the northern regions and to circumpolar science, which is reflected in their discussion of climate-change priorities. They identify the importance of advancing global knowledge and scholarship of climate change while also acknowledging the importance of contributing to local knowledge of climate change for solving sustainability issues [20]. Priorities at the community scale, as identified by northern NGOs, northern Indigenous governments and organizations, territorial governments, and ITK, include how climate change and resource development impact the land, wildlife, and community well-being in their region. Their priorities are to identify local solutions and adaptations to support sustainability, emphasizing the critical need for climate-change research to be relevant and contribute to local decision-making. Climate change and resource development at the local scale are discussed as cross-cutting issues that impact sociocultural and economic sustainability, as well as environmental sustainability, with wide-ranging impacts on cultural activities, food security, community well-being, and traditional and resource economies, among others [57,81,113,152,159,160].

## 6. Conclusions

This examination of northern research-policy documents from across Canada emphasizes the critical importance of sustainability research in addressing policy and decision-making priorities identified by a broad range of organizations. While there are overarching issues commonly identified across the North, local interests can vary greatly, as can local perspectives on what sustainability entails. Understanding and responding to this diversity are important aspects of addressing local needs. Our analysis serves to highlight the importance of coproduction and participatory approaches in aligning sustainability research and policy. We also articulate the impact that including policy and decision-making organizations can have on creating a pathway from research to evidence-informed policy. Ultimately the policy documents analyzed emphasize the need for sustainability research to be conducted in a way that is attentive to the research process, regardless of methodology, including accountable partnerships, institutional and community ethics, and clear and open communication.

Although the need for research that is inclusive of northerners, focuses on northern priorities, and is led and conducted by northerners has been recognized for years, the most recently produced policy documents and current academic literature still discuss challenges and gaps with research meeting specific northern needs. Although in many ways, northern sustainability research has advanced the coproduction of knowledge, northern organizations continue to focus their limited resources on advocating for better research practices, indicating that there are ongoing issues. Arising from our policy analysis are five key recommendations that can support both researchers and northern and Indigenous organizations to strengthen the research-policy interface for Arctic sustainability:

1. **Develop multiscale, inclusive research partnerships**–Research partnerships can be improved at both the individual and organizational scales to focus on northern sustainability priorities. This involves being inclusive of northern leadership, recognizing northern conceptualizations of sustainability, and ensuring benefits for northerners. Better partnerships with individual researchers result in improved projects and initiatives; however, organizational-scale partnerships (e.g., Indigenous or territorial government to university) have the potential to influence research agendas and improve research across multiple projects. The value of multiscale research partnerships has been recognized for some time; however, they can require considerable investments in personnel and financial capacity by all or some of the partners. When accompanied by policy directives, these investments can be impactful.
2. **Ensure distributed benefits**–To ensure that the benefits of sustainability research are distributed across the North, both larger partner organizations and individual researchers need to consider how they can contribute to local research capacity and where their research could have the most impact. Some communities may be overburdened with research, while others do not benefit because they are lacking capacity to attract and engage researchers. Ensuring that benefits reach underserved communities or organizations may require more time invested in building relationships and more financial support for local participation.
3. **Develop dynamic, tailored communications for different audiences**–Policy documents can be an effective means to communicate an organization's research interests; however, they need to be tailored to the appropriate audience. The messages must be clear, concise, and avoid jargon. While sustainability issues are often complex and persistent, local priorities may change regularly. This type of information needs to be updated regularly and can best be communicated through a website, as opposed to a static report. Developing communication materials and policy documents can require dedicated staff or external support.
4. **Expand on research guidance related to policy contributions**–There are numerous documents already available that provide research guidance and several established ethical and permitting processes for research. Organizations considering creating their own research guidance are encouraged to consider how they can fill gaps in existing guidance documents and avoid duplication of existing advice. Useful additions would be organizational-specific advice on connecting research with policy priorities, including clear guidance on how, when, and who to approach at a specific organization. Articulating specific policy and research interests or priorities can also help to attract relevant research partners. This requires time and human resources on the part of the organization; however, it can minimize duplication and maximize effort in developing tailored guidance where needed.
5. **Create more inclusive and accountable processes**–Organizations that have the resources to develop and communicate broad research policies can create more inclusive

processes to ensure transparency and accountability in how they include and represent Indigenous and community voices. Northern voices have been included in both scientific and policy literature in the past; however, their interests are usually communicated through the lens of the authoring organization. Crediting all contributors, including direct quotes, articulating and respecting Indigenous values, and developing more creative approaches to communicating guidance or policy-relevant findings (e.g., artistic graphics, short, plain-language text with appropriate translation, videos, podcasts, etc.) can help to enhance inclusion and accountability. Supporting the inclusion of smaller organizations and community members in policy development will require dedicated funding for their time to contribute. It also requires openness and transparency in ensuring that the input of contributors is clearly communicated and attributed.

While systematic reviews provide insight into research findings from peer-reviewed literature, they rarely capture research or science policies in the grey literature (e.g., government documents, NGO and Indigenous reports and policies). Research-policy documents are an avenue for northern organizations and governments to communicate their research goals and priorities and advocate for research that responds to their needs and contributes to Arctic sustainability. Policy documents provide insight into the motivations of organizations that are actively involved in guiding, legislating, and conducting research. This study provides unique insight into the policy context of northern research, examining issues from the lens of different types of organizations. Although this study did not examine funding-program documentation, our review does demonstrate the influence of funding programs in shaping how research is governed and conducted. Future research into the funding landscape could help to expand our understanding of the role funding plays in linking research policy with how research is conducted. Relationships, partnership, and communication are the foundation of relevant and applicable research, and our analysis shows that these are also essential in ensuring research can inform sustainability policy and decision-making. Policy applications must be considered at the outset of a research project, while ensuring inclusive and accountable research processes throughout, to present evidence that is meaningful in northern and Indigenous contexts. While common sustainability issues were shared amongst the organizations, addressing diverse perspectives and priorities in different regions and at different scales means that sustainability research needs to be embedded in local needs and perspectives. Ultimately, aligning northern research programs and science-policy objectives requires dedicated time and resources for communication and engagement.

**Supplementary Materials:** The following are available online at https://www.mdpi.com/article/10.3390/su132112035/s1, Table S1: Details of policy documents analyzed.

**Author Contributions:** A.D.P.: concept, methodology, analysis, and writing; G.L.: academic supervision, contributions to concept development, review methodology, thematic code structure, paper review and editing; A.O.: contributions to concept development, paper review, and editing. All authors have read and agreed to the published version of the manuscript.

**Funding:** This research was funded by a doctoral scholarship granted by the Social Sciences and Humanities Research Council (SSHRC) and graduate student funding from the Government of Yukon.

**Data Availability Statement:** The data supporting the reported results in the present study will be available on request from the corresponding author or the first author.

**Acknowledgments:** In developing this research project, we held preliminary planning consultations with numerous organizations across the North, which provided initial direction for this paper. We would like to thank representatives of Tr'ondëk Hwëch'in First Nation, Selkirk First Nation, Arctic Institute of Community-Based Research, Council of Yukon First Nations, Wildlife Management Advisory Council—North Slope, Yukon University, Arctic Institute of North America, Yukon Government, Gwich'in Renewable Resources Board, Inuvialuit Regional Corporation, Joint Secretariat – Inuvialuit Settlement Region, NWT Association of Communities, Aurora Research Institute, Government of the Northwest

Territories, Qaujigiartiit Health Research Centre, Nunavut Tunngavik Inc., Nunavut Research Institute, Government of Nunavut, Makivik Corporation, Inuit Tapiriit Kanatami, Crown-Indigenous Relations and Northern Affairs Canada, and Polar Knowledge Canada. We would like to thank Chris Burn for his feedback on an early draft of this paper, Stephanie Sahl for assistance in creating the map of northern regions, and the students of the StraightUpNorth research group (https://straightupnorth.ca) for their advice and feedback during paper development. We would like to thank the three anonymous reviewers whose thoughtful comments helped us improve this article.

**Conflicts of Interest:** The authors declare no conflict of interest.

## Appendix A. Coding Framework

**Table A1.** Coding framework.

| Coding Topics | Parent Codes | Sub-Codes |
| --- | --- | --- |
| Research concepts | Capacity building | Compliance capacity; Cultural competency; Funding stability; Governance capacity; Research capacity; Review capacity; Training |
| | Communication and engagement | Authorship; Clear communication; Consultation-engagement; Language; Local hiring; Local knowledge; Methods of community engagement [1]; Oral history; Plain language; Translation |
| | Context | Academic standards [1]; Advance northern scholarship; Colonial history of research; Community ignored [1]; Cross-cultural context; Increase/decrease in research; Interdisciplinary; Knowledge network; New research paradigm; Research agenda |
| | Definitions | Definition community; Definition data; Definition Indigenous knowledge; Definition north/Arctic [1] |
| | Ethical protocols | Informed consent; Local approval; Local protocol; Withdrawing from research [1] |
| | Indigenous knowledge (IK) | Documenting IK; Generational knowledge [1]; Indigenous research methodologies; Indigenous worldviews; Integrating IK & science; Sacred knowledge; Traditions [1] |
| | Research outcomes | Impact of research; Research benefits; Sustainability [1] |
| | Partnership | Accountability; Building relationships; Co-developing [1]; Community needs [1]; Northern participation; Research agreement; Research network; Researcher needs [1]; Respect; Understanding |
| | Place | Culture camp; Land claims; Mapping; Research burden; Traditional homelands-territories; Traditional place names |
| Research lifecycle | Research process | |
| | Research planning | Identify funding; Identify project priorities & goals; Preliminary engagement; Recruit project team |
| | Research design | Choose methodology; Create timeline; Develop research question; Identify resource requirements; Literature-information search; Write proposal; Proposal review |
| | Preparing for research | Acquire permits; Acquire research license; Ethical review process; Gain local context; Gain local permissions; Plan logistics |
| | Conducting research | Analyze data; Collect data & information; Compliance requirements; Recording information; Storing data; Validate results |
| | Communicating research | Community reporting; Publications; Reporting; Sharing data; Writing results |
| | Implementing research | Evaluation; informing policy-decision |
| Research roles | | Advisor-mentor; Boundary organization [1]; Community; Coordinator; Elder; Funder; Guide; Legislator-regulator; Local researcher; Participant; Partner organization; Principal investigator; Researcher; Reviewer; Student; Support staff-organization; Team member; Translator; User-consumer; Videographer-photographer; Youth |

[1] Emergent code.

## References


1. Government of the Northwest Territories. *Knowledge Agenda: Northern Research for Northern Priorities;* Government of the Northwest Territories: Yellowknife, NT, Canada, 2017.
2. Government of Yukon; Government of the Northwest Territories; Government of Nunavut. *A Pan-Northern Approach to Science;* Government of Yukon; Government of the Northwest Territories; Government of Nunavut: Whitehorse, YT, Canada; Yellowknife, NT, Canada; Iqaluit, NU, Canada, 2016.
3. Government of Yukon. *Science Strategy;* Government of Yukon: Whitehorse, YT, Canada, 2016.
4. Inuit Tapiriit Kanatami. *National Inuit Strategy on Research;* Inuit Tapiriit Kanatami: Ottawa, ON, Canada, 2018.
5. Brunet, N.D.; Hickey, G.M.; Humphries, M.M. Local Participation and Partnership Development in Canada's Arctic Research: Challenges and Opportunities in an Age of Empowerment and Self-Determination. *Polar Rec.* **2016**, *52*, 345–359. [CrossRef]
6. Ogden, A.E.; Schmidt, M.; van Dijken, B.; Kinnear, L. Science in the Yukon: Advancing a Vision for Evidence-Based Decision Making. *Arctic* **2016**, *69*, 210–221. [CrossRef]
7. Petrov, A.N.; BurnSilver, S.; Chapin, F.S.; Fondahl, G.; Graybill, J.; Keil, K.; Nilsson, A.E.; Riedlsperger, R.; Schweitzer, P. Arctic Sustainability Research: Toward a New Agenda. *Polar Geogr.* **2016**, *39*, 165–178. [CrossRef]
8. Riedlsperger, R.; Goldhar, C.; Sheldon, T.; Bell, T. Meaning and means of "sustainability": An example from the Inuit settlement region of Nunatsiavut, Northern Labrador. In *Northern Sustainabilities: Understanding and Addressing Change in the Circumpolar world;* Fondahl, G., Wilson, G.N., Eds.; Springer International: Cham, Switzerland, 2017; pp. 317–336, ISBN 9783319461502.
9. Fondahl, G.; Wilson, G.N. Exploring sustainabilities in the circumpolar North. In *Northern sustainabilities: Understanding and Addressing Change in the Circumpolar World;* Fondahl, G., Wilson, G.N., Eds.; Springer International: Cham, Switzerland, 2017; pp. 1–12, ISBN 9783319461502.
10. Gearheard, S.F.; Shirley, J. Challenges in Community-Research Relationships: Learning from Natural Science in Nunavut. *Arctic* **2007**, *60*, 62–74. [CrossRef]
11. Tondu, J.M.E.; Balasubramaniam, A.M.; Chavarie, L.; Gantner, N.; Knopp, J.A.; Provencher, J.F.; Wong, P.B.Y.; Simmons, D. Working with Northern Communities to Build Collaborative Research Partnerships: Perspectives from Early Career Researchers. *Arctic* **2014**, *67*, 419–429. [CrossRef]
12. Brunet, N.D.; Hickey, G.M.; Humphries, M.M. Understanding Community-Researcher Partnerships in the Natural Sciences: A Case Study from the Arctic. *J. Rural Stud.* **2014**, *36*, 247–261. [CrossRef]
13. Wilson, K.J.; Bell, T.; Arreak, A.; Koonoo, B.; Angnatsiak, D.; Ljubicic, G.J. Changing the Role of Non-Indigenous Research Partners in Practice to Support Inuit Self-Determination in Research. *Arct. Sci.* **2020**, *6*, 127–153. [CrossRef]
14. Carter, N.A.; Dawson, J.; Simonee, N.; Tagalik, S.; Ljubicic, G.J. Lessons Learned through Research Partnership and Capacity Enhancement in Inuit Nunangat. *Arctic* **2019**, *72*, 381–403. [CrossRef]
15. England, J.H.; Dyke, A.S.; Henry, G.H.R. Canada's Crisis in Arctic Science: The Urgent Need for an Arctic Science and Technology Policy; or, "Why Work in the Arctic? No One Lives There". *Arctic* **1998**, *51*, 183–197. [CrossRef]
16. Hik, D.S. A Northern Science Policy for Canada. In Proceedings of the 2030 North: A National Planning Conference, Ottawa, ON, Canada, 1–4 June 2009. Paper 4.
17. Korsmo, F.L.; Graham, A. Research in the North American North: Action and Reaction. *Arctic* **2002**, *55*, 319–328. [CrossRef]
18. Morris, D.W.; Beaulieu, M.; Hamilton, S.; Hik, D.S.; Lemelin, R.H.; Moses, M.M.; Müller, D.K.; Smith, M.A.; Smol, J.P. The Lakehead Manifesto: Principles for Research and Development in the North. *Arctic* **2013**, *66*, iii–iv. [CrossRef]
19. Pigford, A.E.; Hickey, G.M.; Klerkx, L. Towards Innovation (Eco)Systems: Enhancing the Public Value of Scientific Research in the Canadian Arctic. In *Arctic Yearbook;* 2017; Arctic Portal: Akureyri, Iceland; pp. 1–28.
20. Natural Sciences and Engineering Research Council of Canada; Social Sciences and Humanities Research Council of Canada. *From Crisis to Opportunity: Rebuilding Canada's Role in Northern Research;* Natural Sciences and Engineering Research Council of Canada; Social Sciences and Humanities Research Council of Canada: Ottawa, ON, Canada, 2000.
21. Castleden, H.E.; Morgan, V.S.; Lamb, C. "I Spent the First Year Drinking Tea": Exploring Canadian University Researchers' Perspectives on Community-Based Participatory Research Involving Indigenous Peoples. *Can. Geogr. Géographe Can.* **2012**, *56*, 160–179. [CrossRef]
22. McGregor, D.; Bayha, W.; Simmons, D. "Our Responsibility to Keep the Land Alive": Voices of Northern Indigenous Researchers. *Pimatisiwin J. Aborig. Indig. Community Health* **2010**, *8*, 101–124.
23. Yukon Native Brotherhood. *Together Today for Our Children Tomorrow: A Statement of Grievances and an Approach to Settlement by the Yukon Indian People;* Council for Yukon Indians: Whitehorse, YT, Canada, 1973.
24. Castleden, H.E.; Mulrennan, M.E.; Godlewska, A. Community-Based Participatory Research Involving Indigenous Peoples in Canadian Geography: Progress? An Editorial Introduction. *Can. Geogr. Géographe Can.* **2012**, *56*, 155–159. [CrossRef]
25. Ford, J.D.; Couture, N.; Bell, T.; Clark, D.G. Climate Change and Canada's North Coast: Research Trends, Progress, and Future Directions. *Environ. Rev.* **2017**, *26*, 82–92. [CrossRef]
26. Martello, M.L. Global Change Science and the Arctic Citizen. *Sci. Public Policy* **2004**, *31*, 107–115. [CrossRef]
27. Nunavut Tunngavik Inc. *Annual Report on the State of Inuit Culture and Society 2011–2013;* Nunavut Tunngavik Inc: Iqaluit, NU, Canada, 2013.
28. Wong, C.; Ballegooyen, K.; Ignace, L.; Johnson, M.J.; Swanson, H. Towards Reconciliation: 10 Calls to Action to Natural Scientists Working in Canada. *Facets* **2020**, *5*, 769–783. [CrossRef]

29. Association of Canadian Universities for Northern Studies. *Research Excellence in Yukon: Increasing Capacity and Benefits to Yukoners in the Social Sciences, Humanities and Health Sciences*; Association of Canadian Universities for Northern Studies: Ottawa, ON, Canada, 2017.
30. Association of Canadian Universities for Northern Studies. *Research Excellence in the Northwest Territories: Holistic, Relevant and Ethical Research in the Social Sciences, Humanities and Health Sciences*; Association of Canadian Universities for Northern Studies: Ottawa, ON, Canada, 2018.
31. Pedersen, C.; Otokiak, M.; Koonoo, I.; Milton, J.; Maktar, E.; Anaviapik, A.; Milton, M.; Porter, G.; Scott, A.; Newman, C.; et al. ScIQ: An Invitation and Recommendations to Combine Science and Inuit Qaujimajatuqangit for Meaningful Engagement of Inuit Communities in Research. *Arct. Sci.* **2020**, *339*, 1–14. [CrossRef]
32. Mearns, R. Nunavut, Uqausivut, Piqqusivullu Najuqsittiarlavu (Caring for Our Land, Language and Culture): The Use of Land Camps in Inuit Knowledge Renewal and Research. Master's Thesis, Carleton University, Ottawa, ON, Canada, 2017.
33. Sudlovenick, E. *A Serological Survey and Inuit Qaujimajatuqangit of Ringed Seals (Nattiit) in Frobisher Bay, Nunavut*; University of Prince Edward Island: Charlottetown, PE, Canada, 2019.
34. Price, J. *Tukisivallialiqtakka: The Things I Have Now Begun to Understand: Inuit Governance, Nunavut and the Kitchen Consultation Model*; University of Victoria: Victoria, BC, Canada, 2007.
35. Karetak, J.; Tester, F.; Tagalik, S. (Eds.) *Inuit Qaujimajatuqangit: What Inuit Have Always Known to Be True*; Fernwood Publishing: Halifax, NS, Canada; Winnipeg, MN, Canada, 2017.
36. Snook, J.; Cunsolo, A.; Dale, A. Co-Management Led Research and Sharing Space on the Pathway to Inuit Self-Determination in Research. *North. Public Aff.* **2018**, *6*, 52–56.
37. Moffitt, M.; Chetwynd, C.; Todd, Z. Interrupting the Northern Research Industry: Why Northern Research Should be in Northern Hands. *North. Public Aff.* **2016**, *4*, 32–37.
38. Inuit Tapiriit Kanatami; Inuit Circumpolar Council Canada. *Building Inuit Nunaat: The Inuit Action Plan*; Inuit Tapiriit Kanatami; Inuit Circumpolar Council Canada: Ottawa, ON, Canada, 2007.
39. Grimwood, B.S.R.; Cuerrier, A.; Doubleday, N.C. Arctic Community Engagement during the 2007–2008 International Polar Year. *Polar Geogr.* **2012**, *35*, 189–193. [CrossRef]
40. Pulsifer, P.L.; Gearheard, S.F.; Huntington, H.P.; Parsons, M.A.; McNeave, C.; McCann, H.S. The Role of Data Management in Engaging Communities in Arctic Research: Overview of the Exchange for Local Observations and Knowledge of the Arctic (ELOKA). *Polar Geogr.* **2012**, *35*, 271–290. [CrossRef]
41. Gearheard, S.F.; Aporta, C.; Aipellee, G.; O'Keefe, K. The Igliniit Project: Inuit Hunters Document Life on the Trail to Map and Monitor Arctic Change. *Can. Geogr.* **2011**, *55*, 42–55. [CrossRef]
42. Kulkarni, T.; Watkins, J.M.; Nickels, S.; Lemmen, D.S. Canadian International Polar Year (2007–2008): An Introduction. *Clim. Change* **2012**, 1–11. [CrossRef]
43. Hovelsrud, G.K.; Krupnik, I. IPY 2007-08 and Social/Human Sciences: An Update. *Arctic* **2007**, *59*, 341–348. [CrossRef]
44. Bone, R.M. *The Canadian North: Issues and Challenges*, 5th ed.; Oxford University Press: Don Mills, ON, Canada, 2016.
45. Abele, F.; Courchene, T.J.; Seidle, F.L.; St-Hillaire, F. (Eds.) *Northern Exposure: Peoples, Powers, and Prospects in Canada's North*; Institute for Research on Public Policy: Montreal, QC, Canada, 2009.
46. Dacks, G. Implementing First Nations Self-Government in Yukon: Lessons for Canada. *Can. J. Polit. Sci.* **2004**, *37*, 671–694. [CrossRef]
47. Government of Canada. *Northwest Territories Lands and Resources Devolution Agreement*; Government of Canada: Ottawa, ON, Canada, 2013.
48. Government of Canada. Canada's Arctic and Northern Policy Framework. Available online: https://www.rcaanc-cirnac.gc.ca/eng/1560523306861/1560523330587 (accessed on 15 January 2020).
49. Higginbotham, J.; Spence, J. (Eds.) *North of 60: Toward a Renewed Canadian Arctic Agenda*; Centre for International Governance Innovation: Waterloo, ON, Canada, 2016.
50. Larsen, J.N.; Fondahl, G. (Eds.) *Arctic Human Development Report: Regional Processes and Global Linkages*; Nordic Council of Ministers: Copenhagen, Denmark, 2014; ISBN 9789289338813.
51. Rudolph, A.N. *Breaking down Colonial Borders in Inuit Nunaat through Education*; The Gordon Foundation: Toronto, ON, Canada, 2017.
52. Truth and Reconciliation Commission. *Canada's Residential Schools: The Inuit and Northern Experience*; McGill-Queen's University Press: Montreal, QC, Canada; Kingston, ON, Canada,, 2015.
53. Petrov, A.N. Re-Tracing Development Paths: Exploring the Origins and Nature of the 20th Century's Northern Development Paradigms in Russia and Canada. In *Arctic Yearbook*; 2018; Arctic Portal: Akureyri, Iceland; p. 21.
54. Simon, M. Inuit and the Canadian Arctic: Sovereignty Begins at Home. *J. Can. Stud. Détudes Can.* **2009**, *43*, 250–260. [CrossRef]
55. Wilson, G.N. Nunavik and the Multiple Dimensions of Inuit Governance. *Am. Rev. Can. Stud.* **2017**, *47*, 148–161. [CrossRef]
56. Jeffrey, A.; Fiser, A.; Brender, N.; Dowdall, B. *Building a Resilient and Prosperous North: Centre for the North Five-Year Compendium Report*; Conference Board of Canada: Ottawa, ON, Canada, 2015.
57. Special Senate Committee on the Arctic. *Northern Lights: A Wake-up Call for the Future of Canada*; Senate of Canada: Ottawa, ON, Canada, 2019.
58. Ogden, A.E.; Thomas, M.-E. Re: The Lakehead Manifesto. *Arctic* **2013**, *66*, 508. [CrossRef]

59. Ile Royale Enterprises Ltd. *The Canadian Network of Northern Research Operators Strategic Plan 2016–2021*; Ile Royale Enterprises Ltd: Yellowknife, NT, Canada, 2016.
60. Council of Yukon First Nations. History of Land Claims. Available online: https://cyfn.ca (accessed on 23 April 2019).
61. Government of Yukon. *Yukon Science Policy*; Government of Yukon: Whitehorse, YT, Canada, 1986.
62. Government of Yukon. *Scientists and Explorers Act*; Government of Yukon: Whitehorse, YT, Canada, 2002.
63. Government of Yukon. *Guidebook on Scientific Research in the Yukon*; Government of Yukon: Whitehorse, YT, Canada, 2013.
64. Dedats'eetsaa: Tłı̨chǫ Research & Training Institute. Research Agenda. Available online: https://research.tlicho.ca/about/research-agenda (accessed on 2 June 2020).
65. Government of the Northwest Territories. *Knowledge Agenda: Action Plan 2019–2024*; Government of the Northwest Territories: Yellowknife, NT, Canada, 2019.
66. Government of the Northwest Territories. *Building a Path for Northern Science: Government of the Northwest Territories' Science Agenda*; Government of the Northwest Territories: Yellowknife, NT, Canada, 2009.
67. Aurora Research Institute. *Doing Research in the Northwest Territories: A Guide for Researchers*; Aurora Research Institute: Inuvik, NT, Canada, 2019.
68. Nunavut Tunngavik Inc. *Tukisittiarniqsaujumaviit? A Plain Language Guide to the Nunavut Land Claims Agreement*; Nunavut Tunngavik Inc: Iqaluit, NU, Canada, 2004.
69. Government of Nunavut. *Consolidation of Scientists Act*; Government of Nunavut: Iqaluit, NU, Canada, 2013.
70. Rodon, T.; Grey, M. The Long and Winding Road to Self-Government: The Nunavik and Nunatsiavut Experiences. In *Northern Exposure: Peoples, Powers and Prospects in Canada's North*; IRPP: Quebec, QC, Canada, 2009; pp. 317–343.
71. Makivik Corporation. Nunavik Research Centre. Available online: https://www.makivik.org/nunavik-research-centre/ (accessed on 28 March 2020).
72. Nunatsiavut Government. The Path to Self-Government. Available online: https://www.nunatsiavut.com/government/the-path-to-self-government/ (accessed on 10 March 2020).
73. Nunatsiavut Government. Nunatsiavut Research Centre. Available online: https://nunatsiavutresearchcentre.com/about/ (accessed on 28 March 2020).
74. Nunatsiavut Government. Nunatsiavut Government Research Advisory Committee. Available online: https://nunatsiavutresearchcentre.com/ngrac/ (accessed on 28 March 2020).
75. Stuhl, A. The Politics of the "New North": Putting History and Geography at Stake in Arctic Futures. *Polar J.* **2013**, *3*, 94–119. [CrossRef]
76. Government of Canada. *Canada's Northern Strategy: Our North, Our Heritage, Our Future*; Government of Canada: Ottawa, ON, Canada, 2009.
77. Government of Canada. *Building a Strong Arctic Together: Arctic Policy Framework Discussion Guide*; Government of Canada: Ottawa, ON, Canada, 2017.
78. Government of Canada. Polar Knowledge Canada. Available online: https://www.canada.ca/en/polar-knowledge.html (accessed on 22 March 2020).
79. Inuit Tapiriit Kanatami. *Arctic and Northern Policy Framework: Inuit Nunangat*; Inuit Tapiriit Kanatami: Ottawa, ON, Canada, 2019.
80. Inuit Tapiriit Kanatami. What We Do. Available online: https://www.itk.ca/what-we-do/ (accessed on 6 February 2020).
81. Inuit Tapiriit Kanatami. *2016–2019 Strategy and Action Plan*; Inuit Tapiriit Kanatami: Ottawa, ON, Canada, 2016.
82. Inuit Tapiriit Kanatami. *National Inuit Strategy on Research: Implementation Plan*; Inuit Tapiriit Kanatami: Ottawa, ON, Canada, 2018.
83. Inuit Tapiriit Kanatami; Nunavut Research Institute. *Negotiating Research Relationships with Inuit Communities: A Guide for Researchers*; Nickels, S., Shirley, J., Laidler, G.J., Eds.; Inuit Tapiriit Kanatami; Nunavut Research Institute: Ottawa, ON, Canada; Iqaluit, NT, Canada, 2007.
84. Nickels, S.; Knotsch, C. Inuit Perspectives on Research Ethics: The Work of Inuit Nipingit. *Études/Inuit/Studies* **2011**, *35*, 57. [CrossRef]
85. Petrov, A.N.; Hinzman, L.D.; Kullerud, L.; Degai, T.S.; Holmberg, L.; Pope, A.; Yefimenko, A. Building Resilient Arctic Science amid the COVID-19 Pandemic. *Nat. Commun.* **2020**, *11*, 6278. [CrossRef] [PubMed]
86. Vuntut Gwitchin First Nation. *Integrated Community Sustainability Plan*; Vuntut Gwitchin First Nation: Old Crow, YT, Canada, 2009.
87. World Commission on Environment and Development. *Our Common Future*; Oxford University Press: Oxford, UK, 1987.
88. Kealiikanakaoleohaililani, K.; Giardina, C.P. Embracing the Sacred: An Indigenous Framework for Tomorrow's Sustainability Science. *Sustain. Sci.* **2016**, *11*, 57–67. [CrossRef]
89. Valentin, A.; Spangenberg, J.H. A Guide to Community Sustainability Indicators. *Environ. Impact Assess. Rev.* **2000**, *20*, 381–392. [CrossRef]
90. Sheremata, M. Listening to Relational Values in the Era of Rapid Environmental Change in the Inuit Nunangat. *Curr. Opin. Environ. Sustain.* **2018**. [CrossRef]
91. Canadian Polar Commission. *The State of Northern Knowledge in Canada*; Government of Canada: Ottawa, ON, Canada, 2014.
92. International Expert Panel on Science Priorities for the Canadian Arctic Research Initiative. *Vision for the Canadian Arctic Research Initiative: Assessing the Opportunities*; Council of Canadian Academies: Ottawa, ON, Canada, 2008.

93. Rathwell, K.J.; Armitage, D.R.; Berkes, F. Bridging Knowledge Systems to Enhance Governance of the Environmental Commons: A Typology of Settings. *Int. J. Commons* **2015**, *9*, 851–880. [CrossRef]
94. Bowen, G.A. Document Analysis as a Qualitative Research Method. *Qual. Res. J.* **2009**, *9*, 27–40. [CrossRef]
95. Corbin, J.; Strauss, A. *Basics of Qualitative Research: Techniques and Procedures for Developing Grounded Theory*, 3rd ed.; Sage Publications: London, UK, 2012; ISBN 9781412906449.
96. Charmaz, K. *Constructing Grounded Theory*, 2nd ed.; Sage Publications: London, UK, 2014; ISBN 978-0-85702-9133.
97. Karppinen, K.; Moe, H. What we talk about when we talk about document analysis. In *Trends in Communication Policy Research: New Theories, Methods and Subjects*; Intellect: Bristol, UK, 2012; pp. 177–193.
98. Cardno, C. Policy Document Analysis: A Practical Educational Leadership Tool and a Qualitative Research Method. *Educ. Adm. Theory Pract.* **2019**, *24*, 623–640. [CrossRef]
99. Pielke, R.A. *The Honest Broker: Making Sense of Science in Policy and Politics*; Cambridge University Press: Cambridge, UK, 2007.
100. Berrang-Ford, L.; Pearce, T.D.; Ford, J.D. Systematic Review Approaches for Climate Change Adaptation Research. *Reg. Environ. Chang.* **2015**, *15*, 755–769. [CrossRef]
101. Gough, D.; Thomas, J.; Oliver, S. Clarifying Differences between Review Designs. *Syst. Rev.* **2012**, *1*, 28. [CrossRef]
102. Gough, D.; Oliver, S.; Thomas, J. *An Introduction to Systematic Reviews*, 2nd ed.; Sage Publications: London, UK, 2017.
103. Miake-Lye, I.M.; Hempel, S.; Shanman, R.; Shekelle, P.G. What Is an Evidence Map? A Systematic Review of Published Evidence Maps and Their Definitions, Methods, and Products. *Syst. Rev.* **2016**, *5*, 28. [CrossRef] [PubMed]
104. Gwich'in Renewable Resources Board. Research. Available online: http://www.grrb.nt.ca/research.htm#communities (accessed on 16 May 2020).
105. International Arctic Social Sciences Association. Research Principles. Available online: https://iassa.org/about-iassa/research-principles (accessed on 23 June 2020).
106. Aqqiumavvik Society. ᐋᔩᖃᑎᒌᖕᓂᖅ *Aajiiqatigiingniq: An Inuit Research Methodology*; Aqqiumavvik Society: Arviat, NU, Canada, 2020.
107. Qaujigiartiit Health Research Centre. Our Research Model. Available online: https://www.qhrc.ca/our-approach/our-research-model/ (accessed on 16 May 2020).
108. Yukon Research Centre. *Protocols and Principles for Conducting Research with Yukon First Nations*; Yukon College: Whitehorse, YT, Canada, 2012.
109. Association of Canadian Universities for Northern Studies. *Ethical Principles for the Conduct of Research in the North*; Association of Canadian Universities for Northern Studies: Ottawa, ON, Canada, 2003.
110. Van Bibber, M.; George, A. *Doing Good Health Research in Northern Indigenous Communities: A Guide to Research Review*; Arctic Institute of Community-Based Research: Whitehorse, YT, Canada, 2012.
111. Gouvernement du Québec. *Québec Research and Innovation Strategy 2017–2022*; Gouvernement du Québec: Québec, QC, Canada, 2017.
112. First Nations Centre. *Ownership, Control, Access, and Possession (OCAP) or Self-Determination Applied to Research*; National Aboriginal Health Organization: Ottawa, ON, Canada, 2005.
113. Inuit Tapiriit Kanatami. *Submission to the Naylor Panel for Canada's Fundamental Science Review*; Inuit Tapiriit Kanatami: Ottawa, ON, Canada, 2016.
114. Métis Centre. *Principles of Ethical Métis Research*; National Aboriginal Health Organization: Ottawa, ON, Canada, 2010.
115. Inuvialuit Regional Corporation. *Guidelines for Research in the Inuvialuit Settlement Region*; Inuvialuit Regional Corporation: Inuvik, NT, Canada, 2013.
116. Dehcho First Nations. *Traditional Knowledge Research Protocol*; Dehcho First Nations: Fort Simpson, NT, Canada, 2004.
117. Sahtú Renewable Resources Board. Research in the Sahtú Region. Available online: https://www.srrb.nt.ca/index.php?option=com_content&view=article&id=145&Itemid=649 (accessed on 16 May 2020).
118. Council of Yukon First Nations. *Traditional Knowledge Research Guidelines: A Guide for Researchers in the Yukon*; Council of Yukon First Nations: Whitehorse, YT, Canada, 2000.
119. Simon, M. *A New Shared Arctic Leadership Model*; Government of Canada: Ottawa, ON, Canada, 2017.
120. Polar Knowledge Canada. *Strategic Plan*; Government of Canada: Ottawa, ON, Canada, 2020.
121. Government of the Northwest Territories. *Traditional Knowledge Policy: Implementation Framework*; Government of the Northwest Territories: Yellowknife, NT, Canada, 2004.
122. Healey, G.K.; Walker, J.B.; Heron-Herbert, S. Health Research Ethics in Canada's Northern Territories. 2008. Available online: https://www.qhrc.ca/wp-content/uploads/2019/09/AHRN-Ethics-DEC-2008.pdf (accessed on 16 May 2020).
123. Gwich'in Renewable Resources Board. *GRRB Research and Management Priorities 2018–2023*; Gwich'in Renewable Resources Board: Inuvik, NT, Canada, 2018.
124. Ferrazzi, P.; Tagalik, S.; Christie, P.; Karetak, J.; Baker, K.; Angalik, L. *Describing Aajiiqatigiingniq as an Inuit Consensus Methodology in Health Research*; Aqqiumavvik Society: Arviat, NU, Canada, 2020.
125. Ikaarvik. *SciQ: Science and Inuit Qaujimajatuqangit—Research and Meaningful Engagement of Northern Indigenous Communities*; Ocean Wise Conservation Association: Pond Inlet, NU, Canada, 2018.
126. Council of Yukon First Nations. *Ownership, Control, Access and Possession of Data (OCAP) Resolution*; Council of Yukon First Nations: Whitehorse, YT, Canada, 2006.

127. Government of the Northwest Territories. *Working Together towards Relevant Environmental Monitoring and Research in the NWT*; Government of the Northwest Territories: Yellowknife, NT, Canada, 2004.
128. First Nations Information Governance Centre. *OCAP: Ownerhship, Control, Access and Possession—The Path To FN Information Governance*; First Nations Information Governance Centre: Ottawa, ON, Canada, 2014; ISBN 9780987988287.
129. Government of Canada. *Statement on Canada's Arctic Foreign Policy: Exercising Sovereignty and Promoting Canada's Northern Strategy Abroad*; Government of Canada: Ottawa, ON, Canada, 2010.
130. Government of Canada. *Achievements under Canada's Northern Strategy, 2007–2011*; Government of Canada: Ottawa, ON, Canada, 2011.
131. Purple Dog Consulting. *Canadian Arctic Research Licensing Initiative (CARLI): Researcher Consultation*; Purple Dog Consulting: Ottawa, ON, Canada, 2010.
132. First Nations Information Governance Centre. *First Nations Regional Health Survey (RHS)—Best Practice Tools for OCAP Compliant Research*; First Nations Information Governance Centre: Ottawa, ON, Canada, 2011.
133. Perrin, A. Narrowing the Gap in Northern Research. *North. Public Aff.* **2018**, *6*, 38–40.
134. Brockman, A. *Canadian Arctic Research Licensing Initiative: NWT Research Licensing Workshop with Land and Water Regulators*; Government of Canada; Aurora Research Institute: Ottawa, ON, Canada; Inuvik, NT, Canada, 2010.
135. Brockman, A. *Canadian Arctic Research Licensing Initiative: NWT Research Licensing Workshop Regional/Community Reviewers and Health and Social Science Regulators*; Government of Canada; Aurora Research Institute: Ottawa, ON, Canada; Inuvik, NT, Canada, 2010.
136. Irlbacher-fox, S. *Developing a Sustainable Knowledge Economy in the Northwest Territories*; Hotìı ts'eeda; Tłı̨chǫ Government: Behchoko, NT, Canada, 2018.
137. Ford, J.D.; Knight, M.; Pearce, T.D. Assessing the "usability" of Climate Change Research for Decision-Making: A Case Study of the Canadian International Polar Year. *Glob. Environ. Chang.* **2013**, *23*, 1317–1326. [CrossRef]
138. Brunet, N.D. *Assessing Stakeholder Participation in Northern Scientific Research*; McGill University: Montreal, QC, Canada, 2014.
139. Institute for Circumpolar Health Research. *ICHR Strategic Plan 2015–2020*; Institute for Circumpolar Health Research: Yellowknife, NT, Canada, 2015.
140. Council of Yukon First Nations. *Yukon First Nations' Research Protocol Template*; Council of Yukon First Nations: Whitehorse, YT, Canada, 2018.
141. Gwich'in Renewable Resources Board. *Research and Management Interests for the GSA*; Gwich'in Renewable Resources Board: Inuvik, NT, Canada, 2018.
142. Inuit Tapiriit Kanatami; Nunavut Research Institute. Negotiating Research Relationships: A Guide for Communities. *Pimatisiwin* **2003**, *1*, 17–25.
143. Healey, G.K. *Health Research Ethics Workshop and Community Consultation in Rankin Inlet, NU*; Qaujigiartiit Health Research Centre: Iqaluit, NU, Canada, 2007.
144. Gwich'in Tribal Council. *Conducting Traditional Knowledge Research in the Gwich'in Settlement Area: A Guide for Researchers*; Gwich'in Tribal Council: Inuvik, NT, Canada, 2011.
145. Henri, D.A.; Carter, N.A.; Irkok, A.; Nipisar, S.; Emiktaut, L.; Saviakjuk, B.; Ljubicic, G.J.; Smith, P.A.; Johnston, V. Qanuq Ukua Kanguit Sunialiqpitigu? (What Should We Do with All of These Geese?) Collaborative Research to Support Wildlife Co-Management and Inuit Self-Determination. *Arct. Sci.* **2020**, *207*, 1–35. [CrossRef]
146. Brunet, N.D.; Hickey, G.M.; Humphries, M.M. The Evolution of Local Participation and the Mode of Knowledge Production in Arctic Research. *Ecol. Soc.* **2014**, *19*, 69. [CrossRef]
147. Grimwood, B.S.R.; Doubleday, N.C.; Ljubicic, G.J.; Donaldson, S.G.; Blangy, S. Engaged Acclimatization: Towards Responsible Community-Based Participatory Research in Nunavut. *Can. Geogr. Géographe Can.* **2012**, *56*, 211–230. [CrossRef]
148. Government of Yukon; Government of the Northwest Territories; Government of Nunavut. *Developing a New Framework for Sovereignty and Security in the North*; Government of Yukon; Government of the Northwest Territories; Government of Nunavut: Whitehorse, YT, Canada; Yellowknife, NT, Canada; Iqaluit, NU, Canada, 2005.
149. Blackstock, K.L.; Kelly, G.J.; Horsey, B.L. Developing and Applying a Framework to Evaluate Participatory Research for Sustainability. *Ecol. Econ.* **2007**, *60*, 726–742. [CrossRef]
150. Petrov, A.N. Exploring the Arctic's "Other Economies": Knowledge, Creativity and the New Frontier. *Polar J.* **2016**, *6*, 51–68. [CrossRef]
151. Polar Knowledge Canada; Pitquhirnikkut Ilihautiniq/Kitikmeot Heritage Society. *A Guidebook for Research with Nunavut Communities: Iqaluktuuttiaq (Cambridge Bay) Edition*; Polar Knowledge Canada; Pitquhirnikkut Ilihautiniq/Kitikmeot Heritage Society: Cambridge Bay, NU, Canada, 2016.
152. Shirley, J. *C-CIARN North—Nunavut Community Research Needs Survey: Summary Report*; Nunavut Research Institute: Iqaluit, NU, Canada, 2005.
153. Kawerak Inc.; Raymond-Yakoubian, B.; Raymond-Yakoubian, J. Research Processes and Indigenous Communities in Western Alaska: Workshop Report. 2017. Available online: https://kawerak.org/wp-content/uploads/2018/04/Research-Processes-and-Indigenous-Communities-in-Western-Alaska-Workshop-Report.pdf (accessed on 16 May 2020).

154. Kawerak Inc.; Aleut Community of St. Paul Island; Association of Village Council Presidents; Bering Sea Elders Group. Letter to the National Science Foundation's Navigating the New Arctic Program. 2020. Available online: https://kawerak.org/natural-resources/knowledge-sovereignty/ (accessed on 4 October 2021).
155. Omma, E.M.; Scheepstra, A.; Saxinger, G.; Dale, B. Co-Creating Research Projects—Some Personal Experiences from Saami Council and Arctic Researchers. 2020. Available online: https://www.saamicouncil.net/news-archive/co-creating-research-projects-some-personal-experiences-from-saami-council-and-arctic-researchers (accessed on 4 October 2021).
156. Stone, R. Indigenous Alaskans demand a voice in research on warming. *Science* **2020**, *369*, 1284–1285. [CrossRef] [PubMed]
157. Canadian Institutes of Health Research; Institute of Aboriginal Peoples' Health; Institute of Nutrition Metabolism and Diabetes. *Health Research Needs North of 60: Northern Town Hall Meetings, September 2001*; Canadian Institutes of Health Research: Ottawa, ON, Canada, 2002.
158. Knapp, C.N.; Trainor, S.F. Alaskan Stakeholder-Defined Research Needs in the Context of Climate Change. *Polar Geogr.* **2015**, *38*, 42–69. [CrossRef]
159. Lichen Consulting. *Qaujigiartiit Strategic Plan 2018–2028*; Qaujigiartiit Health Research Centre: Iqaluit, NU, Canada, 2018.
160. Yukon Initiating Group. *Canadian Mountain Network Yukon Initiating Group Discussion Paper*; Canadian Mountain Network: Edmonton, AB, Canada, 2017.

*sustainability* MDPI

*Perspective*

# The Arctic Highlights Our Failure to Act in a Rapidly Changing World

**Peter Schlosser [1,*], Hajo Eicken [2], Vera Metcalf [3], Stephanie Pfirman [4], Maribeth S. Murray [5] and Clea Edwards [1]**

1 Global Futures Laboratory, Arizona State University, Tempe, AZ 85281, USA; clea.edwards@asu.edu
2 International Arctic Research Center, University of Alaska Fairbanks, Fairbanks, AK 99775, USA; heicken@alaska.edu
3 Eskimo Walrus Commission, Nome, AK 99762, USA; vmetcalf@kawerak.org
4 College of Global Futures, Arizona State University, Tempe, AZ 85281, USA; spfirman@asu.edu
5 Arctic Institute of North America, University of Calgary, Calgary, AB T2N 1N4, Canada; murraym@ucalgary.ca
* Correspondence: schlosser@asu.edu

**Abstract:** In this perspective on the future of the Arctic, we explore actions taken to mitigate warming and adapt to change since the Paris agreement on the temperature threshold that should not be exceeded in order to avoid dangerous interference with the climate system. Although 5 years may seem too short a time for implementation of major interventions, it actually is a considerable time span given the urgency at which we must act if we want to avoid crossing the 1.5 to <2 °C global warming threshold. Actions required include co-production of research exploring possible futures; supporting Indigenous rights holders' and stakeholders' discourse on desired futures; monitoring Arctic change; funding strategic, regional adaptation; and, deep decarbonization through transformation of the energy system coupled with negative carbon emissions. We are now in the decisive decade concerning the future we leave behind for the next generations. The Arctic's future depends on global action, and in turn, the Arctic plays a critical role in the global future.

**Keywords:** Arctic; COVID-19; Indigenous rights; climate change; co-production; desired futures; adaptation; mitigation; decarbonization; rapid change

**Citation:** Schlosser, P.; Eicken, H.; Metcalf, V.; Pfirman, S.; Murray, M.S.; Edwards, C. The Arctic Highlights Our Failure to Act in a Rapidly Changing World. *Sustainability* **2022**, *14*, 1882. https://doi.org/10.3390/su14031882

Academic Editor: Oran Young

Received: 19 October 2021
Accepted: 2 February 2022
Published: 7 February 2022



## 1. Introduction

Through feedback and coupled processes, the Arctic system amplifies global change signals including global warming. Arctic amplification increases temperature rise by a factor of between 2 and 4 at the pan-Arctic scale [1]. In 2016, a workshop called 'A 5C Arctic in a 2C World' was convened by the Columbia Climate Center to explore the implications of projected temperature increases in an Arctic context [2]. The report from the workshop identified a series of measures, couched in terms of recommendations to the First Arctic Science Ministerial (ASM) [3], to inform and advance effective responses to a rapidly changing Arctic. Five years later, where do we stand on some of these proposed actions? Has progress been made; has the rate of change outpaced responses? Is it time to rethink some of what was proposed?

As participants in the 2016 workshop, we briefly review the trajectory of the Arctic system in a warming world and revisit the recommendations of the 2016 workshop report, referred to in this contribution as the 5C Arctic report [2]. The contribution is not meant to be comprehensive, in particular with respect to research updates, many of which are reflected in recent Intergovernmental Panel on Climate Change (IPCC) reports [1,4] and the annual Arctic Report Cards [5,6]. Rather, we focus on the exploration of actions taken to mitigate warming and adapt to change over the past 5 years. Although 5 years may seem too short a time for the implementation of major interventions, it actually is a considerable time span given the urgency at which we must act if we want to avoid crossing the 1.5 to <2 °C global warming threshold set by the Paris Accord [7] and, more recently, the IPCC Special Report SR15 (the '1.5 degree report') [8]. According to model projections, in order to avoid crossing this threshold, we need to reach net zero carbon emissions by the middle of this century. This means that we have only a decade left to cut emissions by half, which requires cutting emissions by ca. 7 to 8 percent each year starting now, i.e., in 2022.

Discussion of the root causes underlying the failure to address climate change at the global or national scale is beyond the scope of this contribution. We acknowledge that the rapidity of change outpaces timescales over which most societies are able to develop consensus and commitment to action. Some of the discussions at COP 26 illustrate this circumstance through the lens of the underlying value systems including equity and climate justice. At the same time, there are indications that research, focused on understanding how societies—through social learning and effective decision making—can address major threats such as climate change, is starting to bear fruit. Below, we provide examples of such progress on scales relevant to Arctic issues.

## 2. Materials and Methods

In this perspective, we first consider continuing rapid change throughout the Arctic system including secular changes, abrupt changes, and shocks. We then explore the question: are we on the right response trajectory at the required pace? Issues reviewed include the co-production of research exploring possible futures, supporting rights holder and stakeholder discourses on desired futures, the need to monitor Arctic change, and funding strategic and regional adaptation—in contrast to the ad hoc adaptation and self-adaptation that has mostly been the case thus far. We then address the fact that the Arctic's future depends on global action, and, in the inverse, that the Arctic plays a critical role in the global future.

## 3. Results

### *3.1. Continuing Rapid Change throughout the Arctic System*

#### 3.1.1. Secular Changes

Since the 2016 workshop, we have seen continuing trends of the key pressure points on the Arctic system. Cryospheric changes have resulted in major, compounding impacts on ecosystems and environmental system services important to Arctic Indigenous Peoples and humanity as a whole [4]. Specifically, summer sea ice reduction, likely unprecedented over the past millennium, continues unabated [4]. Loss of sea ice is resulting in major shifts with implications for food webs and marine living resources [9]. Winter sea ice loss in the Bering Sea, tied to anthropogenic warming [5,10], has led to a northward shift of fish stocks, with disruption of subsistence and commercial fisheries, compounded by marine heatwaves [11].

At the same time, a consensus has emerged that sea ice loss is reversible with decreasing atmospheric greenhouse gas concentrations [12]. Permafrost thaw and degradation, on the other hand, are irreversible on decadal to centennial timescales and have reached record levels, threatening the release of greenhouse gases from large reservoirs [13]. The increased surface melt of the Greenland Ice Sheet has contributed significantly to accelerated global sea level rise, with Greenland expected to continue as the single largest contributor to global sea level rise in the coming decades [14]. Ice sheet loss mechanisms suggest a greater vulnerability to ocean heat increases and surface melt than previously envisioned, emphasizing the potential for greater uncertainty and underestimation of Greenland's contribution to sea level rise [1,4].

Terrestrial Arctic ecosystems are undergoing substantial changes, ranging from "greening" of the land cover to the increasing importance of wildfires in landscape disturbance and carbon loss [15,16]. These changes are stressing Arctic communities and societal dynamics, including challenges related to transportation, infrastructure, and food security.

For example, residents are facing the ongoing loss of biodiversity and decline in health and populations of critical subsistence species both in marine and terrestrial environments.

These latter threats are tied to health and well-being, in particular through food security. The Utqiaġvik Declaration expressed this in the following way, "Inuit food security is multi-faceted and reflective of interconnecting elements, such as language, child development, mental and physical health, high cost of transportation, economic development, and management. The Arctic's living resources and the ability of our hunters to harvest and process these resources are fundamental to food security and core to Inuit identity, making the health and availability of Arctic wildlife of utmost concern" [17]. This is fundamentally true for all Arctic Indigenous Peoples—hunters, herders or otherwise.

Reflecting on the 6th Assessment Report of the IPCC [1], Inuit Circumpolar Council (ICC) Chair Dalee Sambo Dorough stated, "Inuit have moved beyond 'if' climate change is real to action to protect Inuit Nunaat—our Inuit homeland—including the Arctic land, sea ice and the Inuit way of life. Inuit have been calling for immediate action to contain temperature rise to 1.5 °C, as even this increase will see the reduction in Arctic sea ice, snow cover, and permafrost loss continue. Both the Policy Summary and the Technical Summary note with high confidence that the rate change continues, with sea ice becoming younger, thinner, and more dynamic (very high confidence). Such change has severe consequences for our food security and multiple other aspects of our day to day lives" [18].

The current work indicates major changes in ecosystems, including the appearance of novel contaminants [19] and invasive species [20]. Expanded shipping and fishing in Arctic waters along with wind and ocean currents bringing and accumulating microplastics result in Arctic communities having to deal with increased levels of debris, fishing gear, and microplastics [21].

Concomitant with these changes, adaptation is occurring at both local and regional scales. An example of a new adaptive strategy is vegetable farming [22], such as off-grid containerized agriculture [23]. However, many such approaches and innovations are fragmented, disconnected, and/or still in development stages and hence lose adaptive significance and do not increase overall resilience (e.g., [24]). The lack of strategic adaptation naturally leads to ad hoc self-adaptation of the Arctic system that has the potential to add additional challenges to our capacity to respond to the multi-dimensional and highly interconnected set of changes seen in the Arctic system.

#### 3.1.2. Abrupt Changes and Shocks

Studying past changes manifested in paleo archives, such as ocean sediments or ice cores, reveals that changes in complex systems—and, the Earth system is the ultimate complex system—can occur relatively smoothly. However, they also demonstrate that typically these complex systems also show rapid (abrupt) changes. During 2020 and 2021 we have been reminded by the rapid emergence of COVID-19 how quickly changes can occur—in essence, as shocks to the system [25].

The COVID-19 pandemic created another unsettling stress on Arctic Indigenous communities compounding the already significant challenges presented by the multiple pressures they have experienced in recent decades [26]. Most communities responded with great concern and established recommended social distancing protocols between households, with preliminary research suggesting that such measures were effective relative to regions at lower latitudes [27]. Travel was restricted and involved extended quarantines. As for so many people, this caused separation and a sense of isolation. However, Arctic Indigenous people greatly depend on large, extended families for their social, cultural, nutritional, and emotional well-being. Indigenous food security in small village communities that rely on communal harvests and sharing practices were altered because of social distancing. Significant community gatherings for celebrating, honoring, and healing were interrupted for over a year, which has caused unresolved public health concerns.

COVID-19 alerted us to how vulnerable our highly interconnected Earth system is and how shocks to one part of it will ripple through the entire system, including all environmental and societal components [25]. However, it also showed that society has the capacity to react quickly and offer responses such as testing to diagnose the extent of the impact and vaccination to control the pandemic. Additionally, we learned that measures that are available to minimize adverse effects of particular pressure on our planet or one of its subsystems are frequently not taken up by parts of local, regional, and global communities. And, we learnedthat even when uptake is desired, solutions may not be readily implemented due to wealth, infrastructure, and other disparities. In the end, it also highlights that the deciding factor for how we master challenges to the Earth system is not the availability of (technological) solutions, but our willingness to act. Additionally, in many cases, we are too slow in our response, thereby increasing the challenges for the present and future generations to thrive on a healthy planet.

### 3.2. *Are We on the Right Response Trajectory at the Required Pace?*

As highlighted in the 2019 IPCC Special Report on the Oceans and the Cryosphere [4], governance systems at the pan-Arctic and global scales are challenged by the rapidity and interconnectedness of changes, pushing existing response mechanisms to the brink and straining the limited capacity at the level of Arctic communities and regions. However, there has also been progress on several fronts. While since 2016 we saw several major changes in the political and governance systems with slow or no tendency to action related to the threats faced by global society around the world, we are noting a recent reversal of some of these trends, initiatives, and legislation. Especially when it comes to climate solutions, there is a significant set of actions that are under consideration or have been moved forward.

#### 3.2.1. Co-Production of Research Exploring Possible Futures

The 5C Arctic report recommended: "Enhance and support research in projecting which future states of the Arctic are possible in principle, under which conditions they can be reached, and which impact they would have" [2]. Some progress has been made in recent years recognizing the role of different approaches and knowledge systems—in particular Indigenous knowledge (IK) [28,29]—in describing the plausible range of future states of the Arctic system and the impacts of changes of the present state, from the local to the global scale. However, holistic approaches and mechanisms for bringing relevant knowledge to bear on Arctic issues remain peripheral and under-resourced (see also Table 1).

Earth system models, as well as multi-sector dynamics or integrated assessment models, hold significant promise but are currently not able to capture the full range of plausible futures and associated outcomes in a rapidly changing Arctic. Key processes, such as implications of permafrost degradation, are not yet fully captured, nor are societal and geopolitical drivers of change. In the near term, this shortfall presents challenges in meeting global and Arctic carbon policy and management goals, for example with respect to the impact of underestimated permafrost carbon feedbacks [30]. In this context, participatory scenarios may serve as a useful complement to models and have gained prominence in identifying impacts and vulnerabilities in the Arctic [31,32]. Scenarios draw on both quantitative and qualitative information and expertise, and hold great promise as vehicles for mutual social learning and the formation of communities of practice. A major hurdle that needs to be overcome is how to facilitate participation by Arctic rights holders and a broader range of stakeholders.

Indigenous involvement in research and policy development must include greater equity with the scientific and research sector. Funding should provide IK scholars with an equitable level of participation and should provide support for IK as a system of knowledge and learning. Some examples are Sea Ice and Walrus Outlook (SIWO) (https://www.arcus.org/siwo, accessed 2 September 2021), which is a collaboration of sea ice observations related to walrus migration in spring, and the U.S. National Science Foundation's Navigating the New Arctic initiative and their efforts to recognize the value of proper co-production of knowledge. In Canada, several compelling examples of Indigenous-led monitoring, co-production and co-management have recently been compiled to illustrate the breadth and efficacy of different approaches [33]. Another example is Sea-Ice Monitoring and Real-Time Information for Coastal Environments (SmartICE). This is a partnership among communities, academia, and governments that draws upon Inuit traditional knowledge and state-of-the-art technology to support ice-information needs (https://smartice.org, accessed 5 January 2022). Since its inception as a co-produced research project, SmartICE has developed into a social enterprise with a business model grounded in Inuit values including intergenerational teaching, community building, and meaningful employment for Inuit youth. A final example of funding IK expertise and participation in climate change research is Ikaaġvik Sikukun (https://www.ikaagviksikukun.org, accessed 7 January 2022) in Kotzebue, Alaska. The stated goals include to "...address key questions concerning the mechanisms and impacts of rapid changes taking place in the Arctic while ensuring that our answers incorporate traditional ways of knowing and are relevant to local needs". In the words of ICC Chair Dalee Sambo Dorough, "There is a pressing need for large-scale institutions to be responsive and adaptive to understand and address diverse issues across scales. Such adaptation requires the involvement and use of Indigenous Knowledge to inform research, observation and monitoring programs, as well as governance" [34].

**Table 1.** ICC Priorities in Guiding Research in Inuit Nunaat as laid out in the Utqiaġvik Declaration 2018 (reprinted with permission from Dalee Sambo-Dorough, ICC Chair, [35]).

| |
|---|
| The following actions are required to protect Inuit Nunaat and guide academic institutions, governments, and researchers in the conduct of the Inuit Nunaat research:<br>• Enhance ICC's work with Arctic research efforts, such as the Arctic Council's Arctic Monitoring and Assessment Programme (AMAP) Working Group, Sustained Arctic Observing Network (SAON), International Arctic Science Committee (IASC), the European Union (EU), and during high-level ministerial processes to ensure our views and concerns are addressed on how research in the Arctic should be conducted and to highlight how ethical approaches for research in the Arctic advance Inuit self- determination in research;<br>• Urge ICC to promote the interconnectedness of drivers of change and the interrelated impacts and implications on our health, economy and environment in high level political discussions and decision-making at fora such as the Arctic Council, the EU and UN agencies among other relevant international fora;<br>• Mandate ICC to participate actively in the operationalization of the United Nations "Local Communities and Indigenous Peoples Platform" to create a space to share best practices, relevant climate change programs and policies, and build capacity for Indigenous Peoples to engage in the UNFCCC process |

Just as Science, Technology, Engineering, and Mathematics (STEM) funding aims to grow the next generation of scientists, engineers, and researchers for universities, government agencies, and research institutions, IK should have equivalent initiatives to ensure there is the next generation of IK experts and practitioners in all communities and regions. This is critical for Indigenous communities' full and proper partnership in the co-production of knowledge projects in the Arctic into the future. The scientific/academic/governmental knowledge production industry is firmly established and well-funded to perpetuate itself, but even it must create large funding initiatives to support additional STEM education, post-graduate internships, and postdoctoral opportunities to develop the next generation of scholars, scientists, and researchers to continue their work.

Co-production approaches and adaptation strategies for Arctic Indigenous communities require a balanced approach with resources committed to support the continuity and growth of IK, the Indigenous Peoples' way of knowing and understanding (Table 1). In broad terms, Indigenous education is comprehensive and involves extensive experiential learning, language instruction, and study of cultural practices, beliefs, and spirituality. The learning experience of IK sharing is a necessity for the well-being of Indigenous people, families, and communities and is necessary to ensure that IK remains a critical partner and contributor to humankind's understanding of the Arctic.

### 3.2.2. Rights Holder and Stakeholder Discourse on Desired Futures

The 5C Arctic report recommended: "Design, initiate, and support a platform for a broad stakeholder dialogue on which future state of the Arctic we should strive for, drawing on existing local and regional platforms. The outcomes of the continuing dialogue have to inform decision-making processes in the context of the evolving Arctic trajectory" [2]. It is important to differentiate Indigenous communities and "local" communities, even though the experiences and knowledge of both contexts have important roles in understanding and responding to the challenges of the Arctic. Specifically, the UN Declaration on the Rights of Indigenous Peoples recognizes "the urgent need to respect and promote the inherent rights of indigenous peoples which derive from their political, economic and social structures and from their cultures, spiritual traditions, histories and philosophies, especially their rights to their lands, territories and resources" [35]. This presents a distinctively different status than "local communities", which is the term generally used to describe the inhabitants of a geographic area. This necessity of ensuring the sovereignty of Indigenous communities to maintain connection to their land and waters and to conserve its resources, which have sustained Arctic Indigenous Peoples for millennia, must be recognized. Therefore, Arctic Indigenous People should be designated as "rights holders" instead of stakeholders. Stakeholders for Arctic discussions include many different interest groups, industries, and organizations.

The viewpoint from an Indigenous coauthor of this perspective (V.M.) ties in with the acknowledgment by participants in the 2016 workshop that "There is no 'one Arctic'"—a point echoed by Young in 2021 [36] from a governance perspective. The need for a platform, in the broadest sense, to support discourse on different desirable and achievable future Arctic states is greater than ever. However, this question has already been answered by the Arctic's Indigenous Peoples. Any discourse about Arctic futures raises fundamental ethical questions that remain as vexing as ever in light of competing interests and post-colonial tensions [34]. Some might point to the Arctic Council as a consultative body or the U.N. Framework Convention on Climate Change and its various instruments to effect climate justice as entities that could support discourse.

At the same time, the importance of rights holder perspectives and place-based solutions has already been reflected on above with respect to community-driven adaptation and change management. This calls for a multitude of conversations in a polycentric approach to foster synergy and cross-Arctic communication and mutual support that is nevertheless tied to a particular place.

As an Indigenous Peoples non-governmental organization, the ICC-International has United Nations Consultative Status and is active in its role, including within the United Nations Permanent Forum on Indigenous Issues (UNPFII), which is the central United Nations coordinating body for Indigenous Peoples. ICC-International is also a Permanent Participant at the Arctic Council, one of six Indigenous Permanent Participants (the others being the Arctic Athabaskan Council, the Aleut International Association, the Gwich'in Council International, the Russian Association of Indigenous Peoples of the North, and the Saami Council). The ICC considers the UN and Arctic Council activities as fundamental to its work as an international organization. Each provides a degree of involvement in areas important to Inuit Nunaat, such as human rights and health, environment and climate issues, and food security.

"There is a strong connection between our culture, environment, and our homeland, which transcends national and political boundaries and connects us as one people. We affirm our right to self-determination and through a unified voice and approach are committed to advocate for, and protect the collective interests of, our membership at the international level" [37] (p. 6).

The priority areas presented in the Utqiaġvik Declaration [17] identify what is important to Inuit in the Arctic of the future. Together they represent the understanding that "We continue to rely on the land and ocean for nutrition, social, cultural, and spiritual well-being as well as traditional healing across Inuit Nunaat" [37]. In the words of co-author Vera

Metcalf, "We are only as healthy as our world is—our homeland and waters, air, and all those living with us".

3.2.3. Monitoring Arctic Change

The 5C Arctic report recommended "Complete and sustain the emerging Arctic Observing System, augmented by early warning components and enhanced Arctic system models to closely track key components of the changing Arctic" [2]. The importance of networks of sustained observations of Arctic change has increased further since 2016, driven by a combination of factors. The extent, pace, and effects of Arctic system change are key drivers of response action to mitigate risks stemming from feedback processes and adverse impacts from the local to the global scale. With accelerated and potentially underestimated changes, in particular in the Arctic cryosphere (see above), anticipating major transitions through observations and observation-informed models is increasingly relevant. As articulated in the National Academy of Sciences "The Arctic in the Anthropocene" report, monitoring is a critical alert system for "unknown unknowns" [38]. At the same time, increasing human activities and potentially competing interests in the Arctic have created an urgent need for observations that meet a range of societal information needs. A series of recent reports have examined in more detail the societal (including economic) benefits of Arctic observing activities in response to priorities defined by the Arctic Observing Summits (AOS) and the ASM process [39,40]. Finally, observations will be central to assessing the efficacy of any global scale action meant to stabilize and reduce atmospheric greenhouse gas concentrations.

The SAON initiative has recently made significant progress towards the implementation of more concerted observing efforts. In this context, the focus on widely shared benefits as a driver of the collaborative or coordinated deployment of observing assets has gained traction and is embedded in SAON's Roadmap for Arctic Observing and Data Systems (ROADS) [41].

A remaining challenge is the lack of effective, internationally coordinated funding support mechanisms for such work. One of three themes of the Second ASM was "Strengthening, Integrating and Sustaining Arctic Observations, Facilitating Access to Arctic Data, and Sharing Arctic Research Infrastructure", which has led to the creation of an Arctic Funders Forum [42]. Yet, major bureaucratic hurdles continue to impede effective transnational funding capabilities and capacities—despite the positive outcomes achieved by the Belmont Forum in accelerated international co-sponsorship of environmental research. Individual countries are also challenged to coordinate internally so as to effectively contribute to an internationally supported effort [43]. At the same time, we need to acknowledge that the co-design and co-management of sustained observations and observing networks in partnership with Arctic Indigenous Peoples has some ways to go. There remain significant challenges in bridging knowledge systems and providing opportunities and resources, although the needle is starting to move. In addition to community-driven observing initiatives and individual collaborations at the local scale [44,45], there is encouragement in SAON ROADS calling for Indigenous-led Expert Panels to help identify and prioritize observations with societal benefits in mind [41]. As another example, Canada and the United Kingdom recently agreed to cooperatively fund the Canada–Inuit Nunangat–United Kingdom Arctic Research Program, which is guided by the National Inuit Strategy on Research [46] and requires full Inuit engagement in leadership, design, development and execution. In the words of ICC Chair Dalee Sambo Dorough: "Our desire is that our perception of the Arctic is well understood, that Indigenous knowledge is acknowledged and utilized, that we seize opportunities for co-production of knowledge, and that we create a shift of what observing priorities are—all these steps create opportunities for action. Each of these points could trigger a change of how science is conducted in the Arctic by the simple act of reaching out to our communities, the willingness to co-produce knowledge, while respecting the value of Indigenous knowledge and the ethics related to utilizing it. There are many opportunities throughout observation processes for real action" [47].

3.2.4. Supporting Adaptation to the Changing Arctic

The 5C Arctic report recommendations included "Expedite research on adaptation of the Arctic to ongoing and expected environmental changes and provide resources for implementation of science-based adaptation strategies" and "Ramp up technical and financial support for Arctic societies needing strategic adaptation solutions—including relocation and soft infrastructure support (building codes, zoning, and others)" [2].

Since the release of the 5C Arctic report, a series of key publications by the Arctic Council's Arctic Monitoring and Assessment Programme (AMAP) has provided a kind of roadmap for broad-based adaptation action—with one important caveat. With three regional foci—Bering/Chukchi/Beaufort Region, Barents Region, and Baffin Bay-Davis Strait Region—the Adaptation Actions for a Changing Arctic (AACA) effort examined a broader range of options and approaches for adaptation measures [32]. Significantly, the Bering/Chukchi/Beaufort Report recognized the importance of developing a shared vision for desired Arctic futures by Arctic rights holders and stakeholders ([32]; see also the previous section), pointing to participatory scenario frameworks as a means to approach such visions. However, despite all their strengths, the AACA activities and reports are positioned mostly within an academic/government agency framework with disproportionate space given to a review of the magnitude and impacts of change, and little involvement of Indigenous expertise in the drafting of the reports. This circumstance is reflected in the Arctic Adaptation Exchange that emerged under the U.S. Climate Resilience Toolkit (https://toolkit.climate.gov/tool/arctic-adaptation-exchange, accessed 1 October 2021) as a follow-on to AACA, which is well positioned but underutilized in an adaptation context.

A number of other actions taken at the national and international level may qualify as adaptation measures or point in the direction of response. For example, the implementation of legally binding frameworks for search and rescue and spill response by the Arctic Council—established as a consultative rather than an executive body—is noteworthy [48,49]. Ratification of the Agreement to Prevent Unregulated High Seas Fisheries in the Central Arctic Ocean (CAOFA) [50] in 2021 with all Arctic coastal states, China, Japan, the Republic of Korea, and the European Union as signatories has sent an even stronger message. Specifically, the CAOFA calls for and has set in motion international scientific collaboration and monitoring to establish the state of marine living resources in the central Arctic Ocean, with the potential to develop a joint, ecosystem-based management regime [51]. In a similar vein, the ASM has demonstrated potential as a forum to advance internationally coordinated research responses to pressing Arctic issues. The Third ASM, co-hosted by Iceland and Japan in 2021, drew science ministers and officials from 27 countries, identifying priorities for international collaboration and helping build momentum for the support of concerted action by both Arctic and non-Arctic states. The guiding principles were transparency, inclusivity, and implementing a bottom-up approach to science. ASM3 expanded the ASM process by "attempting to create a more formal consultation process with the wider research community", complemented by "updates and new projects from participating countries, Indigenous Peoples' organizations, and international organizations engaged in Arctic science and education" [52]. Efforts such as the Sustaining Arctic Observing Networks (SAON) initiative, which is supported by both Arctic Council member states and non-Arctic states through membership of SAON co-sponsor International Arctic Science Committee (IASC), have shown promise as vehicles for follow-through on ASM recommendations. However, challenges remain, such as developing instruments for joint, transnational funding of internationally coordinated response action.

The designation of protected areas in light of rapid change can be a mechanism to develop effective response strategies at the local and regional level. Canada's recent designation of the Last Sea Ice—Tuvaijuittuq Marine Protected Area in the High Arctic—or the Northern Bering Sea Climate Resilience Area in the U.S. are examples of such action. Although, many questions remain as to how to best steward and manage such reserves. Different assessments have demonstrated that Indigenous, community-based management and support activities, such as community-driven monitoring, are most effective in ensuring adaptive and sustainable resource management in rapidly changing environments [45,53]. In the Arctic, increasing devolution of regulatory powers to the local, typically Indigenous-managed scale is evident. Examples include regulatory and management authority granted to the Nunatsiavut Government over terrestrial and marine resources in Labrador, Canada, or the Piniakkanik Sumiiffinni Nalunaarsuineq (Pisuna) locally based monitoring system that informs adaptive management approaches in Greenland [44].

However, finding an effective and equitable balance between local and broader interests and perspectives remains a challenge. For example, there may be many different levels of government involved in observing and monitoring various aspects of the Arctic environment but little or no communication among those entities (federal, regional, local, etc.) to ensure consistency in approach, to eliminate redundancies and fill critical gaps, and to leverage resources including human capacity and Indigenous expertise. This disconnect can hinder efforts to adapt to change, such as leading to a mismatch between regulatory frameworks that may control access to critical resources—e.g., country food and observational information—which indicates a need for regulatory flexibility in a rapidly changing environment. The problem is equally applicable from the local to international scales.

The activities of hunting, fishing, herding, gathering, and sharing, then preserving, preparing, carving, sewing, and more must all be kept and supported as times of learning IK to ensure our next generation of IK bearers is prepared to provide the unique way-of-knowing to Arctic research and governance. This is where the Utqiaġvik Declaration priorities of Food Security, Education and Language, Indigenous Knowledge, Sustainable Wildlife Management, and Environment are most directly aimed and where progress toward Indigenous-led resource management will be critically important to the future of the Arctic. The following from the Declaration illuminates this: "Food security is central to Inuit identity and way of life; is characterized by a healthy environment and encompasses access, availability, economics, physical and mental health, Inuit culture, decision-making power and management, and education. Therefore, it will be promoted and endorsed in all aspects of ICC's work" ([17], preamble).

### 3.3. *The Arctic's Future Depends on Global Action*

Arctic warming is mainly driven by actions taken outside the Arctic. One prime example is the atmospheric concentration of carbon dioxide which is responsible for ca. two-thirds of the greenhouse gas effect. It is emitted primarily in mid and low latitudes but affects the globe as a whole due to the rapid mixing of the atmosphere on hemispheric (less than one year) and global scales (several years). Since 2016, $CO_2$ concentrations in the atmosphere have increased steadily at a rapid pace from about 402 ppm (parts per million) (2016 January) to about 418 ppm (2022 January), thereby driving further global and regional warming. These changes imposed onto the Arctic by drivers located in lower latitudes amplify warming in the Arctic—now assumed to exceed 3–6 °C for a 1.5 °C rise in global temperature. An analysis by Carbon Brief following COP26 estimates that current policies concerning carbon emissions will lead to global warming between 2.6–2.7 °C by 2100 (with an uncertainty range of 2 to 3.6 °C) [54]. On the other hand, if countries meet their long-term carbon emission goals, the warming would be kept at about 1.9 °C. There are feedback loops through which the amplified warming of the Arctic affects the lower latitudes, for example through weather patterns influenced by the strength and geographical pattern of the Arctic Vortex [1,55].

The 5C Arctic report recommended: "Deploy measures for deep decarbonization of the global energy system and accelerate the upscaling and deployment of technologies for negative carbon emissions. Unify the efforts for allocating resources to master this historic challenge". Decarbonization of the energy system and a reduction in carbon emissions require an accelerated buildup of renewable energy production. According to the IEA, it is possible to reach net zero carbon emission by 2050 if the right actions are taken now and are sustained [56]. Most of the substitution for fossil fuels will come from solar and wind energy. The report emphasizes that the pathway toward net zero carbon emissions is narrow and requires the "massive deployment of all available clean energy technologies—such as renewables, EVs, and energy efficient building retrofits—between now and 2030. For solar power, it is equivalent to installing the world's current largest solar park roughly every day". The number of ambitious goals by governments and the private sector is growing steadily (e.g., Europe's goal to cut emissions by 55 percent by 2030 to stay on course for carbon neutrality in 2050 [57], China's announcement to stop financing coal-fired power plants abroad [58], Microsoft's plan to be carbon negative by 2030 [59]), but in many cases the pace of implementation still lags behind the stated goals.

In addition to the buildup of renewable energy sources, reaching net zero carbon emissions by 2050 requires a substantial capacity for negative emissions. Negative emissions are one of the major concerns laid out in the 2016 report. There are increasing efforts to upscale negative emissions through nature-based and technological solutions. Examples of nature-based solutions include reforestation (e.g., the American Forests Trillion Trees Movement), the restoration and expansion of wetlands and peatlands, regenerative agriculture, and marine ecosystem augmentation or restoration practices. Technological solutions include mechanisms that extract carbon dioxide from the atmosphere, including those developed by the construction of $CO_2$ extraction units by at least four startups (Climeworks, Carbon Engineering, Carbon Collect, and Global Thermostat.). Although those startups are still in the process of scaling their technology to the next level of tons to hundreds of tons per day, they principally can be scaled to the Gigaton level needed by mid-century to reach the goals of the Paris agreement.

Overall, there is more global action towards the recommendations related to the transformation of the energy system towards carbon neutrality by 2050 listed in the 2016 report. However, it is also clear that the pace towards implementation of the measures outlined in this report has to be accelerated.

### *3.4. The Arctic Plays a Critical Role in the Global Future*

In the past, the Arctic was frequently seen as a part of the Earth system that is largely decoupled, being located 'at the end of the world,' but also playing a minor role in global budgets such as heat and water transport. The Arctic system was thought to be mainly driven by global processes without major feedback on lower latitudes. It is now clear that there are powerful feedback loops from the Arctic to lower latitudes including impacts on weather systems, freshwater budgets in critical areas of the world ocean, and impacts on the global albedo and radiation balance. This recognition has led to improved integration of the Arctic into Earth system models, and it is now also widely accepted that Arctic communities and their knowledge systems have to be included in decision-making processes within the Arctic and beyond its geographical borders.

This changing situation is well captured by Lisa Koperqualuk, ICC Canada Vice-President (International), who stated "Inuit recognized early that safeguarding the Arctic would protect the planet—however, these calls remain unheard. As an observer to the IPCC, ICC advocated for the co-production of knowledge to guide the AR6, which would include Indigenous Knowledge as an important knowledge source" [18].

## 4. Conclusions and Outlook

We see increasing activity in most of the areas addressed in the 2016 report 'A 5C Arctic in a 2C World' [2]. It is also clear that the present pace of response to the challenges facing the Arctic and its communities remains too slow. However, there is still hope that humankind can turn the corner and implement the necessary steps to reach solutions to the challenges caused by the ever-expanding human footprint on our planet. These are solutions that include the Arctic with its amplified response to global drivers.

A key challenge is the fact that the rate and magnitude of change experienced by Indigenous Peoples of the Arctic and other Arctic residents outpace changes at lower latitudes. Yet, policy in many Arctic nations is made in locales and by policymakers far removed from where the greatest impacts are felt. This reality, combined with a history of colonization and focus on extractive resource development, complicates the translation of lessons learned in the Arctic into national or global policy frameworks. This is also reflected in the discourse around climate justice at COP 26 where a rift has expanded between industrialized nations on the one hand, and developing countries and Indigenous Peoples on the other. Part of the work in bridging this rift and supporting climate action involves framing and communication of the issues highlighted in this perspective from an Arctic into a global setting.

Strong Arctic voices can bring the major issues facing the Arctic to the attention of those involved in the decision-making processes that will decide the trajectory of the Arctic and indeed the planet. Although finding these voices is a challenge in itself, and clearly there are many positions concerning which Arctic future is the most desirable, a consensus has to be found concerning the future for which we should aim. For example, the voices of the people of the low-lying island nations had a major impact on the definition of the temperature targets during COP 21 that led to the Paris Agreement [7].

Unfortunately, at this critical time when it is so necessary for the voices of Arctic Indigenous Peoples to be heard, the COVID-19 pandemic's impact on engagement with government managers on policy, and with science on research, has limited Indigenous participation even more. Since most boards, commissions, and workgroups, which are primarily made up of individuals from different communities, postponed travel for face-to-face meetings, engagement, full involvement, and collaboration did not happen. Virtual meetings did not provide the same opportunity for effective participation, especially since Internet-based communication remains too expensive, very slow, and unavailable to most community members—a good illustration of Indigenous People's ongoing disparity in their ability to respond to crises or participate in solution development due to insufficient infrastructure.

We are now in the decisive decade concerning the future we leave behind for the next generations. Every year we fail to meet the milestones towards the targets that would reroute our planet and the Arctic onto a sustainable track into the future will require stronger, more concentrated, and more focused efforts in the years ahead. Additionally, if we hedge the milestones for too many more years, the envisioned targets will be unreachable. The coming few years will give us a clear indication if we are moving towards a future of hope or if we have missed the chance to preserve the 'safe operating space for humanity' in the decades and centuries to come [60].

**Author Contributions:** Conceptualization and methodology, P.S., S.P. and H.E.; writing—original draft preparation, P.S., H.E., M.S.M., V.M. and S.P.; writing—review and editing, H.E., P.S. and C.E.; project administration, C.E. All authors have read and agreed to the published version of the manuscript.

**Funding:** Pfirman is partially funded by the National Science Foundation Award (1928235), ARC-NAV: Arctic Robust Communities-Navigating Adaptation to Variability.

**Institutional Review Board Statement:** Not applicable.

**Informed Consent Statement:** Not applicable.

**Data Availability Statement:** Not applicable.

**Conflicts of Interest:** The authors declare no conflict of interest.

## References

1. IPCC. *Climate Change 2021: The Physical Science Basis. Contribution of Working Group I to the Sixth Assessment Report of the Intergovernmental Panel on Climate Change;* Masson-Delmotte, V.P., Zhai, A., Pirani, S.L., Connors, C., Péan, S., Berger, N., Caud, Y., Chen, L., Goldfarb, M.I., Gomis, M., et al., Eds.; Cambridge University Press: Cambridge, UK, 2021.
2. Schlosser, P.; Pfirman, S.L.; Pomerance, R.; Williams, M.; Ack, B.; Duffy, P.; Eicken, H.; Latif, M.; Murray, M.; Wallace, D. A 5 C Arctic in a 2 C World: Challenges and Recommendations for Immediate Action. 2016. Available online: https://www.researchgate.net/publication/308748753_A_5_C_Arctic_in_a_2_C_World (accessed on 30 September 2021).

3. Joint Statement of Ministers: On the occasion of the first White House Arctic Science Ministerial. Washington, DC, USA, 2016. Available online: https://obamawhitehouse.archives.gov/the-press-office/2016/09/28/joint-statement-ministers (accessed on 12 January 2022).
4. IPCC. *IPCC Special Report on the Ocean and Cryosphere in a Changing Climate*; Pörtner, H.-O., Roberts, D.C., Masson-Delmotte, V., Zhai, P., Tignor, M., Poloczanska, E., Mintenbeck, K., Alegría, A., Nicolai, M., Okem, A., et al., Eds.; 2019. Available online: https://www.c40.org/news/c40-response-ipcc-oceans-ice/?gclid=EAIaIQobChMIxIHz1snl9QIVxKqWCh1N2wHEEAAYASAAEgJwi_D_BwE (accessed on 12 January 2022).
5. Richter-Menge, J.; Druckenmiller, M.L.; Jeffries, M. (Eds.) *Arctic Report Card 2019*; National Oceanic and Atmospheric Administration: Washington, DC, USA, 2019. Available online: https://www.arctic.noaa.gov/Report-Card (accessed on 2 October 2021).
6. Thoman, R.L.; Richter-Menge, J.; Druckenmiller, M.L. (Eds.) *Arctic Report Card 2020*; National Oceanic and Atmospheric Administration: Washington, DC, USA, 2020. [CrossRef]
7. UNFCC. The Paris Agreement. 2015. Available online: https://unfccc.int/process-and-meetings/the-paris-agreement/the-paris-agreement (accessed on 2 October 2021).
8. IPCC. *Global Warming of 1.5°C. An IPCC Special Report on the Impacts of Global Warming of 1.5 °C above Pre-Industrial Levels and Related Global Greenhouse Gas Emission Pathways, in the Context of Strengthening the Global Response to the Threat of Climate Change, Sustainable Development, and Efforts to Eradicate Poverty*; Masson-Delmotte, V.P., Zhai, H.-O., Pörtner, D., Roberts, J., Skea, P.R., Shukla, A., Pirani, W., Moufouma-Okia, C., Péan, R., Pidcock, S., et al., Eds.; World Meteorological Organization: Geneva, Switzerland, 2018.
9. Lewis, K.M.; van Dijken, G.L.; Arrigo, K.R. Changes in phytoplankton concentration now drive increased Arctic Ocean primary production. *Science* **2020**, *369*, 198–202. [CrossRef] [PubMed]
10. Thoman, R.L.; Bhatt, U.S.; Bieniek, P.A.; Brettschneider, B.R.; Brubaker, M.; Danielson, S.L.; Labe, Z.; Lader, R.; Meier, W.N.; Sheffield, G.; et al. The Record Low Bering Sea Ice Extent in 2018: Context, Impacts, and an Assessment of the Role of Anthropogenic Climate Change. *Bull. Am. Meteorol. Soc.* **2020**, *101*, S53–S58. [CrossRef]
11. Prewitt, J.; McFarland, H.R. (Eds.) *Bering Science: Winter 2021 Bering Region Ocean Update, Issue 3 [Newsletter]*; Alaska Ocean Observing System: Anchorage, AK, USA, 2021.
12. Notz, D.; Stroeve, J. The Trajectory Towards a Seasonally Ice-Free Arctic Ocean. *Curr. Clim. Chang. Rep.* **2018**, *4*, 407–416. [CrossRef] [PubMed]
13. Hugelius, G.; Loisel, J.; Chadburn, S.; Jackson, R.B.; Jones, M.; MacDonald, G.; Marushchak, M.; Olefeldt, D.; Packalen, M.; Siewert, M.B.; et al. Large stocks of peatland carbon and nitrogen are vulnerable to permafrost thaw. *Proc. Natl. Acad. Sci. USA* **2020**, *117*, 20438–20446. [CrossRef] [PubMed]
14. Slater, T.; Hogg, A.E.; Mottram, R. Ice-sheet losses track high-end sea-level rise projections. *Nat. Clim. Chang.* **2020**, *10*, 879–881. [CrossRef]
15. Myers-Smith, I.H.; Kerby, J.T.; Phoenix, G.K.; Bjerke, J.W.; Epstein, H.E.; Assmann, J.J.; John, C.; Andreu-Hayles, L.; Angers-Blondin, S.; Beck, P.S.A.; et al. Complexity revealed in the greening of the Arctic. *Nat. Clim. Chang.* **2020**, *10*, 106–117. [CrossRef]
16. Lin, S.; Liu, Y.; Huang, X. Climate-induced Arctic-boreal peatland fire and carbon loss in the 21st century. *Sci. Total Environ.* **2021**, *796*, 148924. [CrossRef]
17. Utqiaġvik Declaration. Inuit—The Arctic We Want. 2018. Available online: https://www.arctictoday.com/wp-content/uploads/2018/07/2018-Utigavik-Declaration.pdf (accessed on 2 October 2021).
18. Inuit Circumpolar Council [ICC]. Inuit Review the IPCC 6th Assessment Report on The Physical Science Basis and Prepare for Crucial COP26 Talks in Glasgow. 2021. Available online: https://iccalaska.org/wp-icc/wp-content/uploads/2021/08/PR-IPCC-6th-Assessment-pfd.pdf (accessed on 2 October 2021).
19. Sonne, C.; Dietz, R.; Jenssen, B.M.; Lam, S.S.; Letcher, R.J. Emerging contaminants and biological effects in Arctic wildlife. *Trends Ecol. Evolution.* **2021**, *36*, 421–429. [CrossRef]
20. Chan, F.T.; Stanislawczyk, K.; Sneekes, A.C.; Dvoretsky, A.; Gollasch, S.; Minchin, D.; David, M.; Jelmert, A.; Albretsen, J.; Bailey, S.A. Climate change opens new frontiers for marine species in the Arctic: Current trends and future invasion risks. *Glob. Chang. Biol.* **2019**, *25*, 25–38. [CrossRef]
21. PAME [Protection of the Arctic Marine Environment]. Desktop Study on Marine Litter Including Microplastics in the Arctic. 2019. Available online: https://oaarchive.arctic-council.org/bitstream/handle/11374/2389/Desktop%20Study%20on%20marine%20litter.pdf (accessed on 30 September 2021).
22. Friedrich, D. Vegetable Farms 'Mushrooming' Across the Arctic. High North News 3 January 2018. 2018. Available online: https://www.highnorthnews.com/en/vegetable-farms-mushrooming-across-arctic (accessed on 30 September 2021).
23. Wilkinson, A.; Gerlach, C.; Karlsson, M.; Penn, H. Controlled environment agriculture and containerized food production in northern North America. *J. Agric. Food Syst. Community Dev.* **2021**, *10*, 127–142. [CrossRef]
24. Olsen, J.; Nenasheva, M.; Hovelsrud, G.K. 'Road of life': Changing navigation seasons and the adaptation of island communities in the Russian Arctic. *Polar Geogr.* **2021**, *44*, 1–19. [CrossRef]
25. Schlosser, P.; Laubichler, M.; Edwards, C.M.; Beschloss, S.; Berman, N.; van der Leeuw, S. COVID-19: The Ultimate Stress Test for Our Global Futures. *Medium* **2020**. [CrossRef]

26. Arctic Council. The Impact of Covid-19 on Indigenous Peoples in the Arctic. 2020. Available online: https://arctic-council.org/news/the-impact-of-covid-19-on-indigenous-peoples-in-the-arctic/ (accessed on 1 October 2021).
27. Petrov, A.N.; Welford, M.; Golosov, N.; DeGroote, J.; Devlin, M.; Degai, T.; Savelyev, A. Lessons on COVID-19 from Indigenous and remote communities of the Arctic. *Nat. Med.* **2021**, *27*, 1491–1492. [CrossRef]
28. Inuit Circumpolar Council [ICC]; Lynge, A.; Stenbæk, M. *Inuit Arctic Policy*; Inuit Circumpolar Conference: Barrow, AK, USA, 2010.
29. Arctic Council. *Meaningful Engagement of Indigenous Peoples and Local Communities in Marine Activities Part 2 Report-Findings for Policy Makers*; Protection of the Arctic Marine Environment (PAME) Working Group: Akureyri, Iceland, 2019.
30. Natali, S.M.; Holdren, J.P.; Rogers, B.M.; Treharne, R.; Duffy, P.B.; Pomerance, R.; MacDonald, E. Permafrost carbon feedbacks threaten global climate goals. *Proc. Natl. Acad. Sci. USA* **2021**, *118*, e2100163188. [CrossRef]
31. Flynn, M.; Ford, J.D.; Pearce, T.; Harper, S.L.; IHACC Research Team. Participatory scenario planning and climate change impacts, adaptation and vulnerability research in the Arctic. *Environ. Sci. Policy* **2018**, *79*, 45–53. [CrossRef]
32. AMAP. *Adaptation Actions for a Changing Arctic: Perspectives from the Bering-Chukchi-Beaufort Region*; Arctic Monitoring and Assessment Programme (AMAP): Oslo, Norway, 2017; 255p.
33. Johnson, N.; Pearce, T.; Breton-Honeyman, K.; Etiendem, D.; Loseto, L. Knowledge co-production and co-management of Arctic wildlife. *Arct. Sci.* **2020**, *6*, 124–126. [CrossRef]
34. Inuit Circumpolar Council [ICC]. ICC Holds Workshops to Develop Circumpolar Protocols for Equitable and Ethical Engagement of Inuit Communities and Indigenous Knowledge. 2021. Available online: https://www.inuitcircumpolar.com/news/icc-holds-workshops-to-develop-circumpolar-protocols-for-equitable-and-ethical-engagement-of-inuit-communities-and-indigenous-knowledge/ (accessed on 30 September 2021).
35. United Nations General Assembly. United Nations declaration on the rights of indigenous peoples. *United Nations* **2007**, *12*, 1–18.
36. Young, O.R. Arctic Futures–Future Arctics? *Sustainability* **2021**, *13*, 9420. [CrossRef]
37. Inuit Circumpolar Council [ICC]. Implementing the Utqiaġvik Declaration 2018–2022. 2019. Available online: https://iccalaska.org/wp-icc/wp-content/uploads/2019/09/InternationalStrategicPlan-9.18.19.pdf (accessed on 30 September 2021).
38. National Research Council. *The Arctic in the Anthropocene: Emerging Research Questions*; The National Academies Press: Washington, DC, USA, 2014. [CrossRef]
39. Dobricic, S.; Ferrario, F.M.; Pozzoli, L.; Wilson, J.; Gambardella, A.; Tilche, A. *Impact Assessment Study on Societal Benefits of Arctic Observing Systems*; European Commission: Brussels, Belgium, 2018. [CrossRef]
40. Strahlendorff, M.; Veijola, K.; Gallo, J.; Vitale, V.; Hannele, S.; Smirnov, A.; Tanaka, H.; Sueyoshi, T.; Nitu, R.; Larsen, J.R. Value Tree for Physical Atmosphere and Ocean Observations in the Arctic. Finnish Meteorological Institute. Report no. 9523360728. 2019. Available online: https://helda.helsinki.fi/bitstream/handle/10138/300768/Value%20tree%20for%20physical%20atmosphere%20and%20ocean%20observations%20in%20the%20Arctic.pdf (accessed on 30 September 2021).
41. Starkweather, S. Sustaining Arctic Observing Networks' (SAON) Roadmap for Arctic Observing and Data Systems (ROADS). *Arctic* **2021**. *in review*.
42. Report of the 2nd Arctic Science Ministerial: Co-operation in Arctic Science–Challenges and Joint Actions. In Proceedings of the 2nd Arctic Science Ministerial, Berlin, Germany, 25–26 October 2018.
43. Murray, M.S.; Loseto, L.; Sankar, R.; Pulsifer, P.; Dawson, J. Canadian Coordination in Support of Sustained Observations of Arctic Change . *Arct. Arct. Obs. Summit Spec. Issue* **2021**, *in press*.
44. Tengö, M.; Austin, B.J.; Danielsen, F.; Fernández-Llamazares, Á. Creating Synergies between Citizen Science and Indigenous and Local Knowledge. *BioScience* **2021**, *71*, 503–518.
45. Eicken, H.; Danielsen, F.; Sam, J.-M.; Fidel, M.; Johnson, N.; Poulsen, M.K.; Lee, O.A.; Spellman, K.V.; Iversen, L.; Pulsifer, P.; et al. Connecting Top-Down and Bottom-Up Approaches in Environmental Observing. *BioScience* **2021**, *71*, 467–483. [CrossRef] [PubMed]
46. Inuit Tapiriit Kanatami. National Inuit Strategy on Research. 2018. Available online: https://www.itk.ca/national-strategy-on-research-launched/ (accessed on 30 September 2021).
47. Sambo Dorough, D. Long-Term Observations Are an Inherent Process in Our Communities-Arctic Council Interview with Dalee Sambo-Dorough/ICC Chair at 2020 Arctic Observing Summit. 2021. Available online: https://arctic-council.org/en/news/long-term-observations-are-an-inherent-process-in-our-communities/ (accessed on 30 September 2021).
48. Arctic Council. *Agreement on Cooperation on Aeronautical and Maritime Search and Rescue in the Arctic*; Arctic Council Secretariat: Tromsø, Norway, 2011.
49. Arctic Council. *Agreement on Cooperation on Marine Oil Pollution Preparedness and Response in the Arctic*; Arctic Council Secretariat: Tromsø, Norway, 2013.
50. *Agreement to Prevent Unregulated High Seas Fisheries in the Central Arctic Ocean*; Ilulissat, Greenland, 2018. Available online: https://ec.europa.eu/oceans-and-fisheries/news/arctic-agreement-prevent-unregulated-fishing-enters-force-2021-06-25_en (accessed on 18 October 2021).
51. Snoeijs-Leijonmalm, P.; Flores, H.; Volckaert, F.; Niehoff, B.; Schaafsma, F.; Hjelm, J.; Hentati-Sundberg, J.; Niiranen, S.; Crépin, A.S.; Österblom, H. Review of the research knowledge and gaps on fish populations, fisheries and linked ecosystems in the Central Arctic Ocean (CAO). *EFICA* **2020**. [CrossRef]

52. Ministry of Education Science and Culture. Knowledge for a Sustainable Arctic: 3rd Arctic Science Ministerial Report. 2021. Available online: https://asm3.org/ (accessed on 2 October 2021).
53. Danielsen, F.; Enghoff, M.; Poulsen, M.K.; Funder, M.; Jensen, P.M.; Burgess, N.D. The Concept, Practice, Application, and Results of Locally Based Monitoring of the Environment. *BioScience* **2021**, *71*, 484–502. [CrossRef] [PubMed]
54. "Analysis: Do COP26 promises keep global warming below 2C?". Available online: www.carbonbrief.org/analysis-do-cop26-promises-keep-global-warming-below-2c (accessed on 12 November 2021).
55. National Research Council. *Arctic Matters: The Global Connection to Changes in the Arctic*; The National Academies Press: Washington, DC, USA, 2015. [CrossRef]
56. IEA. *Net Zero by 2050*; IEA: Paris, France, 2021. Available online: https://www.iea.org/reports/net-zero-by-2050 (accessed on 12 January 2022).
57. European Commission. 2030 Climate Target Plan. 2021. Available online: https://ec.europa.eu/clima/policies/eu-climate-action/2030_ctp_en (accessed on 12 January 2022).
58. Sommer, L. Ahead of Climate Talks, China Vows to Stop Building Coal Power Plants Abroad. NPR, 21 September 2021. 2021. Available online: https://www.npr.org/2021/09/21/1039486454/china-coal-power-climate-change (accessed on 18 October 2021).
59. Smith, B. Microsoft Will Be Carbon Negative by 2030. 2020. Available online: https://blogs.microsoft.com/blog/2020/01/16/microsoft-will-be-carbon-negative-by-2030/ (accessed on 18 October 2021).
60. Steffen, W.; Richardson, K.; Rockström, J.; Cornell, S.E.; Fetzer, I.; Bennett, E.M.; Biggs, R.; Carpenter, S.R.; De Vries, W.; De Wit, C.A.; et al. Planetary boundaries: Guiding human development on a changing planet. *Science* **2015**, *347*, 1259855. [CrossRef] [PubMed]

sustainability

MDPI

*Perspective*

# Sustaining the Arctic in Order to Sustain the Global Climate System

**Daniel Bodansky [1,*] and Rafe Pomerance [2]**

[1] Sandra Day O'Connor College of Law, Arizona State University, Phoenix, AZ 85004, USA
[2] Woodwell Climate Research Center, Falmouth, MA 02540, USA; rafe.pomerance@gmail.com
* Correspondence: Daniel.Bodansky@asu.edu

**Abstract:** The unraveling of the Arctic is bad enough for the Arctic itself, but it will have enormous consequences for the entire planet since the Arctic is a crucial component of the global climate system. Current policies do not provide much hope to prevent these harms. We have committed the earth to too much warming to take a step-by-step approach. We have entered a period of history when planetary management has become unavoidable and must move forward on many fronts simultaneously. Key components of a multiprong approach include decarbonization, focus on short-lived climate forcers, greenhouse gas removal, adaptation, Arctic interventions, and solar climate intervention. This article discusses the last option, which may be the only means of cooling the earth quickly enough to save Arctic ice and permafrost. Scientific research is essential to better understand its feasibility, effectiveness, and safety. However, research is not enough; we need to be ready to respond right away if Arctic or global temperatures need to be lowered quickly. This means we need significant technology research and development so that solar climate intervention technologies are deployment-ready in the relatively near future, perhaps in a decade or two, and could be used should the need arise and should research show that they are effective and safe.

**Keywords:** Arctic; climate engineering; climate intervention; solar radiation management

**Citation:** Bodansky, D.; Pomerance, R. Sustaining the Arctic in Order to Sustain the Global Climate System. *Sustainability* **2021**, *13*, 10622. https://doi.org/10.3390/su131910622

Academic Editors: Stephanie Pfirman, Gail Fondahl, Grete K. Hovelsrud and Tero Mustonen

Received: 18 July 2021
Accepted: 18 September 2021
Published: 24 September 2021

**Publisher's Note:** MDPI stays neutral with regard to jurisdictional claims in published maps and institutional affiliations.



The year 2021 has seen a number of welcome developments on the international climate change front—the reengagement by the United States in the Paris Agreement; the increasing number of pledges by countries to achieve net-zero emissions by mid-century or thereabouts; and the slew of climate change initiatives by investors, industry groups, and other non-state actors. But while these are all essential steps forward in international climate policy, will they be enough to keep the Arctic from unraveling and causing dire harm to the global climate system? We think not and therefore propose a multiprong strategy to sustain the Arctic, including the possibility of solar climate interventions, if shown to be safe and effective.

Each week seems to bring more bad news:

- Arctic sea ice thinned 60% more than previous estimates in the period from 2002 to 2018 [1].
- Warm water from the North Atlantic is hindering sea-ice growth in the Arctic Ocean [2].
- Melting at the bottom of the ice sheet (not just surface melting) accounts for a significant amount of Greenland ice loss [3].
- The majority of marine-terminating glaciers in northwest and central-west Greenland are experiencing accelerating mass loss [4].
- Temperatures in the Russian Arctic exceeded 30 °C in May, more than 20 °C above average for that time of year, accelerating the collapse of Russian infrastructure from thawing permafrost [5].

In a powerful message, a recent report by the Arctic Monitoring and Assessment Program (AMAP) found that the Arctic has warmed three times as much as the global average over the last fifty years, not two or two-and-a-half times as much as previously believed [6]. One study found that we may be nearing the threshold beyond which Greenland ice loss will be irreversible [7].

The unraveling of the Arctic is bad enough for the Arctic itself, but it will have enormous consequences for the entire planet since the Arctic is a crucial component of the global climate system [8]. Its sea ice and spring snow cover reflect an enormous amount of heat back into space, its frozen high-latitude soils contain double the carbon content of the atmosphere [9], and its glacial ice holds enough water to cause over seven meters of sea-level rise [10].

Given its central role in the climate system, "what happens in the Arctic does not stay in the Arctic," as an Arctic official once put it [11]. As Arctic ice melts and spring snow cover shrinks, the Arctic Ocean and land surface become darker and reflect less sunlight back into space, thereby exacerbating global warming [12]. As Arctic permafrost thaws, it releases carbon dioxide and methane, potentially using up much, if not all, of the greenhouse gas budget available to limit global warming to 1.5 °C, according to recent estimates [13]. As Greenland and other smaller Arctic glaciers melt, sea-level rise accelerates, tying the fate of Greenland to the fate of Miami. As the Arctic atmosphere warms, the jet stream likely weakens, radically altering mid-latitude weather patterns, as we are witnessing in real time [14].

Can these harms be prevented? Current policies do not provide much hope.

- *1.5 °C temperature goal*—Global climate policy, as reflected in the Paris Agreement, aims to limit global warning to 1.5 °C at best. However, as the IPCC 1.5 °C Special Report indicates, even if we were able to limit global warming to 1.5 °C, there would still be significant harms [15]. Already, the Great Barrier Reef and most of the world's other coral reefs are dead or dying, and the world is experiencing more extreme weather events, including worsening droughts and wildfires, greater heat extremes, and more intense tropical storms. Additionally, global warming of 1.5 °C means Arctic warming of about 4.5 °C. Moreover, that is the best-case scenario. Limiting warming to 1.5 °C would be huge stretch, given the continuing upward trend in global greenhouse gas emissions and the fact that the atmosphere already contains 419 ppm of $CO_2$, 50% percent above pre-industrial levels. In reality, we will be lucky to limit global warming to 2 °C, a temperature increase that the paleoclimate record suggests would ultimately be accompanied by many meters of sea-level rise [16]. The world could easily be on its way to a warming of 3 °C, which would mean 9 °C of Arctic warming, if the Arctic continues to warm three times faster than the global average.
- *Emissions pathways: overshoot and return*—Most of the emission pathways considered by the IPCC to achieve the 1.5 °C target assume that we will initially overshoot the target and then need negative emissions to bring temperature back down. The problem is that some of the harm from overshoot will be effectively irreversible in meaningful time frames, such as the release of carbon dioxide and methane from thawing permafrost and the disappearance of Greenland and other Arctic—as well as Antarctic and mid-latitude—glaciers.
- *Arctic policy*—Arctic policymaking appears even less promising. The Arctic Council, the principal Arctic-specific governance body, is an informal institution that lacks any regulatory powers and shows no signs of being up to the task of taking significant action. Despite the release of the 2021 Arctic assessment report showing that the Arctic is warming three times faster than the global average, the Arctic Council has not even called worldwide attention to the critical role the Arctic plays for the rest of the world. It has failed to establish a tolerable upper bound for climate change in the Arctic or to answer the question, "What is the Arctic we have to have to sustain the global climate system?" Instead, at its most recent meeting, it merely "not[ed] with concern the serious threats to Arctic ecosystems due to climate change" and "reiterate[d] the need for enhanced action to meet the long-term temperature goal and effective implementation of the Paris Agreement [17]."

What is to be done? There are no perfect options. We have committed the earth to too much warming to take a step-by-step approach. We must realize that we have entered a period of history when planetary management has become unavoidable and move forward on many fronts simultaneously. Key components of a multiprong approach include

1. *Decarbonization*—First and foremost, the world needs to very rapidly decarbonize, as the Paris Agreement suggests. Even if emission reductions are unlikely to take effect quickly enough to completely preserve Arctic ice and permafrost, they are essential in the long run to stabilizing the global climate system. This is climate policy's most urgent task.
2. *Short-Lived Climate Forcers (SLCFs)*—Second, we need to halt emissions of SLCFs, such as methane and black carbon, which are particularly potent contributors to climate change [18].
3. *Greenhouse gas removal*—Third, we need to intensify efforts to remove carbon dioxide and other greenhouse gases from the atmosphere through nature-based and/or technological means (for example, afforestation, soil carbon sequestration, enhanced terrestrial weathering, mineral carbonation, or direct air capture).
4. *Adaptation*—Fourth, to the extent possible, we need to take measures to help the Arctic adapt to the effects of climate change.
5. *Arctic interventions*—Fifth, we need to explore proposals for saving Arctic ice through local interventions to limit summer melt, enhance winter freezing, or stabilize Greenland glaciers.
6. *Solar climate intervention (SCI)*—Sixth, we need to explore the possibility of using technology to reflect more sunlight away from the earth in order to cool it rapidly.

The first four approaches have been written about extensively, are well accepted internationally, and need little defense. By contrast, the last two approaches have received much less attention. One of us has written recently about Arctic interventions [19], so we focus here on the last option, using technology to reflect more sunlight—also referred to as solar climate engineering or solar radiation management (SRM). Solar climate intervention is perhaps the most controversial proposal to address climate change but may be the only means of cooling the earth quickly enough to save the Arctic. Potentially, it could be used to eliminate overshoot on the way to achieving the 1.5 °C Paris target—a huge environmental benefit. Indeed, solar climate intervention could actually lower warming below 1.5 °C, perhaps reaching the equivalent of radiative forcing of 350 ppm of $CO_2$.

Various techniques to reflect sunlight have been suggested. These include injecting aerosols into the stratosphere to scatter incoming sunlight (stratospheric aerosol injection or SAI) and spraying sea salt from ships to provide cloud condensation nuclei and thereby brighten marine clouds (marine cloud brightening or MCB).

Volcanic eruptions provide proof-of-concept that stratospheric aerosols cool the planet. The sulfur aerosols injected into the stratosphere by the eruption of Mount Pinatubo in 1991 cooled the planet by about 0.5 °C. Hence, we dub using solar climate intervention to cool the planet by 1.0 °C a "Two Pinatubo" strategy. Recent modeling suggests that injecting aerosols into the stratosphere in the spring near the Arctic could restore significant amounts of Arctic sea ice and substantially reduce both local and global impacts of climate change [20].

A Two Pinatubo strategy raises many questions. For example,

- How would solar climate interventions affect the Arctic climate, as well as other regional climates and the global climate system?
- What are the advantages and disadvantages of stratospheric aerosol injection in the Arctic as compared to marine cloud brightening, taking into consideration technical feasibility, ability to scale up quickly, effectiveness in reducing temperature, and safety?
- If stratospheric aerosol injection were being considered, which aerosols should be used, how should they be lofted into the stratosphere and dispersed, and at what location(s) and altitude?
- What are the risks of solar climate intervention? What are the benefits?

- How should SCI be governed, either nationally and/or internationally? Who should make decisions and how? What safety, environmental, and other requirements would need to be met before an intervention should be allowed to proceed?

Thus far, insufficient research has been done on solar climate intervention, so we still know too little about its feasibility, effectiveness, and safety. In response, there have been growing calls to develop a research program on solar climate intervention, most recently in a report of the National Academy of Sciences [21].

We strongly support an immediate and focused research program, exploring all potential options to address the Arctic crisis. Given the gravity of the threat and the dwindling time to respond, we believe solar climate intervention research, in particular, warrants considerably higher funding than the modest amounts recommended by the National Academy of Sciences report ($100–200 million over five years).

Scientific research is not enough, however. Given the urgency of the Arctic crisis, we need to be ready to respond right away if Arctic or global temperatures need to be lowered quickly. This means we need significant technology research and development so that solar climate intervention is deployment-ready in the relatively near future, perhaps in a decade or two.

In saying that SCI should be deployment-ready in a decade or two, we want to be absolutely clear: we are not advocating its deployment. Whether it makes sense to deploy SCI will depend on what research shows about its effectiveness and safety. The response to COVID-19 provides a useful analogue. The goal of Operation Warp Speed was to develop deployment-ready vaccines, but the decision to go ahead with Warp Speed did not prejudge whether any of the vaccines that were developed should actually be used. Decisions about use depended on what the clinical trials showed about a vaccine's effectiveness and safety.

When one is dealing with a crisis, one cannot afford to proceed in an incremental, stepwise manner. That is why, if we are to have any chance of saving the Arctic the world needs, we cannot afford to do basic research first and only start developing deployable technologies later, as the NAS report suggests. We need to proceed on all fronts simultaneously. We need to develop the necessary technologies now so that we are ready to deploy them at scale, should the need arise and should the research show that they are safe.

**Author Contributions:** Conceptualization, writing, review and editing: D.B. and R.P. Both authors have read and agreed to the published version of the manuscript.

**Funding:** This research received no external funding.

**Institutional Review Board Statement:** Not applicable.

**Informed Consent Statement:** Not applicable.

**Data Availability Statement:** Not applicable.

**Conflicts of Interest:** The authors declare no conflict of interest.

## References

1. Mallett, R.D.C.; Stroeve, J.C.; Tsamados, M.; Landy, J.C.; Willatt, R.; Nandan, V.; Liston, G.E. Faster decline and higher variability in the sea ice thickness of the marginal Arctic seas when accounting for dynamic snow cover. *Cryosphere* **2021**, *15*, 2429–2450. [CrossRef]
2. Ricker, R.; Kauker, F.; Schweiger, A.; Hendricks, S.; Zhang, J.; Paul, S. Evidence for an increasing role of ocean heat in Arctic winter sea ice growth. *J. Clim.* **2021**, *34*, 5215–5227.
3. Karlsson, N.B.; Solgaard, A.M.; Mankoff, K.D.; Gillet-Chaulet, F.; MacGregor, J.A.; Box, J.E.; Citterio, M.; Colgan, W.T.; Larsen, S.H.; Kjeldsen, K.K.; et al. A first constraint on basal melt-water production of the Greenland ice sheet. *Nat. Commun.* **2021**, *12*, 3461. [CrossRef]
4. Black, T.E.; Joughin, I. Multi-Decadal Retreat of Marine-Terminating Outlet Glaciers in Northwest and Central-West Greenland. *The Cryosphere* **2021**. Available online: https://doi.org/10.5194/tc-2021-164 (accessed on 10 September 2021).
5. Staalesen, A. The looming Arctic collapse: More than 40% of north Russian buildings are starting to crumble. *Arctic Today* **2021**. Available online: https://www.arctictoday.com/the-looming-arctic-collapse-more-than-40-of-north-russian-buildings-are-starting-to-crumble/ (accessed on 10 September 2021).

6. Arctic Monitoring and Assessment Programme (AMAP). *Arctic Climate Change Update 2021: Key Trends and Impacts*; AMAP: Tromsø, Norway, 2021.
7. Boers, N.; Rypdal, M. Critical slowing down suggests that the western Greenland Ice Sheet is close to a tipping point. *Proc. Natl. Acad. Sci. USA* **2021**, *118*, e2024192118. [CrossRef] [PubMed]
8. Moon, T.A.; Overeem, I.; Druckenmiller, M.; Holland, M.; Huntington, H.; Kling, G.; Lovecraft, A.L.; Miller, G.; Scambos, T.; Schädel, C.; et al. The expanding footprint of rapid Arctic change. *Earth's Future* **2019**, *7*, 212–218. [CrossRef]
9. Tarnocai, C.; Canadell, J.G.; Schuur, E.A.G.; Kuhry, P.; Mazhitova, G.; Zimov, S. Soil organic carbon pools in the northern circumpolar permafrost region. *Glob. Biogeochem. Cycles* **2009**, *23*, GB2023. (estimating northern circumpolar permafrost region contains 1672 Pg of organic carbon, compared to 750 PgC in the at-mosph ere). [CrossRef]
10. Aschwanden, A.; Fahnestock, M.A.; Truffer, M.; Brinkerhoff, D.J.; Hock, R.; Khroulev, C.; Mottram, R.; Khan, S.A. Contribution of the Greenland Ice Sheet to sea level over the next millennium. *Sci. Adv.* **2019**, *5*, eaav9396. [CrossRef]
11. 'What Happens in the Arctic, Does Not Stay in the Arctic'–Climate Change in the Arctic Will Have Global Consequences and Cannot Be Ignored. *NATO Parliamentary Assembly News*. 17 May 2017. Available online: https://www.nato-pa.int/news/what-happens-arctic-does-not-stay-arctic-climate-change-arctic-will-have-global-consequences (accessed on 10 September 2021).
12. Pistone, K.; Eisenman, I.; Ramanathan, V. Radiative heating of an ice-free Arctic Ocean. *Geophys. Res. Lett.* **2019**, *46*, 7474–7480. (estimate that an ice-free Arctic Ocean would result in radiative forcing of 0.71 $W/m^2$, the equivalent of one trillion tons of $CO_2$ emissions, speeding global warming by an estimated 25 years). [CrossRef]
13. Natali, S.M.; Holdren, J.P.; Rogers, B.M.; Treharne, R.; Duffy, P.B.; Pomerance, R.; MacDonald, E. Permafrost carbon feedbacks threaten global climate goals. *Proc. Natl. Acad. Sci. USA* **2021**, *118*, e2100163118. (estimating e missions from permafrost thaw of up to 550 Gt of $CO_2$ by 2100, assuming weak climate policies, com-pared to a carbon budget of 340–1000 Gt of $CO_2$-e for a likely chance of remaining below 2 °C and a budget of 290–440 Gt to remain below 1.5 °C). [CrossRef]
14. Overland, J.E.; Wang, M. Resolving future Arctic/midlatitude weather connections. *Earth's Future* **2018**, *6*, 1146–1152. [CrossRef]
15. Intergovernmental Panel on Climate Change. Global Warming of 1.5 °C: An IPCC Special Report. 2018. Available online: https://www.ipcc.ch/sr15/ (accessed on 10 September 2021).
16. Hansen, J.; Sato, M.; Hearty, P.; Ruedy, R.; Kelley, M.; Masson-Delmotte, V.; Russell, G.; Tselioudis, G.; Cao, J.; Rignot, E.; et al. Ice melt, sea level rise, and superstorms: Evidence from paleoclimate data, climate modeling, and modern observations that 2 °C global warming could be dangerous. *Atmos. Chem. Phys.* **2016**, *16*, 3761–3812. [CrossRef]
17. 12th Arctic Council Ministerial Meeting, Reykjavík Declaration 2021, preambular 8, para. 19. Available online: https://oaarchive.arctic-council.org/bitstream/handle/11374/2600/declaration%202021%20web%20EN.pdf?sequence=9&isAllowed=y (accessed on 10 September 2021).
18. Arctic Council. *Expert Group on Black Carbon and Methane—3rd Summary of Progress and Recommendations*; Arctic Council: Tromsø, Norway, 2021.
19. Bodansky, D.; Hunt, H. Arctic Climate Interventions. *Int. J. Mar. Coast. Law* **2020**, *35*, 596–617. [CrossRef]
20. Lee, W.R.; MacMartin, D.G.; Visioni, D.; Kravitz, B. High-latitude stratospheric aerosol geoengineering can be more effective if injection is limited to spring. *Geophys. Res. Lett.* **2021**, *48*, e2021GL092696. [CrossRef]
21. National Academies of Sciences, Engineering and Medicine. *Reflecting Sunlight: Recommendations for Solar Geoengineering Research and Research Governance*; National Academies Press: Washington, DC, USA, 2021.

MDPI
St. Alban-Anlage 66
4052 Basel
Switzerland
Tel. +41 61 683 77 34
Fax +41 61 302 89 18
www.mdpi.com

*Sustainability* Editorial Office
E-mail: sustainability@mdpi.com
www.mdpi.com/journal/sustainability

www.ingramcontent.com/pod-product-compliance
Lightning Source LLC
Chambersburg PA
CBHW061442210926
51850CB00008B/1588
* 9 7 8 3 0 3 6 5 6 9 2 5 3 *